Symmetry and Structure

Symmetry and Structure

(Readable Group Theory for Chemists)

Second Edition

Sidney F. A. Kettle
*University of East Anglia,
Norwich, England*

and

*Royal Military College,
Kingston, Ontario, Canada*

JOHN WILEY & SONS
Chichester • New York • Brisbane • Toronto • Singapore

Other Wiley Editorial Offices

John Wiley & Sons, Inc., 605 Third Avenue,
New York, NY 10158-0012, USA

Jacaranda Wiley Ltd, 33 Park Road, Milton,
Queensland 4064, Australia

John Wiley & Sons (Canada) Ltd, 22 Worcester Road,
Rexdale, Ontario M9W 1L1, Canada

John Wiley & Sons (SEA) Pte Ltd, 37 Jalan Pemimpin #05-04,
Block B, Union Industrial Building, Singapore 2057

Library of Congress Cataloging-in-Publication Data

Kettle, S. F. A. (Sidney Francis Alan)
    Symmetry and structure : readable group theory for chemists /
Sidney F. A. Kettle. — 2nd ed.
        p.   cm.
    Includes bibliographical references and index.
    ISBN 0-471-95547-7 : — ISBN 0-471-95476-4 (pbk.) :

    1. Chemical structure.   2. Symmetry (Physics)   3. Group theory.
4. Chemical bonds.   5. Molecules.   I. Title.
QD471.K47516   1995
541.2´2—dc 20   94-39498
                                                       CIP

British Library Cataloguing in Publication Data

A catalogue record for this book is available from the British Library
ISBN 0 471 95547 7 (cloth)
ISBN 0 471 95476 4 (paper)

Typeset in 10/12pt Times by Mathematical Composition Setters Ltd, Salisbury, Wiltshire
Printed and bound in Great Britain by Biddles Ltd, Guildford and King's Lynn

To: Anna, Maria and Tanya

Contents

Preface to the First Edition

It has been claimed that group theory is little more than applied common sense. If this is so, then it should be possible to present it in a way which avoids explicit use of formal mathematics and, in particular, matrix algebra. Matrix algebra is not a difficult topic and many students will have met it before entering university. To write a book simply to avoid its use would, then, seem a pointless exercise. Yet this is such a book, the reason for its existence lies deeper. In my experience, chemists prefer to think in terms of models and pictures rather than mathematics, they find it easier to describe a model mathematically than to start with a mathematical development and derive from it a picture. For a full understanding of group theory both picture and mathematics are needed and so it is usual to develop them together. Unfortunately, group theory is a sequential subject—each stage depends on preceding stages so that in a text there is a need for constant referral back to earlier pages. Because the mathematical treatment is more precise and comprehensive than the pictorial, the back references are most readily made to mathematical sections. My experience is that for most students the physical picture becomes more and more hazy as the mathematics takes over. This is the reason for the structure of the present book. The subject is presented pictorially—but accurately—so that the student is not referred back to mathematical equations but rather to pictorial explanations. If the text provides the pictures where, then, is the mathematics? The answer is 'in the appendices'. These comprise a significant proportion of the book and, while I hope that they more or less stand on their own as a text on mathematical group theory, they are also integrated with the main text. In this way I hope to have gained something of the best of both worlds! In the text itself I have avoided mathematics by using chemical topics as vehicles for the group theory. For much of the book the vehicle is chemical bonding but in the later chapters other subjects are covered. Throughout, I have tried to choose topic and group theory in a supportive manner; a topic is only included if it enables some development of the group theoretical theme. Conversely, the use of symmetry arguments must tell us something new about the particular topic. I have therefore been more concerned with establishing a firm basis than with its development. Thus, although both ligand field theory and the Woodward–Hoffmann rules are included, neither is treated at any length because no new principles would have immediately emerged in an extension which are not covered elsewhere.

In writing the book I have assumed an elementary knowledge of valence theory—in particular, the shapes of atomic orbitals. However, whenever I have first used terms such as 'orthogonal' and 'normalized' I have given a reminder of their definition.

The content of the book is largely determined by that group theory which should be of utility to an undergraduate although there are a few points at which I have included something simply because I find the topic fun and hope the reader does too! There is no discussion of space groups because for these it is the symmetry operations, and not the associated group theory, with which most students will be concerned. There is no section on the formation of hybrid orbitals because this would not have been in keeping with the presentation of the rather different approach to chemical bonding which forms a major theme in the book. However, this neglect should not be seen as reflecting on the concept of hybrid orbitals (which, in fact, have a secure basis—see D. B. Cook and P. W. Fowler, *Amer. J. Phys.*, **49** (1981), 857), and which appear at several points in the text.

Chapters 1 to 7 of the book contain basic material whereas Chapters 9, 10 and 11 generally cover more advanced topics. Chapter 8 forms a bridge between these two sections. While I hope that the presentation in the last three chapters is one which the average reader will readily, follow, I have taken the opportunity to include aspects which are omitted in many textbooks. Here, too, I have used a chemically relevant topic as a vehicle to introduce some new aspect of group theory.

Those who wish to approach some of the material in this book from a related but different viewpoint may find it helpful to listen to two audiotapes, called 'Symmetry in Chemistry', which I have recorded for the Royal Society of Chemistry, London.

I am grateful to those institutions at which I have been able to write or revise parts of the manuscript—the University of Massachusetts (Amherst), Northwestern University, The University of South Paris and Turin University.

In writing this book I have been helped by a—usually—tolerant family and by the constructive criticism of many colleagues at the University of East Anglia, by Drs H. Fritzer and A. Hutcheon in particular as well as several anonymous reviewers. To Mrs J. Johnson and Mrs M. Livock I owe a debt of gratitude for producing excellent typescripts from sometimes near illegible manuscripts. Defects and errors that remain are, of course, my own responsibility.

Sidney F. A. Kettle
Norwich, April 1984

Preface to the Second Edition

The major difference between the second and first editions of this book is the addition of two chapters dealing with space groups. These were explicitly excluded from the first edition but the increase in the importance of this area of chemistry made its continued exclusion untenable. In introducing it I have followed the principle which guided the pattern of contents of the first edition—that of the 'need to know'. That only facts, concepts and procedures which are of immediate relevance and utility are to be included. It was this principle that delayed the discussion of the allocation of a molecule to the correct point group until Chapter 7 (because it was only then that all of the necessary symmetry operations had been introduced into the discussion). Similarly, in the treatment of space groups, the Hermann–Mauguin notation has been introduced quite separately from the notation used to denote the space groups themselves. To my knowledge, the approach to space groups in these chapters is not to be found in any other chemistry textbook, although it would be familiar to solid state physicists. This approach, combined with an attack on the problem of 'why are there 230 space groups?' offers, I believe, unique insights into space groups, their construction and utilization. Comparison with the first edition will show a myriad of small changes to the other chapters; more questions have been included and a section on 'Further Reading' added. As with the first edition, my hope is that this book will serve to bridge the gap between the qualitative discussions of symmetry found in many general texts and the rather mathematical treatments contained in more specialist books. Above all, I hope that it will make the simplifications and insights afforded by symmetry accessible to a wider range of students and that their confidence in, and the pleasure that they gain, from chemistry will increase as a result. I am indebted to Dr M. Schlegel-Zawadzka for help with the proofreading.

S.F.A.K.

1
Theories in Conflict

1.1 INTRODUCTION

As its title implies, this book is concerned with the symmetry and structure of molecules. Of these, the latter—both in the sense of the geometric and of the electronic structure of molecules—has long been of concern to chemists. We shall be interested in both these aspects and will adopt the viewpoint that the geometric structure of a molecule tells us something about its electronic structure. The connection between the two will be provided by the molecular symmetry, or rather its expression in what is called group theory. Ultimately, however, this book is concerned with the chemical consequences of molecular symmetry, the application of group theory to molecules, and these extend far beyond the problems of chemical bonding. Rather, the problem of chemical bonding will be used as a particularly convenient—and important—way of introducing the concepts of symmetry and then extend to other areas of chemistry the application of the concepts revealed in this way. In an introductory text such as this there will be no attempt to cover all of the uses of symmetry—an objective which it would be difficult to achieve in any text. Rather, some of the more important aspects will be detailed. The aim will be to provide a cover of the basics of the subject sufficient to enable the reader to apply them in other areas. Further, this will be done in a readable, almost entirely non-mathematical manner. Rather detailed but, hopefully still readable, mathematical treatments are reserved for the appendices.

1.2 THE AMMONIA MOLECULE

The ammonia molecule provides a convenient starting point for our study and it will be used to see the problem of chemical bonding in a rather unusual perspective, one that leads to the approach indicated above—the attempt to infer molecular bonding *from* molecular geometry (in contrast to the more common procedure of explaining molecular geometry in terms of chemical bonding). Several approaches to the bonding in the ammonia molecule will first be reviewed, approaches which have been in the chemical literature for many years. The reader may well not be familiar with all of them but should not spend too much time trying to master any new ones—our concern is with generalities, not details. However, references are given to enable the reader to explore any of the approaches in more detail, should he or she so wish.

1.2.1   The atomic orbital model

This model has an historic importance—it is the only description to be found in many pre-1955 texts.[1] Before looking at it, the facts. The ammonia molecule is pyramidal in shape; all three hydrogen atoms are equivalent, the HNH bond angle being 107° (Figure 1.1). The simplest, the oldest, explanation of the

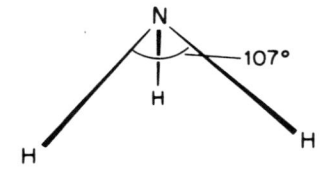

Figure 1.1   The ammonia molecule.

shape follows from the recognition that the ground state electronic configuration of an isolated nitrogen atom is $(1s)^2(2s)^2(2p)^3$, each of the 2p electrons occupying a different p orbital. Each of these 2p electrons may be paired with the electron present in the 1s orbital of a hydrogen atom by placing one hydrogen atom at one end of each 2p orbital so that each nitrogen 2p orbital overlaps with a hydrogen 1s orbital. The result is an ammonia molecule which has the correct, pyramidal shape and which has all of three hydrogen atoms equivalently bonded to the nitrogen (Figure 1.2). However, the angle between any pair of 2p orbitals is 90° so that a bond angle of 90° is predicted

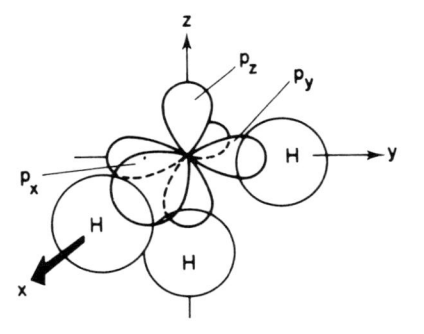

Figure 1.2   N–H bonding in NH_3 envisaged as resulting from the overlap of 2p orbitals of the nitrogen with 1s orbitals of the hydrogens. Because the three nitrogen 2p orbitals have their maximum amplitudes at 90° to each other, bond angles of this value are predicted.

by this model. Agreement with an experimental value of 107° is obtained by postulating the existence of electrostatic repulsion forces between the hydrogen atoms. These repulsions cause the H atoms to repel each other—so that the bond angles increase. If, as seems probable, each N–H bond is slightly polar with each hydrogen carrying a small positive charge, this repulsion is nuclear–nuclear in origin. The consequent modification of the original bonding scheme as a result of this distortion of the bond angle from 90° is not usually considered.

1.2.2 The hybrid orbital model

This is detailed in many post-1955 texts.[2] In this model an alternative description of the bonding in the ammonia molecule is obtained by hybridizing the valence shell orbitals of an isolated nitrogen atom, 2s, $2p_x$, $2p_y$ and $2p_z$ to give four, equivalent, sp^3 hybrid orbitals pointing towards the corners of a regular tetrahedron. Because there are five electrons in the valence shell of the nitrogen atom, three of these hybrid orbitals may be regarded as containing one electron whilst the fourth is occupied by two electrons. As in the previous model, 1s electrons from three hydrogen atoms pair with the unpaired electrons on the nitrogen, now in hybrid orbitals, to give a pyramidal ammonia molecule (Figure 1.3). Again, the three hydrogen atoms are equivalent but the bond angle is predicted to be 109.5°, the angle between the axes of a pair of sp^3 hybrid orbitals. This value is in closer agreement with experiment than that given by the previous model but again some correction is needed if the experimental value is to be reproduced. This correction is usually made by invoking the effects of electron–electron repulsion. It is this electron–electron repulsion which forms the basis of a third model for ammonia and so the way that the 'hybrid orbital' model is modified to give agreement with experiment is contained in the description of the next model.

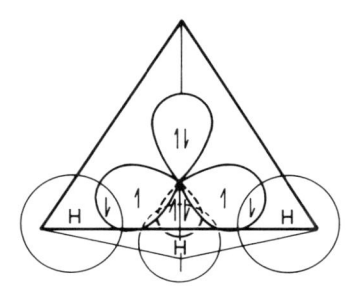

Figure 1.3 N–H bonding in NH_3 envisaged as resulting from the overlap of sp^3 hybrids of the nitrogen with 1s orbitals of the hydrogens. Because sp^3 hybrids have their maximum amplitudes at 109.5° to each other, bond angles of this value are predicted.

1.2.3   The electron-pair-repulsion model

This is the model described in many current texts.[3] The first two models which have been considered seek to explain the structure of the ammonia molecule in terms of the bonding interactions between the constituent atoms. The atoms adopt the arrangement which makes bonding a maximum. In contrast, the present and the next model to be discussed explain the structure not in terms of bonding interactions (although these must exist to hold the atoms together) but by electron repulsion. They recognize that electrons repel each other and regard the structure as being determined by the requirement that the inter-electron repulsion energies are minimized. The first of these models is originally due to Sidgwick and Powell, but was subject to subsequent extensive elaboration and refinement particularly by Nyholm and Gillespie.

In the ammonia molecule there are four electron pairs involving the valence shell of the nitrogen atom. These are the three N–H bonding electron pairs and a non-bonding pair (in the first of the models discussed above these non-bonding electrons were placed in the 2s orbital of the nitrogen; in the second they were placed in an sp^3 hybrid orbital). Because electrons repel each other these four electron pairs would be expected to be as far apart as possible consistent with still being bound to the nitrogen atom (three pairs are also bound to hydrogen atoms). It follows that the preferred orientation of these four electron pairs is that in which they point towards the corners of a regular tetrahedron. Remembering that three of the electron pairs are N–H bonding and that their orientation determines the positions of the hydrogen atoms, a HNH bond angle of 109.5° is predicted, the tetrahedral angle, the same as that given by the second model. It is thought-provoking to recognize that the same bond angle can be predicted either by including bonding interactions or by ignoring them! The refinement of the electron-pair-repulsion model requires the recognition that there are two sorts of electron pairs, those involved in N–H bonding and those which are non-bonding and located on the nitrogen atom. The electron pairs which comprise the N–H bonds are each subject to strong electrostatic attractions from two nuclei, the nitrogen nucleus and that of one of the hydrogen atoms. In contrast, the non-bonding electrons are strongly attracted by one nucleus only, that of nitrogen. It therefore seems reasonable to expect that the centre-of-gravity of the electron density in the N–H bonds will be located at a distance further away from the nitrogen nucleus than that of the lone pair electron density. The recognition of this difference at once leads to a refinement of the model. The accurately tetrahedral arrangement of four electron pairs resulted from the assumption that all the electron pairs were precisely equivalent. In the absence of such equivalence a regular tetrahedral arrangement cannot be expected. It seems reasonable that the repulsive forces occurring between electron density located in two N–H bonds will be less than the electrostatic repulsions between the non-bonding pair of electrons and a N–H bonding pair, simply because the distance between the centres of gravity of electron density will be greater in the former case. It would be expected that this difference in repulsion will lead the molecule to distort accordingly. The conclusion is that the HNH bond angle will be less than 109.5°. Although no quantitative prediction is possible with this simple model the qualitative prediction is in accord with experiment—the bond angle is 107°. These same arguments, applied to the 'hybrid orbital' model (Section 1.2.2), also lead to qualitative agreement with experiment.

1.2.4 The electron-spin-repulsion model

This is a little used model,[4] although it seems to be undergoing a minor resurgence. It differs from the preceding model principally in its recognition that electrons behave as individuals—and so repel each other as individuals—rather than as pairs. It is therefore more appropriate to consider eight electrons associated with the nitrogen atom, four with spin 'up' and four with spin 'down' than to think of there being four electron pairs (with no mention of spin). In the case of eight individual electrons the preferred orientation (in which the electrons are as well separated spatially as possible), would be expected to be one in which the electrons are located at the corners of a cube. A result of detailed quantum mechanics is the recognition that an additional repulsion exists between electrons of like spin, compared with the repulsion between electrons of unlike spin. So, it would be anticipated that an electron of given spin would have as its nearest neighbours at the corners of the cube electrons of the opposite spin. This means that in the cubic orientation of electrons there would be four electrons with spin 'up' defining one tetrahedron and four with spin 'down' defining another (if lines are drawn from one corner of a cube across the face diagonals to other corners and this procedure continued, just four corners are reached. These four corners define a regular tetrahedron. Another regular tetrahedron is defined by the four corners which remain—see Figure 1.4). So far in this model all of the electrons have been associated with the nitrogen atom and we have really been thinking of N^{3-}, with eight valence shell electrons. It follows that when the hydrogen atoms are introduced they must be introduced as bare protons. These protons attract the eight electrons. The attraction between a proton and an electron does not depend upon whether the electron has its spin 'up' or 'down', although, of course, the extra repulsion between electrons of the same spin persists. The net result is that each proton attracts to its locality just one electron with spin 'up' and one with spin 'down'. This attraction brings the two distinct tetrahedral arrangements of electrons into coincidence to give a single tetrahedral arrangement. The conclusion is that two electrons will be associated with each N–H bond and the remaining two will be non-bonding, just the same as we did for the previous model. Clearly, this model also predicts a bond angle of 109.5°, the tetrahedral value. It may be corrected in a manner similar to that described

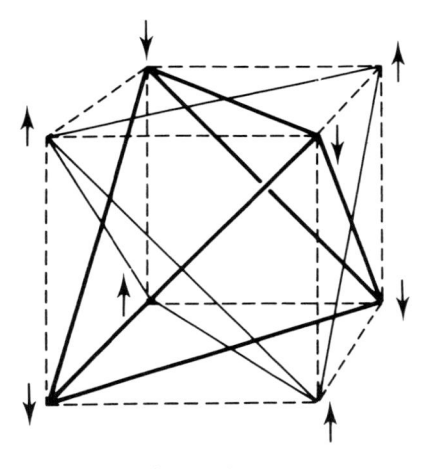

Figure 1.4   The two tetrahedra associated with a cube. Note the association that occurs in Linnett's model between these tetrahedra and the relative spins of the eight electrons placed at the corners of the cube.

above for the electron pair model to give qualitative agreement with experiment.

Although there is considerable overlap between the different models considered above, a survey of them does not lead to any definite conclusion regarding the relationship between the structure of and the bonding in the ammonia molecule. First, they are concerned with a relatively fine point— bond angles. They say nothing about the more important point (in terms of energy) of bond lengths. Second, all start with the supposition that only valence-shell electrons need be considered but then diverge in their explanations. These explanations are not totally distinct but what one model regards as the dominant factor another assumes to be relatively small. The first two models, effectively, say that the geometry is determined by the requirement that bonding interactions be maximized whilst the last two say that it is the consequence of the requirement that non-bonding repulsive forces be minimized. One point that they have in common, however, is the fact that none of them leads to a prediction that the ammonia molecule should be planar.

1.2.5   Accurate calculations

In 1970 Clementi and his co-workers published the results of some very accurate calculations on the ammonia molecule.[5] They were particularly interested in a study of the vibrational motion of the ammonia molecule in which it turns itself inside-out, like an umbrella in a high wind (Figure 1.5). Halfway between the two extremes of this umbrella motion the ammonia

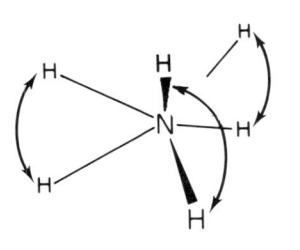

Figure 1.5 The 'umbrella' motion of the ammonia molecule. As the hydrogens move up so the nitrogen moves down (and vice versa) so that the centre of gravity of the molecule remains in the same place.

molecule is planar. The potential energy barrier for the inversion is equal to the difference in total energy between the ammonia molecule in its normal, pyramidal, shape and the planar configuration. In order to obtain a theoretical value for this barrier, Clementi carried out rather detailed calculations for each geometry. The results were very surprising. They showed that the N–H bonding is greater in the planar molecule—there is a loss of bonding of N–H bonding energy of approximately 7.0×10^2 kJ mole^{-1} (167 kcal mole^{-1}) in going from the planar to the pyramidal geometry; this loss is accompanied by a slight lengthening of the N–H bond. Bonding favours a planar ammonia molecule. A comparison of the most stable pyramidal and most stable planar geometries shows that the electron–electron and nuclear–nuclear repulsion energies favour the pyramidal molecule over the planar by about 7.2×10^2 kJ mole^{-1} (172 kcal mole^{-1}). Repulsive forces favour a pyramidal molecule. Note the way that the bonding and repulsive energy changes between the two shapes almost exactly cancel each other. It is the slight dominance of the repulsive forces by 20 kJ mole^{-1} (5 kcal mole^{-1}) which leads to the equilibrium geometry of the ammonia molecule in its electronic ground state being pyramidal.

We are left with a most disturbing situation. There is no doubt that the strongest N–H bonding in the ammonia molecule is to be found when it is planar yet two of the simple models considered earlier in this chapter explained its geometry by the assumption that this bonding is a maximum in the pyramidal molecule! Similarly, the models based on electron–electron repulsion ignored both the fact that nuclear–nuclear repulsion is of comparable importance and the fact that their sum is almost exactly cancelled by changes in the bonding energy. This would not matter so much if there were some assurance that repulsive energies would outweigh the bonding in all molecules (molecular geometries could then reliably be explained using a repulsion-based argument). Unfortunately, no such general assurance can be given. This can be seen if the discussion of the ammonia molecule is extended to include some related species.

The molecules NH_3, PH_3, NH_2F, PH_2F, NHF_2, PHF_2, NF_3 and PF_3 all have similar, pyramidal, structures and would be treated similarly in all simple models. But calculations by Schmiedekamp and co-workers[6] have shown that the first four owe their pyramidal geometry to the dominance of repulsive forces (bonding is stronger when they are planar) but the last four are pyramidal because the bonding is greatest in this configuration and dominates the repulsive forces (which now favour a planar arrangement)! Although this last sentence is marginally stronger than strictly permitted by the calculations, there is no doubt about the general conclusion. Although these eight compounds all have the same structure they do not all have it for the same reason, because of the close competition between repulsive and bonding forces. At present there are no rules to enable the prediction of which will win the competition in a particular case.

Although simple explanations of molecular shape such as those described earlier in this chapter are very useful to the chemist—and are widely and fruitfully used—they can be considered only as guides because they are not infallible. They are more *aides-mémoire* than correct explanations. It is for this reason, and because it happens to be particularly convenient for our purpose, that in this book the opposite strategy of using the experimentally determined shape of a molecule to infer details of the electronic structure of the molecule *in that shape* will be adopted. Few attempts will be made to explain why a molecule has a particular shape, although there will be many points at which the *consequences* of a particular geometry and its changes will become the focus of attention.

Problem 1.1   Consider each of the models for the structure of the bonding in the ammonia molecule detailed above and for each indicate the importance (if any) that it places on (a) electron–nuclear bonding forces, (b) electron–electron repulsion forces and (c) nuclear–nuclear repulsion forces.

Problem 1.2   Show that each of the models described in Sections 1.2.1 and 1.2.4 predicts that the water molecule is non-linear (the bond angle is actually 104.5°).

Problem 1.3   Hazard a guess at whether it is bonding or non-bonding forces which lead to NCl_3 having a pyramidal shape.[7]

NOTES AND REFERENCES

1. See for example, p. 65, *Inorganic Chemistry*, by Barry-Barnett and Wilson (Longmans Green, London, 1953).
2. See for example, p. 159 of *Valency and Molecular Structure* by Cartmell and Fowles (Butterworths, London, 1956).
3. The most recent account of the VSEPR model is to be found in *The VSEPR Model of Molecular Geometry* by R. J. Gillespie and I. Hargittai [Prentice-Hall (Allyn and Bacon), New York, 1991].

4. It is described by Linnett in *The Electronic Structure of Molecules* (Methuen, London, 1964).
5. A. Rauk, L. C. Allen and E. Clementi, *J. Chem. Phys.*, **52** (1970), 4133.
6. A. Schmiedekamp, S. Skaarup, P. Pulay and J. E. Boggs, *J. Chem. Phys.*, **66** (1977), 5769.
7. An answer will be found in a paper by K. Faegri and W. Kosmus in the *Journal of the Chemical Society (Faraday Transactions 2)*, **73** (1977), 1602 (be prepared for a surprise).

2

The Symmetry of the Water Molecule

An investigation into the consequences of molecular symmetry will begin in this chapter. In contrast to most texts, which develop the mechanics of group theory before its application, in this chapter and those following the two will go hand in hand. No ideas will be introduced until they are needed and immediately applicable.† This approach enables a non-mathematical discussion but means that aspects which appear early in most texts appear late in this. It is not until Chapter 7, for instance, that there is a general discussion of the symmetry of molecules, the allocation of the correct point group to a molecule. This is because it is not until then that the reader will have met all of the elements of symmetry needed. Following the discussion in the previous chapter it would be appropriate to develop our arguments with particular reference to the ammonia molecule. Unfortunately, this problem is, for the moment, too difficult and it will be deferred until Chapter 6. Instead, the water molecule will be considered in great detail. In so doing, an approach will be developed which will subsequently be extended to more complicated species (one of which is ammonia). With each increase in complexity of the molecule under study, so the power and scope of the group theoretical methodology available to the reader will increase.

2.1 SYMMETRY OPERATIONS AND SYMMETRY ELEMENTS

When we say that a molecule has high symmetry we usually mean that within the molecule there are several atoms which have equivalent positions in space. Thus, the tetrahedral symmetry of the methane molecule is manifest in the fact that the four hydrogen atoms are equivalent (Figure 2.1). Suppose that you have before you a model of the methane molecule which is so well constructed that no minor blemishes serve to distinguish one hydrogen atom from another. If you were to momentarily close and then open your eyes you would have no means of telling whether someone had rotated the model so that, although each hydrogen atom had been moved, the final position of the model was indistinguishable from its starting position. Such questions provide a convenient approach to symmetry and is the approach which will be followed in this book.

---

† It has been suggested that the book is written around a 'need to know' requirement.

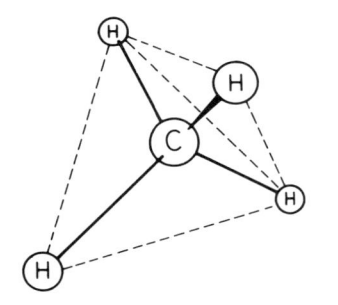

Figure 2.1   The methane molecule, shown in perspective. The important point is that all the hydrogens are equivalent.

The symmetry of a molecule is characterized by the fact that it is possible, hypothetically at least, to carry out *operations* which, whilst interchanging the positions of some (or all) of the atoms, give arrangements of atoms which are indistinguishable from the initial arrangement. Note the phrase 'hypothetically at least'. There is no requirement that all symmetry operations must always be physically possible in the way that a rotation is. As will be seen shortly, reflection in a mirror plane is an important symmetry operation. But if an atom lies in such a mirror plane then, presumably, a physical reflection in the mirror plane would mean one side of the atomic nucleus interchanging with the other and this is scarcely physically possible.† Of the operations that will be met in the following chapters only the rotation operations are physically possible. The others—such as the operation of reflection in a mirror plane or inversion in a centre of symmetry—are not. Those operations which cannot physically be carried out are called 'improper rotations' in contrast to the 'proper rotations' which are physically possible. A more precise definition of improper rotations will be given later (pp. 14 and 145). The distinction between physically possible and impossible operations is not important because our concern is with a mathematics, the mathematics of group theory and what is permissible in this mathematics, and not with ball-and-stick models and what is possible for them.

It is helpful to consider a particular example and in this chapter, the symmetry of the water molecule will be the subject of study. This molecule and symmetry have the advantage of simplicity, both geometrically and mathematically. Indeed, it is possible to gain familiarity with almost all of the important aspects of the application of group theory to molecules using it. Even in the later chapters of the book it will occasionally prove useful to return to the water molecule and take advantage of the simplicity which it offers. Our first task will be to obtain a list of those symmetry operations which turn the water molecule into a configuration indistinguishable from the initial one.

---

† The operation of time reversal—scarcely physically possible—is important in some aspects of theoretical chemistry, although it is not one which will be considered in this book. It has the effect (mathematically) of converting an electron with α spin into one with β.

The most evident symmetry operation which turns the water molecule into itself is the act of rotation by 180° about an axis which bisects the HOH angle and lies in the molecular plane. Figure 2.2 shows the water molecule before, in the middle of, and after completion of this operation. Apart from the arrows, which have been added for clarity, the first and third diagrams are indistinguishable. The effect of the operation is to interchange the two hydrogen atoms. We say that 'the two hydrogen atoms are symmetry-related' or 'they are symmetrically equivalent'. A rotation operation is denoted by the letter C (which may be conveniently thought of as derived from the symbol $\circlearrowleft$). Because it takes two successive rotations to return each atom to its original position, the rotation operation is called a twofold rotation operation and is denoted C_2, pronounced 'see two'. The same symbol, C_2, is used to denote the rotation operation and the axis about which the rotation occurs, although the distinction between the two is rather important. Some authors distinguish between an axis and the corresponding operation by writing the latter in bold type, thus C_2. The use of bold type is very useful if one is developing the

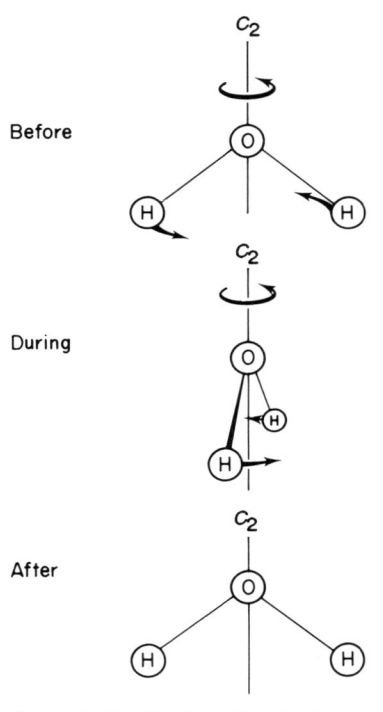

Figure 2.2   The conversion of the H_2O molecule into an arrangement which is indistinguishable from the original by a rotation of 180° ($360/2 \equiv C_2$). In general, it is not possible to give symmetry operations the sort of physical reality which is attempted here.

mathematics of symmetry theory, group theory. In the present book, however, it will always be clear from the context whether an axis or operation is being discussed and bold type will not be used (because its use tends to make the subject look more daunting than need be the case). Twofold rotation operations are not the only ones which can exist, threefold (C_3), fourfold (C_4), fivefold (C_5) and sixfold (C_6) rotation operations are quite common in chemistry; there are examples later in the book.

In defining rotation axes (and the corresponding operations) it is necessary to require that the rotation which is repeated several times in order to return each atom to its original position is always carried out in the same sense (clockwise or anticlockwise). For a C_n axis, where n rotations in the same sense are required to reproduce the starting arrangement, each operation involves a rotation of $360°/n$ about the C_n axis. The language that is used is to say that the larger the value of n, the higher the rotational symmetry of the axis.

The twofold rotation operation is not the only manifestation of symmetry in the water molecule. If the plane defining the water molecule were to be replaced by an infinitely thin mirror, as shown in Figure 2.3, then reflection in this mirror plane would have the effect of turning the water molecule into a configuration indistinguishable from the original one. This operation has the effect of turning the 'front' of the two hydrogen atoms and of the oxygen atom into the 'back' and vice versa. Mirror planes and the operation of reflection in them are both denoted by sigma — σ — (the operation sometimes being distinguished by bold type) and, just as for rotation axes, various subscripts are used. In the present case the subscript v is appended to give the symbol σ_v. This subscript arises because when, as is the convention, the axis of highest symmetry (C_2 in the present case) is arranged so as to be vertical, as in Figure 2.3, then the mirror plane is also vertical. The subscript v on σ is the initial letter of vertical. Thus, more jargon: a σ_v mirror plane is vertical with respect to the axis of highest symmetry (this axis always lies in the σ_v mirror plane).

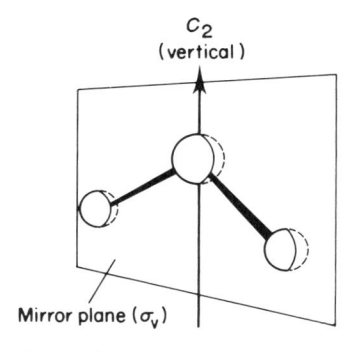

C_2
(vertical)

Mirror plane (σ_v)

Figure 2.3 The mirror plane of symmetry in the molecular plane of the H_2O molecule. The plane should be thought of as infinitely thin and serving to reflect one 'side' of the molecule into the other.

Other subscripts on σ which will be met are h (for horizontal) and d (for dihedral). They will be discussed in detail later in the book.

The C_2 and σ_v symmetry operations do not exhaust the symmetry possessed by the water molecule. Another feature of this symmetry is the existence of a second mirror plane. This mirror plane, which lies perpendicular to the molecular plane, is shown in Figure 2.4. Like the first, the second mirror plane contains the twofold axis (indeed, the line of intersection between the two mirror planes defines the twofold axis). It follows that, like the first mirror plane, the second is denoted σ_v. However, its effect on the molecule is quite different to that of the first—it has the effect of interchanging the two hydrogen atoms, for instance—and so it is necessary to distinguish between them. This is done by adding a prime to the symbol for the second mirror plane, thus: σ_v'. Had the water molecule possessed a third type of vertical mirror plane (which it does not!) then this would have been denoted σ_v'' and so on. Note that a σ_v is an *improper* symmetry operation—although its effect can be seen, it cannot physically be carried out.

In this section it has sometimes been found convenient to talk of rotation axes and mirror planes almost as if they were physical objects. Collectively, they are called *symmetry elements*. Examples of other symmetry elements— such as a centre of symmetry—will be met later in this book. Although in the preceding paragraph it was convenient to talk about a symmetry operation and the corresponding symmetry element almost at the same time, it cannot be emphasized too strongly that in this book our real concern is with *symmetry operations*. Symmetry elements are introduced simply to enable the corresponding operations to be more readily understood, notwithstanding the fact that symmetry elements appear to have more physical reality than do symmetry operations.

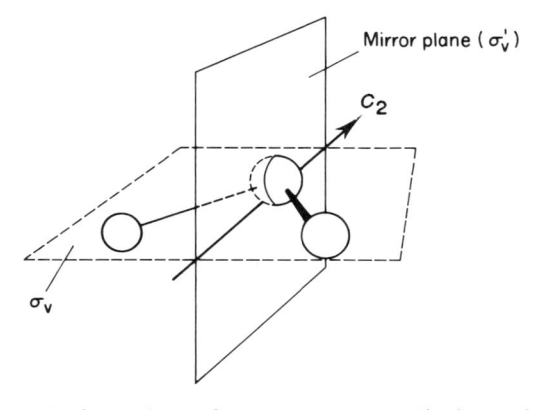

Figure 2.4 A second mirror plane of symmetry, perpendicular to the first, in the H_2O molecule. The line of intersection of the two mirror planes is the twofold rotation axis.

Problem 2.1   By a comparison of the terms 'symmetry element' and 'symmetry operation' as applied to threefold (C_3), fourfold (C_4), fivefold (C_5) and sixfold (C_6) rotations suggest why it is important to recognize the difference between these terms.

As can be shown by an abortive search, no other rotation axes or mirror planes exist in the water molecule, so that it would seem that the three symmetry operations which we have recognized define the symmetry of the water molecule. This, however, is not strictly so. We have already seen that the application of the C_2 operation twice over regenerates the original molecule, with each atom restored to precisely its original position. It is easy to see that the same is true of the σ_v and σ_v' operations. That is, the end result of carrying out any of these symmetry operations twice is the same as that of leaving the molecule alone. The implication of this is that we should formally recognize the possibility that one way of turning a molecule into a configuration indistinguishable from the original is simply to leave the molecule alone. This, so called, *identity operation* will be denoted by the letter E (some books use I). No matter how much or how little symmetry a molecule possesses the identity operation always exists for it.† It is the set of four symmetry operations E, C_2, σ_v and σ_v' which completely defines the symmetry of the water molecule. So, one way of talking about the symmetry of the water molecule would be to give this list. However, rather than give a complete listing of symmetry operations (which for some high symmetry molecules could be rather tedious) this information is compressed into a shorthand symbol which for the set of operations of the water molecule is C_{2v} (pronounced 'see two vee'). One talks of the water molecule as 'having C_{2v} symmetry' or we talk of 'the symmetry operations of the C_{2v} point group' (by which is meant E, C_2, σ_v and σ_v').

The last phrase contained two new words, 'point' and 'group'. The word 'group' arises from the fact that the set of operations satisfy all of the requirements of mathematical group theory. These are covered in detail in Appendix 1. Here, it is sufficient to give an important and relevant example. Apply any two of the symmetry operations to the water molecule one after the other. The result is always equivalent to the effect of applying just *one* of the operations of the group (which may be different from the two that were used). Thus, as will be seen in detail later, for the case of the water molecule, following the σ_v operation by C_2 gives the same result as the application of the σ_v'. Indeed, this combination method is sometimes a useful method of making sure that all of the symmetry operations of a particular molecule have been found. There is a limit to the process, however. Eventually all of the symmetry operations that turn a particular molecule into itself will have been obtained. The successive application of two members of the set of operations will always

---

† It is sometimes convenient to regard the identity operation as a C_1 rotation, i.e. a rotation of 360°. This interpretation highlights the fact that although there is an infinite number of choices of C_1 axis there is only one distinct $C_1(E)$ *operation*.

produce a result which is equivalent to the application of another member of the set. Sets which are closed in this fashion are called *groups*. Our interest is in *groups* of symmetry operations (although there are many other types of group). For any group there has to be a specified method of combining the group elements—for symmetry operations it is applying them one after another. Other types of groups may have very different methods of combination. The complete, formal, definition of a 'group' requires some mathematics and is reserved for Appendix 1 but it may help to give two more examples. Consider the three numbers 1, 0, −1. Do these form a group under the operations of addition or subtraction? While it is clear that all three numbers can be interrelated by these operations, it is equally clear that when the operation (+1) is applied to the number 1 the number 2 is generated. Similarly, (−1) applied to the number −1 gives the number −2. Clearly, 1, 0, −1 do not comprise the entire group because 2 and −2 have to be included. In similar fashion it can be seen that 3, −3 and, indeed, all integers between ∞ and −∞ (plus and minus infinity) have to be included. This is an example of an infinite group. Groups such as this are of importance in the description of the translational symmetry found in crystal lattices, a topic which will be dealt with in Chapter 12 although the detailed group theory is not included in that chapter.

As a second example consider the rectangular table shown in Figure 2.5. The top of the table has been divided into quarters and two of these are coloured black and two white. Were there no such colouration, the table would have the same symmetry as the water molecule, as shown in Figure 2.6. However, the presence of the coloured sections means that the σ_v and σ_v' operations are no longer symmetry operations unless they are each combined with quite a new type of symmetry operation, that of changing colour, black into white and white into black. If these (reflection and colour change) operations are labelled (σ_v) and (σ_v'), then the operations E, C_2, (σ_v) and (σ_v') form a group.†

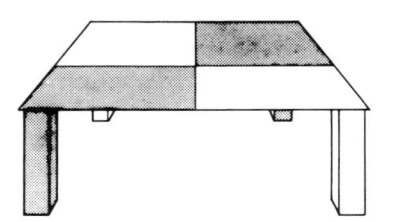

Figure 2.5 A table showing black and white colour-change symmetry. The legs of the table reduce the symmetry so that it is not necessary to compare the top surface of the table with the bottom.

---

† This may seem a contrived example and perhaps it is. However, it is not too far from real applications in chemistry. Suppose we have a molecule which contains an unpaired electron. What is the symmetry relationship between a molecule with spin 'up' and one with spin 'down'? One has to invent the operation of 'spin change', similar to that of 'colour change'. Of course, the footnote on page 11 provides an alternative approach.

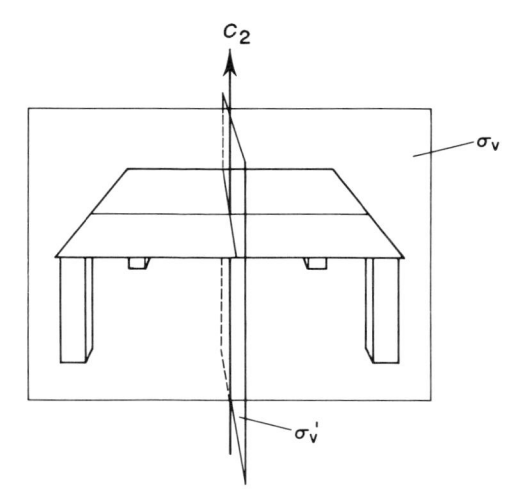

Figure 2.6   The C_{2v} symmetry of the table of Figure 2.5 when uncoloured.

As has been mentioned, a geometrical feature corresponding to a symmetry operation is called a symmetry *element*. Thus, corresponding to a rotation operation is a rotation axis; corresponding to a reflection operation is a mirror plane. Rotation axes and mirror planes (and also other similar things, such as a centre of symmetry) are examples of symmetry elements. For all molecules it is true that all the symmetry elements which they possess pass through a common point in the molecule (in the case of the C_{2v} point group, perhaps confusingly, they pass through an infinite number of common points along the C_2 axis). This is the reason that all such groups (of operations) are called *point groups*. Put another way, there is always at least one point which is left invariant (unchanged) by all of the operations of a point group. This point may or may not be a point at which an atom is located.

Problem 2.2   Explain carefully what is meant by each noun and each adjective in the phrase 'the symmetry operations of the C_{2v} point group'.

2.2   MULTIPLIERS ASSOCIATED WITH SYMMETRY OPERATIONS

From the way that they have been defined above, it is evident that the effect of each of the symmetry operations of the C_{2v} point group when applied to the water molecule, considered as a whole, is to turn the molecule into itself. An alternative way of putting this is to say that the effect of each of the symmetry operations on the molecule is equivalent to multiplication by the number 1. That is, the effect of each of the operations can be represented as shown in the table below.

Symmetry operation	Effect of the operation on the water molecule (considered as a whole)
E	1
C_2	1
σ_v	1
σ_v'	1

The apparently pointless exercise of representing by the number 1 the effects of the behaviour of the water molecule under the symmetry operations begins to acquire some significance when we ask whether all quantities associated with the water molecule are, like the water molecule itself, turned into themselves by the operations of the C_{2v} point group? It will be seen that they are not. Consider, for example, the oxygen $2p_y$ orbital shown in Figure 2.7. In Figure 2.8 are pictured the effects of the symmetry operations of the C_{2v} point group on this orbital. It is evident that, whilst the identity and σ_v operations have the effects of regenerating the original orbital, the C_2 and σ_v' operations have the effect of reversing the phases of the lobes. It follows that while the result of the application of the E and σ_v operations may be represented by the multiplication factor 1, the effects of the C_2 and σ_v' operations have to be represented by the multiplicative factor -1. That is, the association between symmetry operations and multiplicative factors is:

Symmetry operation	Effect on the oxygen $2p_y$ orbital
E	1
C_2	-1
σ_v	1
σ_v'	-1

Having obtained one such set of numbers the question at once arises of how many such sets can be found. Would any combination of 1 and -1 be acceptable? The answer is 'no'. To explore this further, consider the effect of the symmetry operations of the C_{2v} point group on the $2p_x$ and $2p_z$ orbitals of the oxygen. The $2p_x$ orbital is shown in Figure 2.9, where the fact that the positive lobe is located above the plane of the page and negative lobe beneath this plane is indicated by the perspective of the diagram. In order to avoid completely obscuring the negative lobe behind the positive, the water molecule

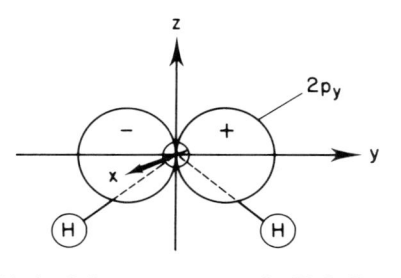

Figure 2.7 The $2p_y$ orbital of the oxygen atom in H_2O. By convention, the y axis is taken to lie in the plane of a planar molecule.

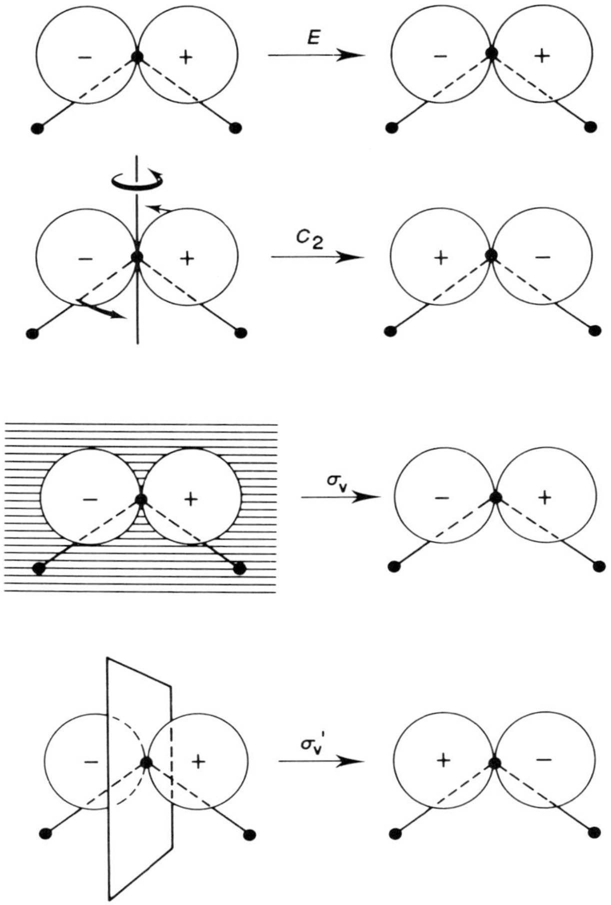

Figure 2.8 The effects of the symmetry operations of the C_{2v} point group on the oxygen $2p_y$ orbital in the water molecule. The point of importance is the relative phases of the orbital 'before' (left) and 'after' (right).

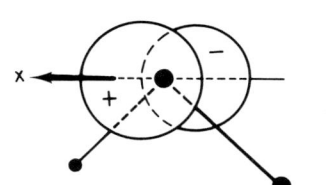

Figure 2.9 The $2p_x$ orbital of the oxygen atom in H_2O. By convention, the x axis is taken to be perpendicular to the plane of a planar molecule.

is viewed from a slightly skew position. Figure 2.10 shows the effects of the four symmetry operations of the C_{2v} point group on the oxygen $2p_x$ orbital. It is evident that, whilst the application of the E and σ_v operations result in the phases of the lobes of the orbital being unchanged, the application of the C_2 and σ_v' operations leads to a reversal of these phases. In this case the numbers representing the effects of the symmetry operations are shown below.

Symmetry operation	Effect on the oxygen $2p_x$
E	1
C_2	-1
σ_v	-1
σ_v'	1

A third set! Before proceeding, a note of warning is necessary. It is a generally accepted convention that the axis of highest rotational symmetry in a molecule (C_2 in the case of the water molecule) is called the z axis. Although the direction of the z axis is therefore uniquely specified for most molecules by this convention it is seldom true that the same can be said for the x and y axes. In this book we are following a convention which has been suggested by Mulliken but is not always followed—that of requiring that a planar molecule lies in the yz plane. So, the reader may find that, in the case of the water molecule, what we have called the x axis some authors will call the y (so that the zx plane, rather than the yz, is the molecular plane). Had the x and y axes been interchanged then, of course, the sets of numbers to which they give rise in the above discussion would also be interchanged.

We now return to the problem of the symmetry properties of the orbitals of the oxygen atom and consider the $2p_z$ orbital. This orbital is shown in Figure 2.11 and its behaviour under the symmetry operations of the group in Figure 2.12. It is evident from this latter figure that, although the symmetry operations may have the effect of turning one side of the orbital into the other, this change is always accompanied by the retention of the *phase* of the lobes of the orbital so that the number representing the effect of each operation is 1 as shown below.

Symmetry operation	Effect of the oxygen $2p_z$ orbital
E	1
C_2	1
σ_v	1
σ_v'	1

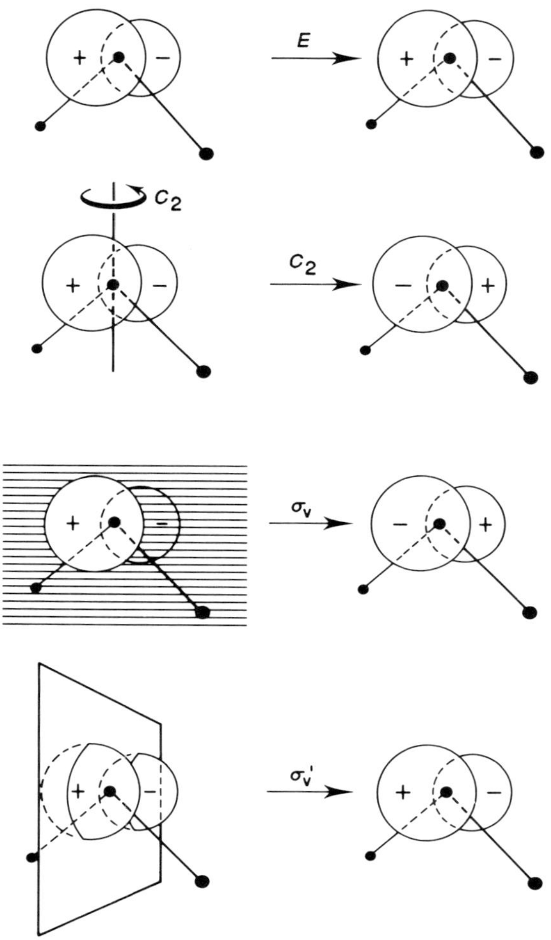

Figure 2.10 The effects of the symmetry operations of the C_{2v} point group on the oxygen $2p_x$ orbital in the water molecule. The point of importance is the relative phases of the orbital 'before' (left) and 'after' (right).

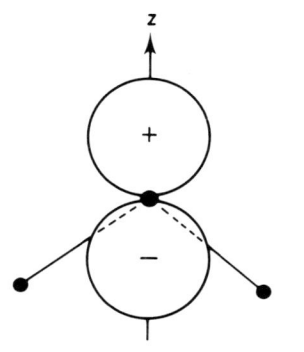

Figure 2.11 The $2p_z$ orbital of oxygen in H_2. By convention, the z axis is taken to lie along the axis of highest rotational symmetry of a molecule (there are departures from this rule for molecules of very high symmetry).

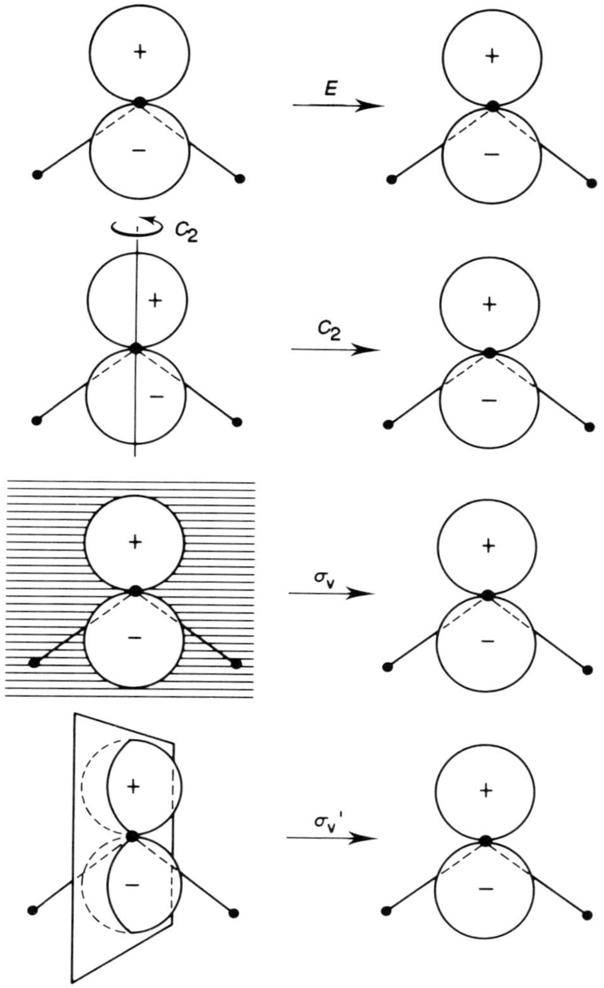

$$\xrightarrow{E}$$

$$\xrightarrow{C_2}$$

$$\xrightarrow{\sigma_v}$$

$$\xrightarrow{\sigma_v'}$$

C_2

Figure 2.12 The effects of the symmetry operations of the C_{2v} point group on the oxygen $2p_z$ orbital in the water molecule. The point of importance is the relative phases of the orbital 'before' (left) and 'after' (right).

This set of numbers if the same as that obtained earlier as a description of the symmetry properties of the whole molecule. The conclusion is that although it is possible for quantities associated with the water molecule to give rise to the same set of numbers as the molecule itself, other alternatives are possible (such as those found for the $2p_y$ and $2p_x$ oxygen orbitals).

We now come to the key point in the argument which is being developed. This is that the differing symmetry properties of, for example, the $2p_y$, $2p_x$ and $2p_z$ orbitals of the oxygen atom in the water molecule (i.e. the fact that their symmetry operations differs), may be *represented* by the sets of numbers which have been obtained. Quantities which have different symmetry properties give rise to different sets of numbers. Evidently, the next question which has to be considered is whether the sets of numbers which have already been generated comprise a complete list of the different types of symmetry behaviour which may be shown by quantities (such as atomic orbitals) associated with the water molecule. The answer is 'no'; there is just one further type of symmetry behaviour (i.e. set of numbers) which have yet to be obtained. Consider the symmetry properties of the $3d_{xy}$ orbital of the oxygen atom. Although this orbital is not commonly included in elementary discussions of the electronic structure of the oxygen-containing compounds (because it is not a valence-shell orbital) it does none the less exist and would be included in most sophisticated calculations of the electronic structure of such molecules. It is shown in Figure 2.13. Note that where the product of coordinate axes xy is

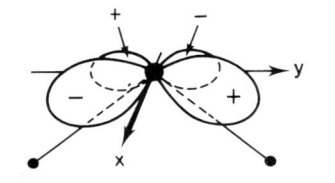

Figure 2.13   The $3d_{xy}$ orbital of oxygen in H_2O. Note that the phases of the lobes of the orbital are those of the product xy.

positive, the phase of the $3d_{xy}$ orbital is also positive (this matching of phases is implicit in the use of the xy subscript). The effects of the symmetry operations of the C_{2v} point group on this orbital are shown in Figure 2.14. This figure shows that the effect of the identity (E) and of the C_2 operations is to regenerate the original orbital with unchanged phases. In the case of the σ_v and σ_v' operations, however, the phase of each lobe of the orbital is reversed. The appropriate multiplicative factors representing the effects of the operations are therefore as shown at the top of the next page.

Four sets of numbers have now been generated. They are collected together in Table 2.1, although the order in which they are presented is not that in which they were obtained.

Against each row of numbers is shown the orbital which leads to its generation, the (O)s indicating that the orbitals considered were those of the

Symmetry operation	Effect on the oxygen $3d_{xy}$ orbital
E	1
C_2	1
σ_v	-1
σ_v'	-1

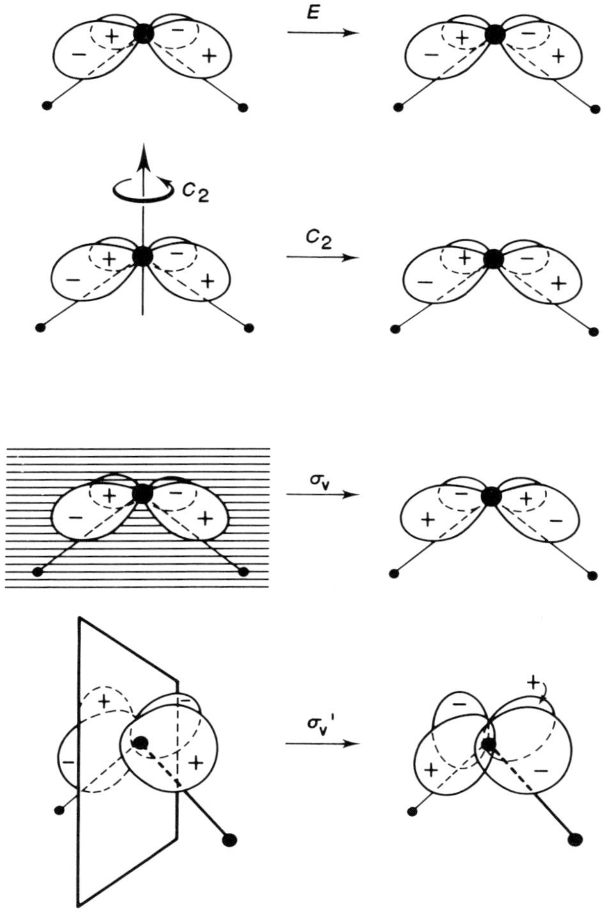

Figure 2.14 The effects of the symmetry operations of the C_{2v} point group on the oxygen $3d_{xy}$ orbital in the water molecule. The point of importance is the relative phases of the orbital 'before' (left) and 'after' (right).

Table 2.1

E	C_2	σ_v	σ_v'	
1	1	1	1	$2p_z(O)$
1	1	-1	-1	$3d_{xy}(O)$
1	-1	1	-1	$2p_y(O)$
1	-1	-1	1	$2p_x(O)$

oxygen atom. The surprising—and important—thing is that it is impossible to find an atomic orbital of the oxygen atom which will generate a set of numbers other than one of those given in this table. A test of this assertion can be obtained by considering the transformation of the 2s and the other 3d orbitals of the oxygen ($3d_{z^2}$, $3d_{x^2-y^2}$, $3d_{zx}$ and $3d_{yz}$). Besides providing a partial proof of the assertion, this exercise will provide the reader with experience which will prove invaluable as the discussion develops.

Problem 2.3 Show that $2s(O)$, $3d_{z^2}(O)$ and $3d_{x^2-y^2}(O)$ individually generate the same set of numbers as $2p_z(O)$; $3d_{zx}(O)$ as $2p_x(O)$ and $3d_{yz}(O)$ as $2p_y(O)$. In each case use the coordinate axis set shown in Figure 2.7 and operation labels indicated in Figure 2.4 (unless one is consistent, unexpected results can be obtained, as explained in the text).

The assertion that one cannot add to the table of numbers given above suggests that this set of numbers has properties beyond those which might be expected from the way in which they were derived. This is so. Indeed, sets of such numbers will provide the basis for the discussion contained in almost all of the remainder of this book. As an illustration of the unexpected properties of these numbers, we make what might appear to be a digression to discuss in more detail the effects of applying two of the symmetry operations of the water molecule in succession. In fact, it is a discussion which touches at the fundamentals of the subject—although this would only become evident after a thorough study of Appendix 1.

2.3 GROUP MULTIPLICATION TABLES

Earlier in this chapter it was asserted that the effect of the successive application of symmetry operations of a group was always equivalent to the effect of some single operation of the group. This will now be investigated in detail for the C_{2v} point group by considering, in turn, each operation and the effect of following it with each of the four symmetry operations of the group considered in turn. It will be helpful to focus attention on a particular molecule. The water molecule is inconvenient for this particular purpose (because of the apparent equivalence of the effects of applying different symmetry operations, a phenomenon encountered several times already in this chapter) and instead the ethylene oxide, $(CH_2)_2O$, molecule will be considered as an example. This molecule is shown in Figure 2.15. In this figure the hydrogen atoms have been labelled with the suffixes a, b, c or d—so that in order to study the effects of the symmetry operations on this molecule all that has to be done is to see how these labels are rearranged. The effects of the operations of the C_{2v} point group on these labels are shown in Figure 2.16, a figure that should be studied carefully until the reader is fully conversant with it. It will be noted that each symmetry operation gives rise to a different final arrangement of labels—a feature which would not have been found with H_2O.

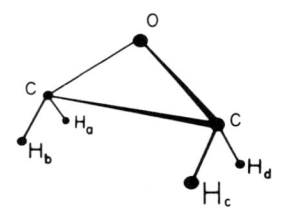

Figure 2.15   The ethylene oxide molecule C_2H_4O.

Because the identity operation does not change the distribution of the labels at all, it is evident that any operation preceded or followed by the identity operation gives rise to the same final arrangement as that operation on its own. It can immediately be concluded that:

$$E \text{ followed by } E \equiv E$$
$$E \text{ followed by } C_2 \equiv C_2$$
$$E \text{ followed by } \sigma_v \equiv \sigma_v$$
$$E \text{ followed by } \sigma_v' \equiv \sigma_v'$$
$$C_2 \text{ followed by } E \equiv C_2$$
$$\sigma_v \text{ followed by } E \equiv \sigma_v$$
$$\sigma_v' \text{ followed by } E \equiv \sigma_v'$$

Less trivial is the result of the successive application of pairs of operations from the set C_2, σ_v and σ_v'. It has been pointed out earlier in this chapter that any one of these operations followed by itself gives rise to the initial arrangement and so the sequence is equivalent to the identity operation. That is,

$$C_2 \text{ followed by } C_2 \equiv E$$
$$\sigma_v \text{ followed by } \sigma_v \equiv E$$
$$\sigma_v' \text{ followed by } \sigma_v' \equiv E$$

The remaining combinations of operations are illustrated in Figure 2.17 and the reader may, by comparison with Figure 2.16, determine which single operation

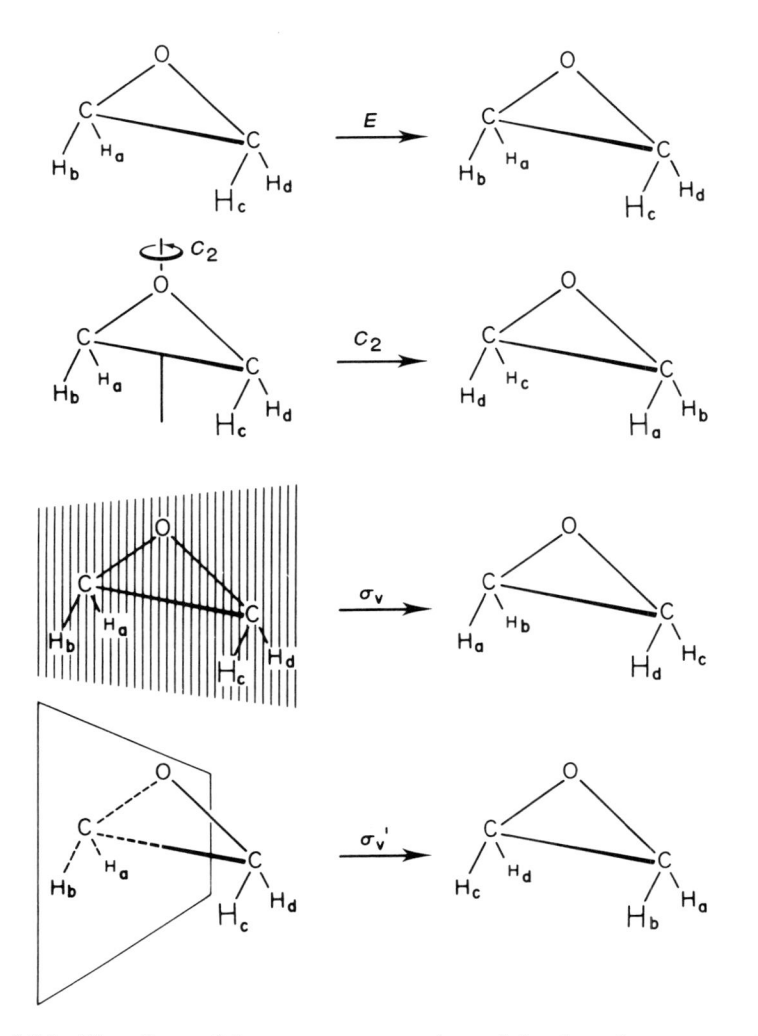

Figure 2.16 The effects of the symmetry operations of the C_{2v} point group on the four hydrogen atoms of ethylene oxide. The point of importance is the relative pattern of the hydrogen atoms 'before' (left) and 'after' (right).

is equivalent to each combination. The conclusion is that

$$C_2 \text{ followed by } \sigma_v \equiv \sigma_v'$$
$$C_2 \text{ followed by } \sigma_v' \equiv \sigma_v$$
$$\sigma_v \text{ followed by } C_2 \equiv \sigma_v'$$
$$\sigma_v \text{ followed by } \sigma_v' \equiv C_2$$
$$\sigma_v' \text{ followed by } C_2 \equiv \sigma_v$$
$$\sigma_v' \text{ followed by } \sigma_v \equiv C_2$$

These results are collected together in Table 2.2.

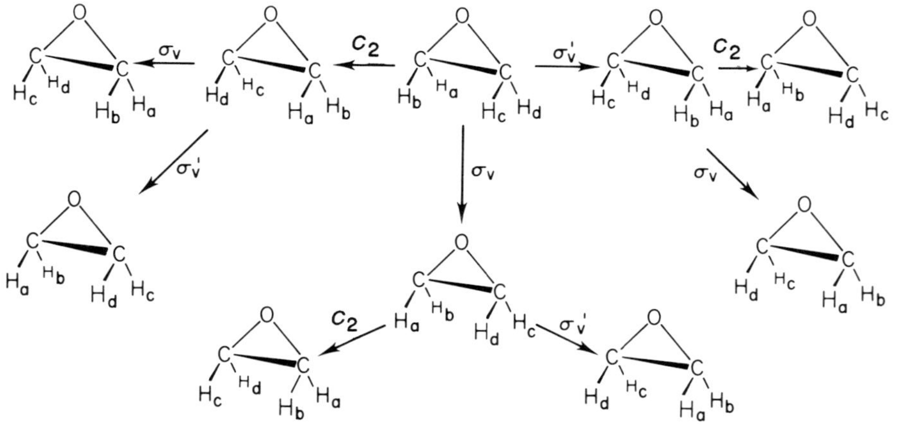

Figure 2.17 The effects of two successive operations of the C_{2v} point group on the four hydrogen atoms of ethylene oxide. The single operation which corresponds to each combination of operations shown here may be determined by comparison with the patterns shown on the right-hand side of Figure 2.16.

Problem 2.4 Check the entries in Table 2.2.

It is usual in the mathematical theory of groups to refer to the law of combination of group elements as 'multiplication' (although only rarely does the operation have anything to do with ordinary arithmetical or algebraic multiplication). So, in the present case, where two symmetry operations combine by being applied in succession, they are said to 'multiply'. Thus we say 'C_2 multiplied by σ_v is equal to σ_v''. Table 2.2 is therefore referred to as the *multiplication table* for the operations of the C_{2v} point group.†

Table 2.2

C_{2v}		First operation			
		E	C_2	σ_v	σ_v'
	E	E	C_2	σ_v	σ_v'
Second	C_2	C_2	E	σ_v'	σ_v
operation	σ_v	σ_v	σ_v'	E	C_2
	σ_v'	σ_v'	σ_v	C_2	E

---

† In Table 2.2 the results of the multiplication are independent of the order in which the operations are applied—the table is symmetric about its leading diagonal. This is not a general property of multiplication tables but it is one that makes that in Table 2.2 a particularly simple case to start with.

Problem 2.5   Construct a multiplication table for the operations E, C_2, (σ_v) and (σ_v') of the coloured table (Figure 2.5 and page 17) and thus demonstrate that the operations form a group (those wishing to pursue this a little further will find that the discussion towards the end of Section A.1.2 of Appendix 1 has some relevance).

Problem 2.6   By forming the group multiplication table show that 1, i, $-i$, -1 form a group under the operation of multiplying them together in the usual meaning of the word 'multiplication'. The numbers 1 and -1 are ordinary numbers and $i = \sqrt{-1}$ (it is perhaps more useful to note that $i^2 = -1$ and $-i^2 = 1$).

Problem 2.7   Before using ethylene oxide as the example for Section 2.3 the author considered urea, $CO(NH_2)_2$, as an alternative. He decided against it because although urea could have a structure with four hydrogen atoms arranged similarly to the pattern shown in Figure 2.15, in the crystal the molecule is planar, with two pairs of equivalent hydrogen atoms. Show that in both of these arrangements, planar and non-planar, urea has C_{2v} symmetry.

2.4   CHARACTER TABLES

The reason for the digression in the previous section was to illustrate some of the properties of the sets of numbers contained in Table 2.1. This is done by combining Tables 2.1 and 2.2 in the following way. Choose any row of Table 2.1, say, the second. Abstracting this row from the table, the association between symmetry operations and numbers is that shown below

$$
\begin{array}{cccc}
E & C_2 & \sigma_v & \sigma_v' \\
1 & 1 & -1 & -1
\end{array}
$$

Turning now to Table 2.2, everywhere in this table that the operation E is listed, replace it by the number with which it is associated in the chosen row of Table 2.1. That is, it is replaced by the number 1. Similarly, wherever C_2 appears in Table 2.2 it is replaced by 1, whilst both σ_v and σ_v' are replaced by -1. When these replacements have been made, Table 2.3 is obtained:

Table 2.3

	1	1	-1	-1
1	1	1	-1	-1
1	1	1	-1	-1
-1	-1	-1	1	1
-1	-1	-1	1	1

The interesting—and important—thing about this table is that, if it is looked upon simply as a table in which numbers multiply each other, arithmetically, then the products are all correct. It is left to the reader to demonstrate that this statement is true no matter which row of numbers is selected from Table 2.1. Indeed, it is possible to approach the topic of the symmetry of the water molecule by first obtaining Table 2.2 and then inviting the student to attempt to find as many sets of numbers as possible that give an arithmetically correct table when substituted consistently into Table 2.2. Apart from the trivial set in which each operation is represented by the number 0, only those sets of numbers contained in Table 2.1 will be found to substitute correctly. This is a result that could not have been anticipated from the way that the sets of numbers were obtained. This is the first hint of the fundamental nature of the set of numbers of Table 2.1; more will be met in the next chapter.†

Problem 2.8   Show, by an abortive search for alternatives, that only the sets of numbers contained in Table 2.1 substitute into Table 2.2 to give an arithmetically correct multiplication table (it is recommended that the reader decide in advance the number of sets that they are prepared to consider).

Because of the close relationship between the multiplication of the operations of the C_{2v} point group (given in Table 2.2) and the multiplication of the numbers in the rows of Table 2.3, each set of numbers may be regarded as representing (i.e. behaving in an analogous way to) the set of symmetry operations.‡ We shall speak of each row of Table 2.1 as being a *representation* of the symmetry operations. Further, we shall call them 'irreducible representations' (the significance of the word 'irreducible' will not become evident until the next chapter, when the concept of a reducible representation will be introduced). In the discussion that follows it will often be necessary to refer to the individual rows in Table 2.1 and it is convenient to circumvent the need to write each one out in full by giving each a label. The labels commonly used are those shown in Table 2.4.

Thus, the set of numbers given at the beginning of this section, $(1\ 1\ -1\ -1)$, would be referred to as 'the A_2 irreducible representation of the C_{2v} point group'. This sounds rather awkward when first encountered but it is the sort of phrase which occurs over and over again in the subject. Because the association between the symmetry operations and irreducible representations given in

---

† This discussion explains why only the numbers 1 and -1 appear in Table 2.1. Had the number 2 been associated with the E operation, for instance, then the product of multiplying E with E, in the group theoretical sense, to give the answer E—and so 2 by substitution—would not be arithmetically correct (arithmetic would call for the number 4).

‡ Note that the word 'multiplication' in this sentence does not have quite the same meaning when applied to symmetry operations as it does when it refers to numbers.

Table 2.4

C_{2v}	E	C_2	σ_v	σ_v'	
A_1	1	1	1	1	$2p_z(O)$
A_2	1	1	-1	-1	$3d_{xy}(O)$
B_1	1	-1	1	-1	$2p_y(O)$
B_2	1	-1	-1	1	$2p_x(O)$

Table 2.4 is unique to the C_{2v} point group, this is indicated by including the group label in the top left-hand corner of the table.

There is some system about the choice of the labels A_1, A_2, B_1 and B_2 in Table 2.4. The A's are distinguished from the B's by the fact that they have numbers of $+1$ for the C_2 operation whereas the B's have -1 (in the general case, A's have characters of $+1$ for rotation about the axis of highest symmetry while B's have a character of -1). A_1, by convention, is the so-called 'totally symmetric' irreducible representation and has $+1$ for all of its numbers. It is called totally symmetric because *all* the operations of the group turn something of A_1 symmetry into itself. Every group has a totally symmetric irreducible representation. Although it may not be labelled A_1 it is always the first A listed (it could be something like A_g or A' for instance).

The system distinguishes A's from B's and A_1 from A_2. This is really the end, although the distinction can be extended to the B's by noting that irreducible representations with the suffix 1 are symmetric (character $+1$) under the σ_v operation whereas those with suffix 2 are antisymmetric (character -1). However, in the case of the B's this distinction is marred by the fact that the distinction between σ_v and σ_v' is somewhat arbitrary—interchange the use of these labels and the labels B_1 and B_2 would have to change too. In practice this means that it is advisable to check the notation used by each author—one worker's notation may not be the same as the next. As long as one is consistent in the notation used for a particular problem there is no ambiguity about the final answer obtained. Similar considerations apply to many of the groups commonly used in chemistry.

Just as the set of operations, $(E, C_2, \sigma_v, \sigma_v')$ may be represented by any of the irreducible representations A_1, A_2, B_1, B_2, so, too, individual symmetry operations, such as C_2, are characterized, in each irreducible representation by a particular number (which, in general, varies from one irreducible representation to the next). These individual numbers are termed *characters* and tables such as Table 2.4 are called *character tables*. As has already been indicated, character tables are of prime importance for the topics discussed in this book. The unexpected properties of the sets of numbers in Table 2.1 become the unexpected properties of character tables. It is these 'unexpected properties' (which are actually fundamental and far from accidental) which are at the heart of their value in chemistry. In the next chapter, when the bonding in the water molecule is discussed, the existence, and to some extent the origin, of these properties will become clear. Because of their importance, this chapter concludes with some further comments on character tables in general and that of the C_{2v} point group in particular.

On the right-hand side of Table 2.4 the oxygen orbital is indicated which was used to generate a particular irreducible representation. Functions which have the property of generating an irreducible representation are commonly listed alongside character tables in this way. Such functions are called 'basis functions'. It has been seen that the transformations of the oxygen $2p_y$ orbital under the operations of the C_{2v} point group lead to the B_1 set of characters—the oxygen $2p_y$ orbital is a basis function for the generation of the B_1 characters. This would normally be said a little more formally: 'the oxygen $2p_y$ orbital is a basis for the B_1 irreducible representation of the C_{2v} point group'. Alternatively, and more simply, 'the oxygen $2p_y$ orbital has B_1 symmetry in the C_{2v} point group'.†

Problem 2.9 Compare the multiplication table obtained in answer to Problem 2.5 with Table 2.2. Thus, or otherwise, construct the character table for the group of Problem 2.5.

Problem 2.10 Repeat Problem 2.5 using the group of Problem 2.3. In this case it will probably be necessary to use the 'or otherwise' method and it may only be possible to generate an incomplete character table. The complete character table has the same set of characters as one that will be met in Chapter 11 (Table 11.1).

Problem 2.11 Show that the dipole moment of the water molecule forms a basis for the A_1 irreducible representation of the C_{2v} group. (*Hint*: the dipole moment must, by symmetry, be directed along the z axis.)

Problem 2.12 Show that the translation of the entire water molecule along the y axis forms a basis for the B_1 irreducible representation of the C_{2v} point group. *Hint*: represent the translation as an arrow and consider the transformations of this arrow and the direction in which it points. The y direction is given in Figure 2.7.

---

† There is a subtle point here. When the *symmetry* is the focus of attention, an upper case (capital) letters is used. However, when the *orbital* is the focus of attention it is denoted by a lower case symbol: thus, 'the b_1 orbital'. In this book the use of lower case symbols will largely be confined to diagrams.

Problem 2.13 Repeat Problem 2.12 but now consider a translation along the x axis and show that it transforms as B_2.

One final word, one which is not important at a first reading but which is included to help the reader understand the logic behind the sequence of the chapters in this book and to explain a word that will be used from time to time. The C_{2v} point group is an *Abelian* point group. Abelian groups have multiplication tables which are symmetric about their leading diagonal (top left to bottom right)—inspection of Table 2.2 shows that this is true for the C_{2v} group. That is, the result of multiplying two operations is independent of the order in which they are multiplied—of which operation comes first and which comes second. It is this, together with the fact that each operation multiplied by itself gives the identity, that makes the C_{2v} group a particularly simple one to work with. An alternative (but equivalent) definition of an Abelian point group is to regard such point groups as those for which the character tables contain only numbers like 1 and -1.† The character tables of Abelian groups never contain numbers such as 3, -3, 2, -2 and 0. The reason why at the beginning of this chapter consideration of the ammonia molecule was deferred is that the character table of its point group contains the numbers 2 and 0 as well as 1 and -1. As will be seen in Chapter 8 (Table 8.2), this originates in the fact that the result of multiplying some of the elements of its group *does* depend on the order in which they are taken.

2.5 SUMMARY‡

In a molecule the axis of highest symmetry is conventionally chosen to be the z axis; recommendations for the choice of x and y exist (p. 20). The concern of this book is with point group symmetry operations (pp. 11,14), which are named according to a conventional nomenclature (pp. 12, 13). These operations form a group (p. 15 and Appendix 1). In the present (and the next two) chapters the discussion is restricted to Abelian point groups (p. 33). In such groups, individual quantities—such as atomic orbitals on a central atom (pp. 18 *et seq.*, 31)—that are transformed into themselves under the operations of the point group may have these transformations described by characters (p. 31). (An example of an Abelian group which shows a more complicated behaviour will be met in Chapter 11.) A complete collection of characters is called a character table (p. 31). Each row of characters is called an irreducible representation (p. 30); each of the individual quantities used to generate them

---

† In Chapter 11 it will be seen that they can also contain complex numbers, such as i and $-i$, which are such that some power of them equals 1 (thus $i^4 = 1$, for instance).
‡ Page numbers refer to the page in the chapter on which a full discussion commences. Sometimes in the summaries words are used in a way that should be evident from the context but which will be discussed in detail in later chapters, e.g. 'isomorphism'.

are said to be the *basis* for the irreducible representation that it generates (p. 32). Characters multiply together in a way that is isomorphous (p. 30) to the way that the operations of the point group multiply (p. 28 but see Appendix 2). Irreducible representations are given labels in a systematic, but not always unambiguous, way (p. 31).

3

The Electronic Structure of the Water Molecule

In Chapter 2 it was shown that it is possible to obtain sets of numbers (characters)—called irreducible representations—by a study of the transformation properties of the atomic orbitals of the oxygen atom in the water molecule; the atomic orbitals served as bases for the generation of irreducible representations. Atomic orbitals are not the only things which may serve as bases. So, in Chapter 2 some problems were concerned with the transformational properties of other quantities such as the dipole moment of the water molecule and it was seen that these serve as bases. In the following chapters a variety of bases will be met; for instance, when studying the vibrations of a molecule the small displacements of individual atoms will be used as bases. Sometimes, the set of numbers—the representation generated by the transformation properties of a basis set—appear in the character table. This is when an irreducible representation is generated. More commonly, however, the representation generated does not appear in the character table. In such cases the representation is a *reducible* one. One of the representations encountered in the present chapter is a *reducible representation*; by studying it, a method of breaking up a reducible representation into a sum of irreducible representations will be obtained. However, to be able to do this it is necessary to recognize more of the special properties of the irreducible representations than those met in Chapter 2. Again, these will be developed with reference to the character table of the C_{2v} point group.

3.1 THE ORTHONORMAL PROPERTIES OF IRREDUCIBLE REPRESENTATIONS

As indicated in Chapter 2, the sets of characters in the C_{2v} character table have properties beyond those which might reasonably be expected from the way that they were derived. One set of these properties proves to be of great importance. Consider any irreducible representations of the C_{2v} point group (Table 3.1) and multiply its individual characters by the corresponding characters of any other irreducible representation. Then sum the products of characters which have been obtained. So, consider as an example the A_2 and B_1 irreducible representations. The sum of the products of characters is:

$$[1 \times 1] + [1 \times (-1)] + [(-1) \times 1] + [(-1) \times (-1)] = 0$$

Table 3.1

C_{2v}	E	C_2	σ_v	σ_v'	
A_1	1	1	1	1	$2s(O), 2p_z(O), \psi(A_1)$
A_2	1	1	−1	−1	
B_1	1	−1	1	−1	$2p_y(O), \psi(B_1)$
B_2	1	−1	−1	1	$2p_x(O)$

In this case, and for *all* others in which the characters of two *different* irreducible representations of the C_{2v} point group are multiplied together, the sum is zero. If, however, instead of multiplying the characters of two different irreducible representations, the characters of an irreducible representation are squared and the answers summed, then a different result is obtained. For the B_2 irreducible representation:

$$[1 \times 1] + [(-1) \times (-1)] + [(-1) \times (-1)] + [1 \times 1] = 4$$

The sum of products is equal to four. The same answer would have been obtained no matter which of the irreducible reducible representations had been chosen. Four is also the number of operations in the C_{2v} point group. This is no accidental coincidence. As mentioned in Chapter 2, every character table contains as its first row a series of 1's, the characters of the totally symmetric irreducible representation. If these are squared and added then the result simply counts the number of operations in the group. So, for this irreducible representation at least (and for any other which has characters of either $+1$ or -1), the answer will always be the number of operations in the group, irrespective of how many operations there are. Because the number of operations in a group turns out to be an important quantity, it is given a name—it is called the *order* of the group. Thus, 'the C_{2v} point group is of order four'.

If, instead of choosing a row of the character table for the calculations of the above paragraph, the columns had been selected a similar result would have been obtained. The sum of the products of the characters in two columns is equal to zero when the characters come from two different columns. If the *same* column is chosen, i.e. the characters squared, then the sum of squares is equal to the order of the group. These results are known as the character table *orthonormality* relationships; a more general form of them will be discussed in Chapter 5 where it will be shown that they may be used to derive character tables as an alternative to the procedure used in Chapter 2. It is in large measure the existence of these relationships which enable symmetry considerations to simplify many problems in the physical sciences. They will be used frequently in this book. The word 'orthonormal' is a composite of the words 'orthogonal' and 'normal' and embodies both. Orthogonal here means independent. When two things are orthogonal it means that one behaves—and can be discussed—

without automatically requiring a change to the other. Thus, all the wavefunctions associated with an atom are orthogonal to each other. In the present case, we can talk of different irreducible representations quite independently of each other. Normal or normalized means 'weighted equally'—and equal weighting usually means being given unit weight. This concept is most easily seen for two one-electron wavefunctions of an atom. Each wavefunction is normalized if, when we (mathematically) ask the question 'How many electrons does each wavefunction describe?' we obtain (mathematically) the answer '1'. If we obtained the answer '1' for the first wavefunction but some different answer, say '1.83' for the second we would say that the second was not normalized and we would have to modify it with a multiplicative scale factor so that we did, indeed, get the answer '1'. This scaled wavefunction would also then be said to be normalized. Later in this chapter we shall be effectively normalizing irreducible representations when we divide the number 4 (obtained by simple arithmetic) by the order of the C_{2v} point group (the total number of operations in the group), which is also 4, to give the number 1. As implied above, the orthonormality relationships are best expressed mathematically and this is done in Appendix 2.

Problem 3.1   Check that each of the irreducible representations of Table 3.1 is orthonormal.

3.2   THE TRANSFORMATION PROPERTIES OF ATOMIC ORBITALS IN THE WATER MOLECULE

In this chapter it will be shown that C_{2v} character table may be used to greatly simplify a discussion of the bonding in the water molecule. As usual, this bonding will be treated as arising from the interaction of orbitals located on the oxygen atom with those on the two hydrogen atoms. For simplicity, the discussion will largely be confined to the valence shell atomic orbitals of these atoms. That is, we shall consider the oxygen 2s, $2p_z$, $2p_x$ and $2p_y$ orbitals together with the two hydrogen 1s orbitals. The transformation properties of the oxygen orbitals have already been discussed (Section 2.2 and Problem 2.3) and symmetry labels placed on them. The results are summarized on the right-hand side of Table 3.1; the hydrogen 1s orbitals will be discussed shortly. It will prove convenient to use phrases like 'orbitals of A_1 symmetry, by which, in the present example, is meant the 2s and $2p_z$ orbitals of the oxygen together with any orbitals of this symmetry which may subsequently be discovered (one arises from the hydrogen 1s orbitals). In a similar way, the $2p_y$ orbital of the oxygen will be referred to as 'an orbital of B_1 symmetry', by which is meant that the characters of the B_1 irreducible representation describe its transformations under the operations of the C_{2v} point group.

So far in this book only the transformational properties of individual orbitals have been considered. However, it could happen that it becomes necessary to consider all three 2p orbitals of the oxygen atom together, as a set. What if the z axis had been chosen to lie in some arbitrary direction in the molecule rather than along the twofold axis (and, similarly, no symmetry constraints had been placed on the x and y axes)? A $2p_z$ orbital pointing in an arbitrary direction would not be turned neatly into itself by all of the symmetry operations of the group. The behaviour of the $2p_x$ and $2p_y$ would similarly be complicated. In fact, any one of them, after being rotated or reflected, would have to be described as a linear combination, a mixture, of all three of the starting orbitals. We would have had to treat the orbitals as a set. Evidently, a careful choice of direction for coordinate axes can simplify symmetry discussions!† Had we persisted in choosing arbitrary (but, of course, mutually perpendicular) directions for our axes the final result would have been the same—we would have ended up with 2p orbitals transforming as A_1, B_1 and B_2. However, the work involved would have been more difficult and it will not be until Appendix 2 has been read that the reader will be equipped to prove the assertion that has just been made.

When we turn to the two hydrogen 1s orbitals in the water molecule and attempt to place symmetry labels on them we are confronted with a similar problem. Should they be considered as individuals or as a pair? The answer to this question is simple (and covers the case of oxygen 2p orbitals oriented along arbitrary axes). Whenever one (or more) operations of the point group has the effect of interchanging or mixing orbitals (or, as sometimes happens, a little of both) then all of the orbitals which are scrambled must be considered together, as a set. This statement applies not just to atomic orbitals; it also holds for other quantities. For instance, in Chapter 9 it will be seen that the small atomic displacements used in the study of molecular vibrations are often scrambled by symmetry operations and these displacements have to be treated as a set. We return now to the specific problem of the transformation of the two hydrogen 1s orbitals in the water molecule. In Figure 3.1 the behaviour of these two orbitals (which will be denoted h_1 and h_2) under the symmetry operations of the C_{2v} point group is shown. For the C_2 and σ'_v operations the two hydrogen 1s orbitals interchange but under E and σ_v each remains itself. Something which remains unchanged under an operation gives rise to a character of 1 (the numbers which were introduced as multiplicative factors in Chapter 2 will now be referred to as characters). So, when two things remain unchanged it is both reasonable and correct to conclude that each makes a contribution of 1 to the character. An aggregate character of 2 is obtained. For both the E and σ_v symmetry operations characters of 2 are therefore obtained. However, for the C_2 and σ'_v operations a situation is encountered which has not previously been met, because

---

† Throughout this book the author will be making educated choices of coordinate axes which simplify the subsequent discussion. The reader may find it amusing to try to catch him making these simplifications. It is also a helpful exercise.

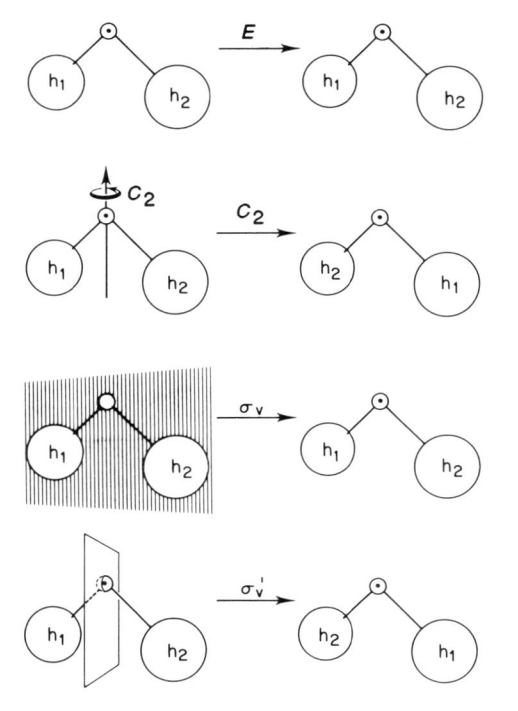

Figure 3.1   The behaviour of the two hydrogen 1s orbitals of H_2O under the symmetry operations of the C_{2v} point group. The point of interest is the permutation of the labels h_1 and h_2.

When the transformation of several things is being considered together the character which they together generate under a symmetry operation is the sum of the characters which they generate as individuals.

the orbitals h_1 and h_2 interchange under these operations. The fact that h_1, for instance, *disappears* from its original position has to be described. The only evident way of doing this is by using a multiplicative factor (character) of zero. The same is true for h_2. We can generalize this result as shown below.

Symmetry operations which lead to all of the members of a set interchanging with each other give rise to a resultant character of zero. If a symmetry operation results in some members interchanging while others remain in the same position, it is only the latter which makes non-zero contributions to the character.

This discussion contains no explicit recognition of the fact that h_1 and h_2 *interchange* under the C_2 and σ_v' operations—only of their disappearance from their original positions. The correct way of describing the situation is by matrix algebra and is described in detail in Appendix 2. The present discussion avoids the use of matrix algebra and so it is not able to provide a description of the fact that h_1 and h_2 interchange. The more detailed treatment given in Appendix 2 contains a proof of the two important rules given in the boxes above and shows that the transformation of h_1 and h_2 under C_2 and σ_v leads to each contributing zero to the aggregate character.†

The set of characters which has just been obtained is:

E	C_2	σ_v	σ_v'
2	0	2	0

This set has properties which are rather different to those of corresponding sets which we obtained in Chapter 2 (and which we called irreducible representations). For instance, when these characters are substituted for the corresponding symmetry operations in the group multiplication table (Table 2.2), the multiplication table obtained is not arithmetically correct.‡

Problem 3.2 Substitute characters for the corresponding operations in Table 2.2 using the correspondence

E	C_2	σ_v	σ_v'
2	0	2	0

and check that the table obtained is not arithmetically correct.

Problem 3.3 Show that the 1s orbitals of the four hydrogen atoms of the ethylene oxide molecule discussed in Section 2.3 and, in particular, Figures 2.15, 2.16 and 2.17, form a basis for the following representation of the C_{2v} point group

E	C_2	σ_v	σ_v'
4	0	0	0

3.3 A REDUCIBLE REPRESENTATION

Although the set of characters generated at the end of the previous section is not identical to any of the irreducible representation of the C_{2v} point group, it *is* equal to a sum of two of them. If the corresponding characters of the A_1 and

---

† The zeros encountered in the text are there shown to be the diagonal elements of the appropriate transformation matrix.
‡ It is very significant that when the transformation matrices of Appendix 2 are substituted, rather than characters, then the multiplication table obtained *is* correct provided that the rules of matrix multiplication are applied to the matrices.

B_1 irreducible representations are added together, the same set of characters is generated as those obtained using h_1 and h_2 as a basis:

	E	C_2	σ_v	σ_v'
A_1	1	1	1	1
B_1	1	-1	1	-1
$A_1 + B_1$	2	0	2	0

That is, the representation which was generated by the transformations of h_1 and h_2 can be decomposed into a sum of irreducible representations. A representation which can be reduced to a sum of other representations is, reasonably enough, called a *reducible representation*. The use of the name 'irreducible representation' for the representations appearing in the character table should now be clear. These representations cannot be reduced further, they are irreducible. There are similarities between reducible representations and irreducible representations, but there are also important differences.† A most important connection between reducible and irreducible representations is found in the orthonormality relationships (Section 3.1). These relationships provide a systematic way of reducing a reducible representation into its irreducible components. These relationships were introduced by multiplying the characters of two different irreducible representations together. What if, instead, one of the selected representations were reducible and only one were irreducible? Multiply the individual characters of the reducible representation generated by the transformation of h_1 and h_2 by the corresponding characters of one of the irreducible representations of the C_{2v} character table and sum the products. Choose, for example, the A_2 irreducible representation:

	E	C_2	σ_v	σ_v'
	2	0	2	0
A_2	1	1	-1	-1
	2	$+0$	-2	$+0 = 0$

For the B_2 irreducible representations the answer zero would also have been obtained. For the A_1 and B_1 irreducible representations, however, non-zero answers result. For example, for the B_1:

	E	C_2	σ_v	σ_v'
	2	0	2	0
B_1	1	-1	1	-1
	2	$+0$	$+2$	$+0 = 4$

---

† One of these has already been seen—for the C_{2v} group the characters of a reducible representation do not multiply arithmetically to give a multiplication table in which there is a consistent correspondence between the numbers it contains and the operations in the corresponding group multiplication table.

It is not difficult to understand why A_2 and B_1 give different results. We know that the reducible representation is a sum of the A_1 and B_1 irreducible representations. So, in the above procedure we were forming products of A_2 and of B_1 with $(A_1 + B_1)$. That is, in the latter, the B_1 case, we were forming products between B_1 and A_1 and between B_1 and B_1 simultaneously. But the first of these gives a sum which is equal to zero, whilst the second gives a sum of four—as was seen in Section 3.1. That is, non-zero answers are obtained by the above procedure when from the character table there is selected an irreducible representation which is contained in the reducible. It is not surprising that an answer of 0 was obtained in the A_2 case, because A_2 is not contained in the reducible representation. This recognition leads to a general method for reducing reducible representations into their irreducible components. This is the method that has just been used for the B_1 reducible representation but, because it is so important, it is worthwhile to repeat it in detail, applied to the A_1 case.

The steps involved are:
Write down the reducible representation

E	C_2	σ_v	σ_v'
2	0	2	0

Write down the characters of the selected (here, A_1) irreducible representation

1	1	1	1

Multiply the characters in the same column

2	0	2	0

Add these products together and then divide the sum by the order of the group

$$4/4 = 1$$

We conclude that our reducible representation contains the A_1 irreducible representation. Had it contained A_1 twice the final answer would have been 2—and so on.

Problem 3.4 Use the method described above to reduce the following reducible representations of the C_{2v} point group.

	E	C_2	σ_v	σ_v'
(a)	2	0	-2	0
(b)	3	-1	-1	-1
(c)	3	1	-1	1
(d)	3	1	1	-1

Problem 3.5 Reduce the following reducible representations of the C_{2v} point group and for each check your answer by adding together the characters of the irreducible representations to regenerate those given below (there is an aspect of the irreducible representations in this

problem which distinguishes them from those in Problem 3.4 and which makes this check worthwhile).

	E	C_2	σ_v	σ_v'
(a)	3	-1	-3	1
(b)	4	0	0	-4
(c)	6	-4	-2	0
(d)	9	1	1	1
(e)	11	1	-3	-1

3.4   SYMMETRY-ADAPTED COMBINATIONS

What is the significance of the fact that the reducible representation generated by the transformation of the two hydrogen 1s orbitals h_1 and h_2 may be reduced into a sum of A_1 and B_1 irreducible representations? As has been seen, an irreducible representation such as A_1 describes the transformation properties of a *single* orbital as too does the B_1 irreducible representation.† The significance therefore has to be that it is possible to derive from the orbitals h_1 and h_2 *one* orbital, the transformations of which are described by the A_1 irreducible representation and a second orbital which transforms as B_1. Evidently, the next step is to investigate the form of these orbitals—to find out what they look like. There is a systematic method of carrying out this task but it will not be introduced until Chapter 4 (when it can be given a wider applicability than is possible here). For the present example a rather simpler argument will suffice.

The reader will recall that in a discussion of the electronic structure of the hydrogen molecule, H_2, two hydrogen 1s orbitals combine to give bonding and antibonding combinations. If, hypothetically, the oxygen atom is removed from a water molecule a hydrogen molecule is left, albeit with a rather stretched H–H bond. It would be reasonable to expect that the combinations of hydrogen 1s orbitals in this stretched H_2 molecule would be related to the correct combinations of hydrogen 1s orbitals in the water molecule. With neglect of overlap between the two atomic orbitals, the bonding and antibonding combinations of hydrogen 1s orbitals in the H_2 molecule have the form

$$\psi(\text{bonding}) = \frac{1}{\sqrt{2}}(h_1 + h_2)$$

$$\psi(\text{antibonding}) = \frac{1}{\sqrt{2}}(h_1 - h_2)$$

where the same labels for the hydrogen atomic orbitals have been used as in the water molecule. Consider the transformation of these bonding and antibonding

---

† This is perhaps best seen in the context of the contents of the first box in Section 3.2, applied to the identity operation, E. The character under E simply counts the number of objects under consideration.

combinations under the operations of the C_{2v} point group. The transformations of the bonding combination are shown in Figure 3.2 where, to emphasize the fact that it is a single orbital which is drawn, the component hydrogen 1s orbitals have been joined together with a thick line. Clearly, under all of the operations of the C_{2v} point group this orbital is transformed into itself. That is, the combination $1/\sqrt{2}(h_1 + h_2)$ is of A_1 symmetry in the C_{2v} point group.

The transformation properties of the antibonding combination of hydrogen 1s orbitals in the hydrogen molecule are shown in Figure 3.3. In this figure, again to emphasize that it shows the transformations of a single orbital, the two parts of the orbital are linked. The characters generated by the action of the operations of the C_{2v} point group on this orbital are:

$$\begin{array}{cccc} E & C_2 & \sigma_v & \sigma_v' \\ 1 & -1 & 1 & -1 \end{array}$$

so that, as expected, it is of B_1 symmetry.

The bonding and antibonding 1s molecular orbitals of the hydrogen molecule have the symmetries A_1 and B_1, respectively, in the C_{2v} point group. At the beginning of this section it was shown that there are A_1 and B_1 combinations of hydrogen 1s orbitals in the water molecule but it was not possible to say what they looked like. Clearly, the molecular orbitals of H_2

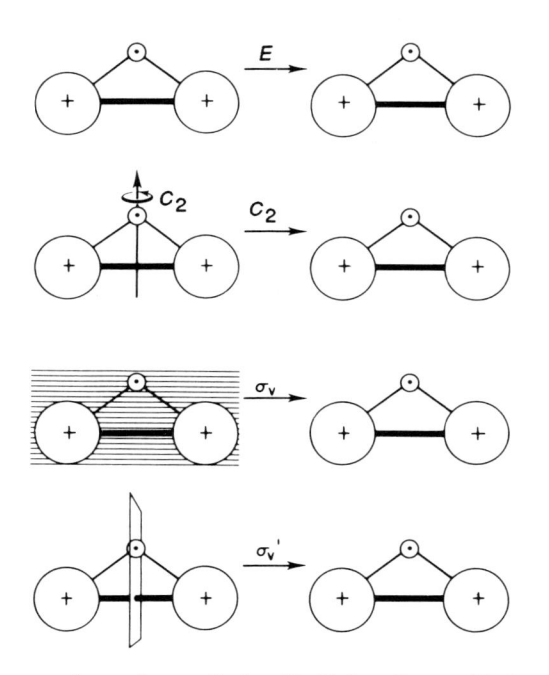

Figure 3.2 The transformations of the H–H bonding orbital of H_2 under the symmetry operations of the C_{2v} point group.

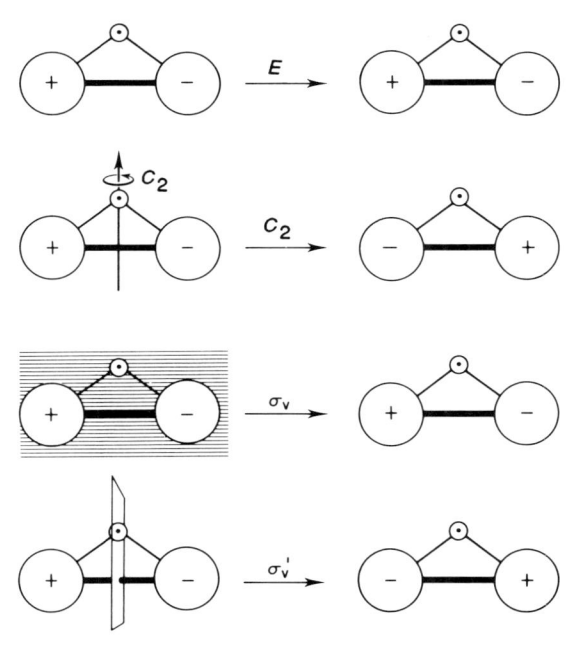

Figure 3.3 The transformations of the H–H antibonding orbital of H_2 under the symmetry operations of the C_{2v} point group. The point of interest is a comparison of the phases of this orbital 'before' (left) and 'after'.

provide the answer. In the water molecule, rather than treating the two hydrogen 1s orbitals separately, it is necessary to work with two combinations, one of A_1 symmetry and one of B_1. The A_1 combination is:

$$\psi(A_1) = \frac{1}{\sqrt{2}}\,(h_1 + h_2)$$

and the B_1 combination is:

$$\psi(B_1) = \frac{1}{\sqrt{2}}\,(h_1 - h_2)$$

The argument used to obtain $\psi(A_1)$ and $\psi(B_1)$ was based on plausibility rather than mathematical rigour. As indicated above, a more rigorous method will be developed in Chapter 4.

3.5   THE BONDING INTERACTIONS IN H₂O AND THEIR ANGULAR DEPENDENCE

The two linear combinations of hydrogen 1s orbitals in the water molecule which transform as the A_1 and B_1 irreducible representations have now been obtained. Although the mathematical form of these orbitals is one which neglects overlap between h_1 and h_2, this neglect in no way affects their symmetry species.

We now come to a vital point in our argument. It involves as the key step an assertion which, for the moment, the reader is asked to take to some extent on trust. A proof will be given in Chapter 10 although a partial justification is included here. The assertion is that:

Interactions between orbitals transforming as different irreducible representations are always zero.†

That is, in a discussion of the bonding in a molecule the argument can be broken up into smaller, separate, discussions, one for each irreducible representation. This is an enormous simplification; the more the molecular symmetry, the greater its value. In the case of the water molecule, for example, the only orbital of B_2 symmetry is the $2p_x$ orbital of the oxygen. There is no hydrogen 1s combination of this symmetry and so the oxygen $2p_x$ orbital does not interact with any other orbital in the molecule. That is, it is a non-bonding orbital located on the oxygen atom. This conclusion required virtually no work to obtain, yet it gives us chemically useful information on one of the orbitals of the water molecule. Symmetry arguments are useful! Incidentally, as a little thought about their transformation properties should confirm, when two (or more) orbitals of the same symmetry species interact, the final molecular orbitals are all of the same symmetry species as the initial orbitals.

Figure 3.4 provides some justification for the assertion that consideration of bonding interactions can be confined to those between orbitals of the same symmetry species. It shows the overlap between the 2s orbital of the oxygen—of A_1 symmetry—and the B_1 combination of hydrogen 1s orbitals, $\psi(B_1)$. It is evident that, although these orbitals overlap with one another, the overlap *integral* is zero since the regions of positive overlap are exactly cancelled by the regions of negative overlap. The zero overlap integral between $2p_x$ and the A_1 or B_1 combinations of hydrogen 1s orbitals $\psi(A_1)$ and $\psi(B_1)$— very relevant to the conclusion that $2p_x$ is a non-bonding orbital—can similarly be demonstrated. The basis functions associated with the C_{2v} character table—Table 3.1—provide a list of all the orbitals which interact with one another. The orbitals of A_1 symmetry which must be discussed are the 2s and $2p_z$ orbitals on the oxygen, each of which interacts with the hydrogen 1s combination $\psi(A_1)$. There are no orbitals of A_2 symmetry and only two of B_1 symmetry—the $2p_y$ orbital of oxygen and a hydrogen 1s

† This statement concerns one-electron terms in the Hamiltonian. An analogous statement may be made which covers the two-electron terms. The general form of such statements will become evident in Chapter 10.

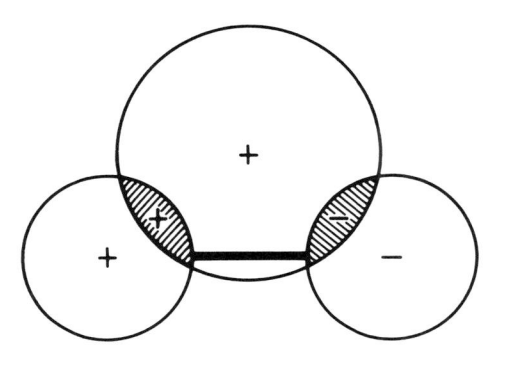

Figure 3.4   The zero overlap *integral* between an orbital of A_1 symmetry (oxygen 2s) and an orbital of B_1 symmetry (a linear combination of hydrogen 1s orbitals).

combination $\psi(B_1)$. We shall first consider the B_1 interactions qualitatively but in some detail.

The interaction between the $2p_y$ orbital of the oxygen and the B_1 combination of hydrogen 1s orbitals will lead to bonding and antibonding molecular orbitals. A schematic representation of the overlap between $2p_y$ and $\psi(B_1)$, together with the form of the resultant bonding and antibonding molecular orbitals is shown in Figure 3.5. The bonding molecular orbital is an out-of-

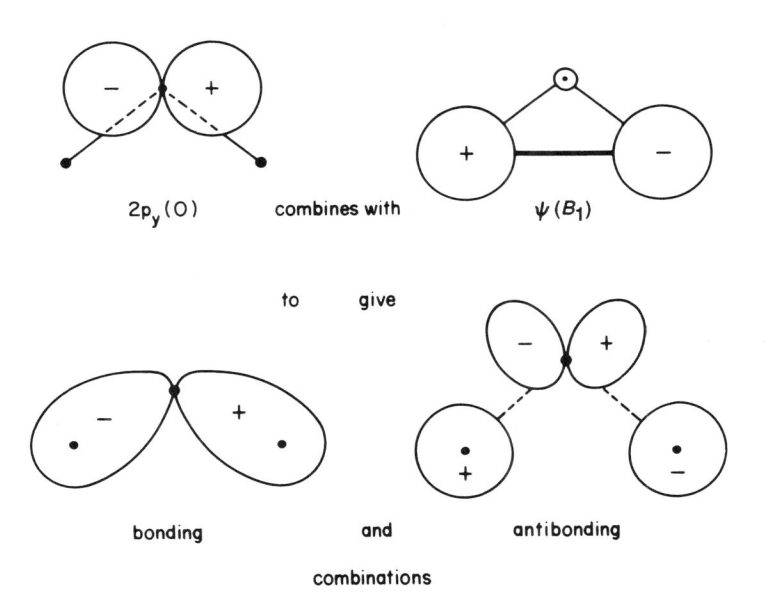

Figure 3.5   Interaction between orbitals of B_1 symmetry, leading to bonding and antibonding combinations.

phase combination of $2p_y$ and $\psi(B_1)$ while the antibonding molecular orbital is an in-phase combination (if this pattern seems strange, compare the relative phases of $2p_y$ and $\psi(B_1)$ in Figure 3.5).

The B_1 bonding orbital will surely be occupied by two electrons in the water molecular and so contribute to the molecular stability. There is an important point which must be made concerning this bonding molecular orbital. Consider, qualitatively, the dependence of the molecular stabilization derived from this orbital upon the HOH bond angle, θ. For the (hypothetical!) case of very small θ, shown in Figure 3.6, the lobes of $\psi(B_1)$ overlap with $2p_y$ in a way that leads to a relatively small value for the overlap integral between them; the overlap integral decreases as the bond angle decreases. When θ is very small, then, the interaction between the two orbitals of B_1 symmetry, which varies roughly as the overlap integral, will be small and the B_1 bonding molecular orbital will make little contribution to the molecular stability. As is qualitatively evident from Figure 3.6 (and is confirmed by more detailed calculations) as θ increases (keeping the O–H bond length constant) the interaction between $2p_y(O)$ and $\psi(B_1)$ smoothly increases with θ and reaches a maximum for a bond angle of 180°. It is legitimate to conclude that, were this interaction the only thing determining the geometry of the water molecule, then H_2O would be a linear molecule. Clearly, there is more to come!

The interaction between the three orbitals of A_1 symmetry is a more difficult problem simply because there are three orbitals to consider, not two. In the case of the B_1 interaction, the final molecular orbitals were mixtures of the two starting. Similarly, we would expect the final A_1 molecular orbitals to be mixtures of $2s(O)$, $2p_z(O)$ and $\psi(A_1)$. Although $2s(O)$ and $2p_z(O)$ were introduced as separate functions they will be mixed (i.e. contribute to the same molecular orbitals) by virtue of their mutual interaction with $\psi(A_1)$ (induced by their mutual overlap with the hydrogen 1s orbitals). Because of their original separation, it is unlikely that this mixing is very large. So, in the water molecule there will probably be an oxygen 2s orbital mixed with a small amount of oxygen $2p_z$ together with a second orbital which is largely $2p_z$ mixed with a little of 2s. Both interact with the same hydrogen 1s orbital

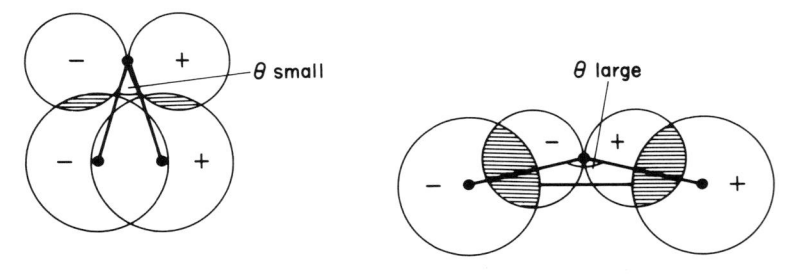

Figure 3.6   Variation of overlap integral of the B_1 orbitals with variation in the bond angle in H_2O.

combination. The question immediately arises as to how many of the three resulting molecular orbitals will be bonding and occupied. Answers of guaranteed accuracy to this question will come either from experiment or from very accurate calculations. At a more approximate level it is usually safe to assume that interactions between orbitals will change orbital energies, but not dramatically. If, for the moment, H$_2$O is regarded as a composite of an oxygen atom and H$_2$ then in the oxygen atom the 2s orbital will certainly be occupied. In H$_2$, the bonding combination $\psi(A_1)$ will certainly be occupied. So, it seems entirely probable that in H$_2$O there are two molecular orbitals of A_1 symmetry which are filled with electrons and contribute to the bonding. It turns out that this is a correct description and that the composition of the orbitals is indeed that assumed; we shall briefly return to this model later. First, however, it is helpful to explore an alternative, simpler but less accurate, description of the A_1 bonding molecular orbitals.

Consider the question 'is it possible to obtain two combinations of the 2s(O) and 2p$_z$(O) orbitals such that one does, and the second does not, interact with $\psi(A_1)$?'. If this is possible then the problem has been reduced to the simplicity of the B_1 case considered earlier; the interaction between two orbitals, not three. The simplification is possible, and the general way that it may be achieved is indicated in Figure 3.7. Figure 3.7 shows, schematically, that if in-phase and out-of-phase combinations of 2s(O) and 2p$_z$(O) are taken then one of the resulting mixed (hybrid) orbitals is directed towards the hydrogen atoms whilst the second combination is largely located in a region remote from them. This second combination would be essentially non-bonding and we may, as a first approximation, ignore its interaction with $\psi(A_1)$. It is convenient to choose 2s(O) and 2p$_z$(O) combinations which simplify the pictorial representation of the problem and, therefore, to assume that they are sp hybrids of the form:

$$\frac{1}{\sqrt{2}} \, [2s(O) + 2p_z(O)] \text{ non-bonding}$$

$$\frac{1}{\sqrt{2}} \, [2s(O) - 2p_z(O)] \text{ involved in bonding}$$

and it is these which are shown, qualitatively, in Figure 3.7.

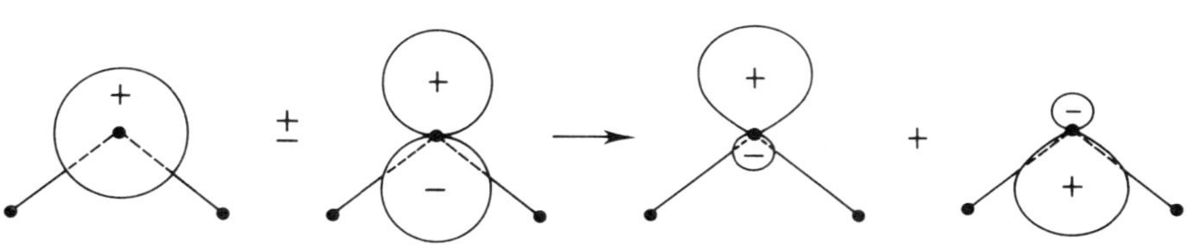

Figure 3.7 In-phase and out-of-phase combinations of oxygen 2s and 2p$_z$ orbitals give two (sp) hybrid orbitals.

The problem has now been simplified so that it is analogous to that discussed earlier for the case of the B_1 interactions; we have only the interaction between two orbitals, $\psi(A_1)$ and the second given above, to consider. These two orbitals will combine to give in-phase and out-of-phase combinations which are, respectively, bonding and antibonding molecular orbitals. These orbitals are shown schematically in Figure 3.8. Although excluded from participation in the bonding, the non-bonding orbital given above will almost certainly be occupied. It may be identified with a lone-pair orbital in the water molecule, the second lone pair orbital being $2p_x$, which, as has been seen, is of B_2 symmetry.

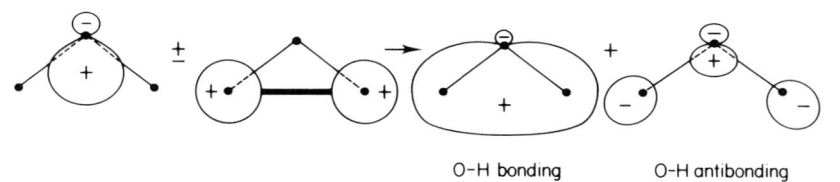

O-H bonding   O-H antibonding

Figure 3.8 Interaction between orbitals of A_1 symmetry, leading to bonding and antibonding combinations.

To complete the picture of the bonding in the water molecule, consider the relationship between the stabilization resulting from the interactions between the various orbitals of A_1 symmetry and the value of the HOH bond angle. For this discussion it proves convenient to consider the interactions involving the $2s(O)$ and $2p_z(O)$ orbitals separately. It is evident, from Figure 3.9, that the magnitude of the interaction between $2s(O)$, and the hydrogen 1s combination $\psi(A_1)$ does not depend upon the HOH bond angle. Because the oxygen orbital is spherically symmetrical the overlap integral is independent of the bond angle. So, this interaction does not vary with bond angle—and this is why it is convenient to consider the $2s(O)$ and $2p_z(O)$ orbitals separately.

The $2p_z(O)-\psi(A_1)$ interaction is shown schematically in Figure 3.10. It is evident from Figure 3.10 that when $\theta = 180°$ there is a zero overlap integral

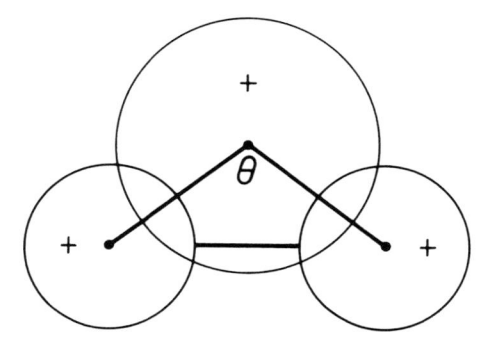

Figure 3.9 An (A_1) overlap integral which does not depend on bond angle (cf. Figure 3.6) because all orbitals concerned are spherical.

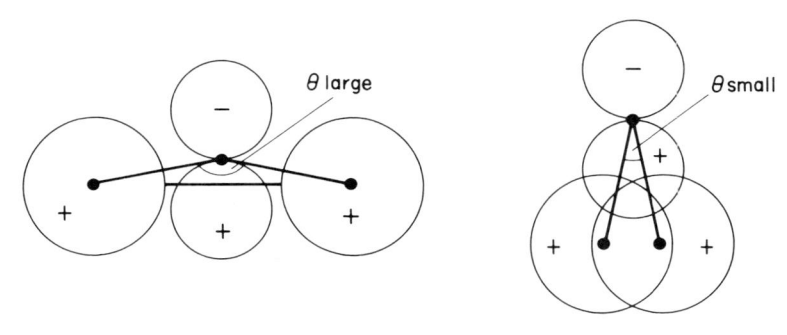

Figure 3.10   An (A_1) overlap integral which varies with bond angle (cf. Figure 3.9) because one of the orbitals involved is non-spherical.

and so no interaction between $2p_z$ and $\psi(A_1)$. This result has its origin in molecular symmetry. The symmetry of a linear water molecule is no longer C_{2v} but that of a different point group (called $D_{\infty h}$). In this latter group the $2p_z(O)$ orbital and $\psi(A_1)$ are of different symmetries; it follows that they do not interact. Evidently, the interaction between $2p_z(O)$ and $\psi(A_1)$ increases smoothly as the H–O–H bond angle decreases from a value of $180°$ and reaches a maximum at the physically unrealistic value $\theta = 0°$. Because the interaction of $\psi(A_1)$ with $2s(O)$ is independent of bond angle it is the interaction of $\psi(A_1)$ with $2p_z(O)$ which determines the angular variation of the interaction of $\psi(A_1)$ with mixtures of $2s(O)$ and $2p_z(O)$ orbitals; this interaction will tend towards a maximum at $\theta = 0°$.

In summary, of the bonding interactions in the water molecule, those of A_1 symmetry favour a bond angle $\theta \to 0°$ and that of B_1 symmetry leads to a stabilization which maximizes as $\theta \to 180°$. The two interactions are opposed and the observed bond angle represents a compromise. It is easy to see that removal of an electron from the B_1 bonding molecular orbital would reduce the tendency towards a large bond angle; this theme is developed later in the chapter. Incidentally, this example provides an illustration of an assertion made in Chapter 1; that symmetry arguments enable us to understand molecular structure rather than to predict it. Had the bond angle in water between $170°$ the discussion above could have been suitably modified (an angle of $180°$ would only have presented problems because the symmetry would no longer have been C_{2v}). Our discussion has also enabled us to conclude that there is only one unambiguously non-bonding orbital. This, a pure $2p_x$ atomic orbital of the oxygen atom, is of B_2 symmetry. A second entirely non-bonding lone pair does not exist in the isolated water molecule (although most simple descriptions of the bonding in the molecule include one). However, in a rather less accurate model, a second non-bonding orbital of A_1 symmetry can be introduced. The physical evidence for two lone pairs, and it is this evidence which provides the motivation for the simple pictures of bonding in the water molecule, comes largely from structural data. Thus, in ice each oxygen is roughly tetrahedrally surrounded by four hydrogens, two close and two distant. It seems reasonable that each of the distant hydrogens should be associated, by hydrogen bonding, with a lone pair. However, it must not be forgotten that attaching two more protons to the water molecule, even loosely, will modify its electronic structure. So, for instance, each of the A_1 molecular orbitals of an isolated water molecule will also be involved in the bonding of these additional protons (just as is the case in methane). Further, the reader should be able to show that the 1s orbitals of the distant hydrogens give rise to a combination of B_2 symmetry, so that even the—genuinely on a symmetry model—non-bonding $2p_x$ orbital of H_2O may become weakly involved in bonding in ice.

> **Problem 3.6**   The individual oxygen atoms in ice are surrounded by a distorted tetrahedron of hydrogen atoms. That is, they resemble the carbon atom of Figure 2.1 but two of the oxygen–hydrogen bonds are longer than the other two. The closely bonded pair are those discussed in Section 3.3. Show that the transformation of the 1s orbitals of the more distant hydrogen atoms give rise to a reducible representation with $A_1 + B_2$ components.

> **Problem 3.7**   In a discussion of the bonding in H_2S (bond angle 93°) the valence shell orbitals on the sulfur are 3s, 3p and 3d. Inclusion of the 3d orbitals would increase the number of possible interactions with $\psi(A_1)$ and $\psi(B_1)$ which would have to be considered. List all of the sulfur valence shell orbitals which could interact with each of them (use Table 2.4 and the results of Problem 2.3).

3.6   THE MOLECULAR ORBITAL ENERGY LEVEL DIAGRAM FOR H_2O

The discussion has now reached the point at which it is possible to obtain a schematic molecular orbital energy level diagram for the water molecule. Rather than work with the first model presented above, the more accurate, we shall consider the approximate. This is because it is the model that is the more compatible with relatively simple ideas about the bonding in H_2O—it is the one most likely to be produced by simply following chemical intuition. Notwithstanding its approximate nature, it gives a good prediction of the *relative* ordering of molecular orbital energies and so gives hope that similar approximate models will also be of value for other molecules. We proceed by presenting a schematic energy level scheme in Figure 3.11 and then detail the arguments used in its derivation. Before doing this note that, in contrast to the discussion in the text, lower case symbols have been used in Figure 3.11. Strictly, as has been briefly mentioned earlier in a footnote, *lower case symbols are used to describe wavefunctions.* Any wavefunctions—they could be vibrational wavefunctions, for instance, and this usage will be met in Chapter 9. Thus, a one-electron wavefunction reasonably enough, characterizes a single electron. However, an orbital can be occupied by two electrons, each with the same (three-dimensional, spatial) wavefunction. The distinction in usage is a rather fine one and many, but not all, authors use lower case symbols for both orbitals and wavefunctions. The convention seems to be that in diagrams such as Figure 3.11 lower case symbols are used but in the associated text one is as likely to meet either 'the orbital of A_1 symmetry' as 'the a_1 orbital'. In this book, upper case symbols will largely be used in the text and lower case symbols largely confined to diagrams. One important context in which lower case symbols are (almost!) invariably used in the literature is when orbital occupancies are specified. So, Figure 3.11 shows an orbital occupancy, starting with the lowest energy first and labelling orbitals of the same symmetry sequentially $(1a_1, 2a_1, 3a_1 ...)$ which is written as

$$1a_1{}^2 \, 1b_1{}^2 \, 2a_1{}^2 \, 1b_2{}^2$$

Here the superscripts indicate the number of electrons in each orbital, two in each case.

We now return to the details of Figure 3.11 and explain the arguments leading to the orbital energy sequence shown there. First, we have used a nodal plane criterion, which experience shows to be reliable.[1] In the simple model of the water molecule there are just two bonding molecular orbitals, one of A_1 and one of B_1 symmetry. Figures 3.5 and 3.8 reveal that although for both A_1 and B_1 bonding molecular orbitals the oxygen–hydrogen interactions are entirely bonding, in the B_1 case the hydrogen–hydrogen interactions are

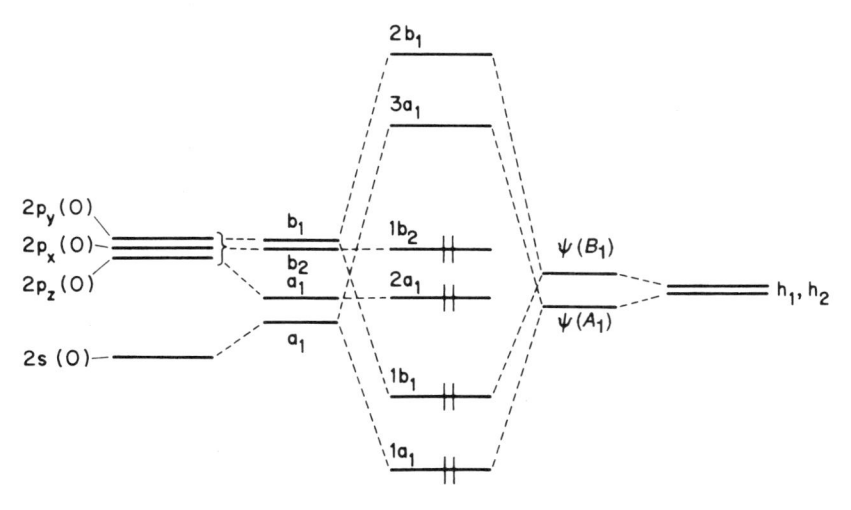

Figure 3.11   A schematic molecular orbital energy level diagram for H_2O.

antibonding— $\psi(B_1)$ corresponds to the antibonding combination of 1s atomic orbitals in the hydrogen molecule. The corresponding component in the A_1 bonding molecular orbital is bonding (Figure 3.8) and so it is reasonable to anticipate that this orbital will be more stable than the B_1. That is, the orbital with the smallest number of nodal planes is usually expected to be the most stable. Second, to obtain the most probable order of the two non-bonding molecular orbitals (of A_1 and B_2 symmetries) note that the 2s orbital of the isolated oxygen atom is of a lower energy than the 2p's. It seems reasonable, therefore, that $\psi(A_1)$, which contains a 2s component, should be of lower energy than the B_2 non-bonding orbital which is pure $2p_x$ (the same argument is also relevant to the A_1 and B_1 bonding molecular orbitals and reinforces the previous conclusions about their relative order). There is another point which must be made in connection with Figure 3.11. In this figure the interaction between the hydrogen 1s orbitals h_1 and h_2 is shown as removing the degeneracy of these two orbitals. This splitting corresponds to the separation between the bonding and antibonding molecular orbitals of H_2 (much reduced in the present case because of the large separation between the two hydrogen atoms). On the other hand, the mixing of the 2s and 2p orbitals of the oxygen combination $\psi(A_1)$ is shown as bringing these two closer together in energy. This is because if the combinations have the idealized sp hybrid forms which they were given earlier then they would be precisely equivalent (although differently orientated in space). It follows that if we were to work out energies associated with these two hybrids then we would expect to obtain the same result for each. The conclusion is, therefore, that when orbitals on the same atom are mixed to give general— not idealized—hybrid orbitals these hybrids should be regarded as having energies intermediate between those of their components.

In the water molecule there are eight valence electrons available to be allocated to the four lowest molecular orbitals shown in Figure 3.11 (the electron configuration of the oxygen atom is $1s^2 2s^2 2p^4$ and contributes six valence electrons; each hydrogen is $1s^1$ and contributes one). It follows that the lowest four orbitals, two non-bonding and two bonding molecular orbitals, are occupied.

3.7 COMPARISON WITH EXPERIMENT

Is there any experimental test of the model that has just been developed? The most pertinent test would be the observation of individual orbital energy levels. Such data are provided by photoelectron spectroscopic measurements, in which electrons are ejected from individual molecules by high energy monochromatic radiation in a high vacuum. The difference between the (measured) kinetic energy of an ejected electron and the energy of the incident photons is the energy required to remove the electron from the molecule. A variety of electron energies results, corresponding to a variety of molecular ionization energies. These ionization energies correspond very closely to the usual definition of orbital energy. An orbital energy is defined as the energy required to remove an electron from a molecule subject to the restriction that the orbitals of the other electrons in the molecule are unchanged. Evidently, this is a theoretical, rather than practical, definition—some readjustment of the orbitals of the residual electrons would be expected. Fortunately, however, the effect of these readjustments is usually rather small. It is therefore possible to use the ionization energies given by photoelectron spectroscopy to test our model.

The photoelectron spectrum of water shows four peaks (of energies 12.62 eV, 13.78 eV, 17.02 eV and 32.2 eV. Qualitatively, then, the photo-electron measurements support the energy scheme given in Figure 3.11. There are four different ionization energies arising from valence-shell electrons. A detailed analysis of the photoelectron spectrum can also give some idea of the symmetry species of the molecular orbital from which a particular electron is photo-ejected. In the case of the water molecule the 12.62 eV peak is probably associated with ionization from a B_2 orbital, the 13.78 and 32.2 eV peaks with ionization from A_1 orbitals and the 17.02 eV peak with ionization from a B_1 orbital, in agreement with the qualitative predictions of Figure 3.11. Agreement with the more accurate model is even better. The ionization from the most stable orbital (that at 32.2 eV) is from a largely 2s(O) orbital; ionizations from the other orbitals are from orbitals which have considerable 2p(O) contri-butions and so are relatively close together in energy.

With the present-day computers it is possible to carry out accurate calcula-tions on molecules with fewer than, perhaps, fifty electrons. The water molecule with a total of only ten electrons should, therefore, be amenable to quite precise theoretical investigation. All of the accurate calculations which have been performed on this molecule lead to roughly the same orbital energies and demonstrate the presence of molecular orbitals of A_1 symmetry at about 14 and 30 eV, a B_1 at about 17 eV and one of B_2 symmetry at about 12.5 eV. The agreement between these data and the photoelectron results is very good. That with Figure 3.11 is as good as could be expected.

It is particularly encouraging to find that the qualitative symmetry based arguments which have been used to discuss the electronic structure of the water molecule should give results which are in excellent qualitative agreement both with those obtained by experiment and those obtained by detailed calculations. Hopefully, the same techniques may be applied to other molecules and similar qualitatively accurate results obtained. It is obvious that as molecular complexity increases the difficulty in arriving at an unambiguous energy level scheme will also increase. However, the symmetry of the water molecule is not particularly high. It is not unreasonable to hope that for larger, but higher symmetry, molecules the increase in molecular complexity will be compen-sated for by the increase in molecular symmetry and so the methodology will remain applicable.

3.8   THE WALSH DIAGRAM FOR TRIATOMIC DIHYDRIDES

We are now in a position to reconsider in more detail a problem which we first encountered in Chapter 1—that of the significance which can be placed upon the observation that the bond angle in the electronic ground state of the water molecule is 104.5°. As shown in Section 3.5, the bonding interactions responsible for the stability of the water molecule are maximized at quite different bond angles. The stabilization resulting from the B_1 interaction maximizes at a bond angle of 180° whereas that from the A_1 interactions maximizes at a small bond angle. The observed bond angle represents a compromise, showing that interactions involving both $2p_y(O)$ and $2p_z(O)$ are of importance. However, the total bonding is unlikely to show a strong dependence on bond angle because although a change in θ will reduce the stabilization resulting from interactions involving orbitals of one symmetry species it will increase the stabilization accruing from the other. Only if water had been found to be linear could it have been concluded that the major contribution to the molecular bonding resulted from interactions between those orbitals which we have identified as being of B_1 symmetry. Conversely, only if the bond angle were very small, say 60°, would it then have been valid to conclude that most of the stabilization resulted from the A_1 interaction. A diagram showing this behaviour schematically is given in Figure 3.12 where,

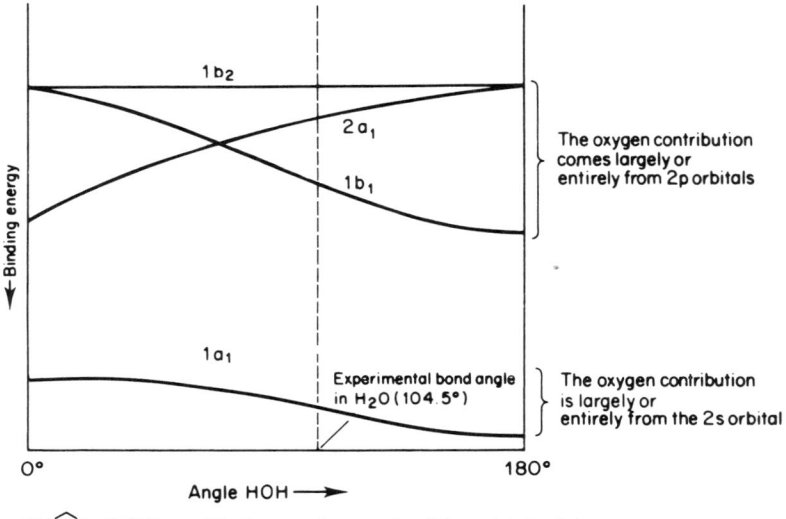

Figure 3.12   A Walsh diagram for H_2O.

again, orbitals have been denoted by lower case symbols. In this diagram there is also recognition of the fact that at the $\theta = 180°$ limit the orbital which has, qualitatively, been called the A_1 non-bonding orbital loses its 2s component and becomes a pure 2p non-bonding orbital. This latter orbital is therefore degenerate with (i.e. has the same energy as) the non-bonding orbital which we have labelled B_2. Conversely, the lower, bonding, A_1 orbital loses its 2p component and becomes a pure 2s orbital at 180°—and so is of lower energy in this limit. The non-bonding B_2 orbital remains unchanged in energy as the bond angle changes (actual calculations show that it increases slightly in energy at the 180° limit but as this is caused by the effects of electron repulsion we have not included it in Figure 3.12). Figure 3.12 is specifically drawn for H_2O—but its general form is applicable to all MH_2 molecules for M atoms which have similar valence shell orbitals to oxygen. Diagrams of this type were first introduced by Walsh[2] and are therefore commonly known as Walsh diagrams.

It is possible to directly relate the observed geometries of first row MH_2 molecules (in electronic ground and excited states) to the occupancy of the highest B_2 and A_1 orbitals in Figures 3.11 and 3.12. Occupancy of the former, which is non-bonding, would be expected to have little effect on bond angle but the lower the occupancy of the latter the larger we expect the bond angle to be (and vice versa). Table 3.2 details relevant data and shows that when there are two electrons in the highest A_1 orbital an angle of about 103° results; one

Table 3.2

Molecule	Orbital occupancy		Bond angle
	A_1	B_2	
BH_2 (excited)	0	1	180°
BH_2	1	0	131°
CH_2 (triplet excited)	1	1	136°
CH_2 (singlet excited)	1	1	140°
NH_2^+ (excited)	1	1	144°
BH_2^-	2	0	100°
CH_2	2	0	102°
CH_2^-	2	1	99°
NH_2	2	1	103°
OH_2^+	2	1	107°
OH_2	2	2	105°
NH_2^-	2	2	104°

Data from E. Wasserman, *Chem. Phys. Letters*, **24** (1974), 18, and Y. Takahata, *Chem. Phys. Letters*, **59** (1978), 472 (note that in this latter paper B_1 and B_2 are interchanged compared with the usage in this chapter).

electron in this orbital leads to a bond angle of about 140°. When this orbital is empty a linear molecule results.

Problem 3.8   The following species have been the subject of theoretical investigations but their bond angles have yet to be determined experimentally. Predict approximate values for their bond angles

$$FH_2^{3+}, \quad BeH_2^{3-}, \quad CH_2^{2-}, \quad BH_2^{3-}, \quad NH_2^{+}$$

3.9   SIMPLE MODELS FOR THE BONDING IN H_2O

This chapter is concluded by investigating the relationship between the picture of the electronic structure of the water molecule which has been developed in the chapter and that given by models such as those discussed for ammonia in Chapter 1. The first model is that in which the oxygen atom in the water molecule is regarded as being tetrahedrally surrounded by electron pairs, two bonding and two non-bonding. The concordance between this model and the general pattern of energy levels shown in Figure 3.11 is easy to show. The bonding electron pairs (which have been called β_1 and β_2) and the lone pair electrons (labelled l_1 and l_2) are shown schematically in Figure 3.13. It is easy to show that the transformations of the two bonding orbitals β_1 and β_2, considered as a pair, generate the reducible representation:

E	C_2	σ_v	σ_v'
2	0	2	0

which has A_1 and B_1 components. These, of course, are precisely the same symmetries as possessed by the bonding molecular orbitals shown in Figure 3.11. It is also easy to show that the transformations of the lone pair orbitals, l_1 and l_2, generate the reducible representation:

E	C_2	σ_v	σ_v'
2	0	0	2

which is also easily shown to be the sum of the A_1 and B_2 irreducible representations. These, again, are the symmetries of what have been identified as the lone pair orbitals in Figure 3.11.

Problem 3.9   Generate the two reducible representations discussed above. Use Figure 3.13.

It is interesting to consider the 'tetrahedral oxygen' model for H_2O in more detail. Let us start with the lone pair orbitals. Since they are localized on the oxygen atom we conclude that they must be derived from the valence shell

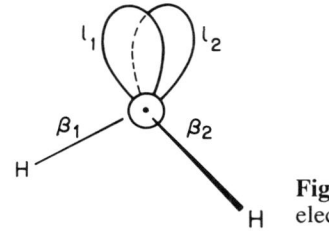

Figure 3.13 The tetrahedral arrangement of bonding electron pairs (β) and lone pairs (l) in H_2O.

atomic orbitals of the oxygen atom. But the only B_2 valence shell oxygen atomic orbital is $2p_x(O)$, so the orbital of this symmetry which is a combination of lone pair orbitals must be identical to that obtained from our symmetry based discussion. Similarly, the A_1 combination must be some mixture of $2s(O)$ and $2p_z(O)$, again in qualitative agreement with our earlier result. Study of the O–H bonding orbitals leads to a similar agreement with the model we have developed in this chapter. Because β_1 and β_2 are bases for A_1 and B_1 irreducible representations we conclude that $2s(O)$, $2p_z(O)$ and $\psi(A_1)$ may contribute to the A_1 combination. Similarly, $2p_y(O)$ and $\psi(B_1)$ contribute to the B_1 combination—conclusions identical to those reached earlier. Further, the statement that β_1 and β_2 are *bonding* means that the A_1 and B_1 combinations must be in-phase, bonding, combinations of oxygen and hydrogen orbitals. We conclude that the simple two bonding, two non-bonding orbital picture of the water molecule may readily be reinterpreted in the language which we have used in this chapter. Further, by using the energy-level criteria based on the number of nodal planes and the relative energies of atomic orbitals, discussed in the present chapter, the qualitative energy level diagram shown in Figure 3.11 can again be derived.

The second simple model of the water molecule to be considered is one that is sometimes quickly discarded. This is the model in which only the 2p orbitals of the oxygen atom are considered as being involved in O–H bonding, the 2s orbital being, implicitly, regarded as non-bonding. In this model, the oxygen 2p orbitals which lie in the plane of the water molecule overlap with the hydrogen 1s orbitals. The relevant oxygen 2p orbitals, which because it is a choice in accord with that made in setting up the group theoretical model, will be taken to be $2p_z(O)$ and $2p_y(O)$, are orientated so as to point directly at the hydrogen atoms. This is shown in Figure 3.14; a bond angle of 90° is predicted. The oxygen $2p_x$ orbital, which is perpendicular to the plane of the molecule, is non-bonding, precisely the role which we found it to play. The two non-bonding orbitals are, therefore, $2s(O)$ and $2p_x(O)$, which are of A_1 and B_2 symmetries, just as found above (their relative energies—one very stable A_1 orbital and one high energy B_2 orbital—also agree with the symmetry-based model). It is easy to show that the two bonding orbitals which result from the overlap of the $2p_z(O)$ and $2p_y(O)$ orbitals (those in the plane of the molecule) with the hydrogen 1s orbitals give rise to a reducible representation with A_1 and B_1

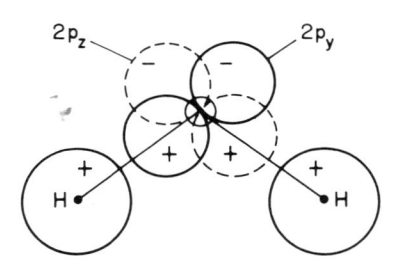

Figure 3.14   Bonding in H_2O with 2p as the only oxygen orbitals involved. Note that the axis labels do not follow the convention adopted elsewhere in this book.

components. Arguments analogous to those developed above for β_1 and β_2 in the previous model demonstrate that the qualitative forms of the corresponding A_1 and B_1 molecular orbitals are those deduced earlier in this chapter. It is perhaps pertinent to comment that this particular simple model gets closer to the results of accurate quantum mechanical calculations than does any other. It predicts a low-lying non-bonding 2s orbital, one pure 2p non-bonding oxygen orbital and two bonding molecular orbitals involving oxygen 2p orbitals. Yet this is a model which fell into disuse in the 1950s—perhaps it is time for its revival.

Problem 3.10   Show that the $2p_z(O)$ and $2p_y(O)$ orbitals shown in Figure 3.14 transform together as $A_1 + B_1$.

Problem 3.11   In the text a variety of alternative arguments, all leading to the same conclusion, have been used to arrive at the general form of Figure 3.11. Select and rehearse a single set of arguments leading to this figure.

3.10   A *RAPPROCHEMENT* BETWEEN SIMPLE AND SYMMETRY MODELS

It is useful at this point to review the development of the arguments in this book. In Chapter 1 it was concluded that simple models of molecular bonding cannot be expected to be infallible predictors of molecular geometry. However, in the present chapter it has been shown, at least for the case of the water molecule, that these simple models may usefully be reinterpreted. It is possible to recast them and to show that the bonding descriptions which they present are equivalent, qualitatively, to a symmetry-based description. Symmetry-based descriptions show that there can be different relationships between molecular geometry and the contribution to the bonding from the various bonding molecular orbitals. The stabilization resulting from one interaction may be independent of geometry, others may be very sensitive to geometry. Different interactions may make a maximum contribution to molecular stability at quite different patterns of bond angles. Although not evident from the discussion so far, these conclusions have a general validity.

It is here that the circle closes. Simple pictures of molecular bonding are perhaps more reliable predictors of relative energies of molecular orbitals than they are of molecular geometries (although more used for the latter rather than the former). When a simple picture fails to give a correct molecular orbital energy level pattern it is usually because there are some interactions in the molecule involving orbitals other than those considered in the simple model. In such cases the simple models are, none the less, usually good starting points for a detailed discussion. Finally, we note that, despite their apparent differences, when there is a variety of simple approaches to the bonding in a molecule, they usually lead to the same qualitative energy level diagram. Again, the only exceptions occur when different models include different interactions, but here the differences are themselves illuminating.

What is the particular attraction of a symmetry-based approach which leads us to refer all other models to it? A computational advantage has already been mentioned—interactions are non-zero between wavefunctions of the same symmetry species so that the size of the problem, the number of interacting orbitals, is reduced. There is another important reason. Whenever excited electronic states or ionized species are considered it becomes essential to use a symmetry-based approach. This is because it is the only one which allows a simple connection between the discussion of the ground and excited (or ionized) states of a molecule. One illustration will make the point. Suppose an electron in the water molecule is excited from a bonding orbital to some high-lying, non-bonding, orbital and suppose that the excited electron comes from a single O–H bond. According to all of the simple models of the bonding in the water molecule those electrons associated with one bond are not associated with the other and so such an excitation would seem entirely possible. In the excited state the two O–H bonds would differ—one has only one bonding electron while the other has two. This is in contradiction to the observation that in all stable excited states of the water molecule the O–H bonds are equivalent (excluding unstable states from which dissociation into H + OH occurs). For a symmetry-based description, in which the bonding electron comes from a molecular orbital spread equally over both hydrogen atoms, the observed equivalence of the two hydrogen atoms in excited or ionized states follows naturally. A symmetry-based description is thus to be preferred because it can be applied to both ground and excited states.†

---

† A paper which makes a direct connection between the models considered above and the photoelectron spectrum of the water molecule is J. Simons, 'Why Equivalent Bonds Appear as Distinct Peaks in Photoelectron Spectra', in *J. Chem. Educ.*, **69** (1992), 522.

Problem 3.12 A student was heard to complain that 'symmetry arguments make difficult problems even harder by adding another complicating consideration'. Write a one-page document assessing this point of view.

3.11 SUMMARY

The irreducible representations which appear in character tables are orthonormal (p. 35)—each component is independent of the others and carries equal weight. This property enables reducible representations (p. 41) to be reduced systematically to their irreducible components (p. 42). In the context of molecular bonding this enables the interactions between orbitals of each symmetry type to be discussed separately (p. 46). Such discussions, together with simple nodal-plane criteria (p. 54), enable qualitative molecular-orbital energy level diagrams to be constructed (p. 53) and the angular variation of each bonding interaction assessed (p. 51). This latter information may be conveniently represented as a Walsh diagram (p. 56). A symmetry analysis of simple pictures of molecular bonding reveals that they have similarities with each other and with the symmetry-based approach (p. 58).

REFERENCES

1. E. B. Wilson, *J. Chem. Phys.*, **63** (1975), 4870.
2. A. D. Walsh, *J. Chem. Soc.* (1963), 2250.

4

The D$_{2h}$ Character Table and the Electronic Structures of Ethene (Ethylene) and Diborane

The present chapter has several objectives, in addition to those indicated by its title. First, to introduce a new symmetry group and its character table. The group has been chosen because it is related to the C_{2v} group with which the reader is now familiar. It is not the simplest group that could have been used to discuss the bonding in ethene and diborane—the simplest would be the group D_2, a group which will be met shortly—but this discussion itself is only part of the objective of the present chapter. Use of the more complicated, D_{2h} (pronounced 'dee two aich') group will enable a start to be made on an exploration of the relationships between groups and the corresponding character tables. A second objective is to present and to use the rather important technique of *projection operators*. Despite their somewhat unattractive name these provide a very simple method of obtaining functions transforming as a particular irreducible representation, otherwise as difficult as it is necessary and important.

4.1 THE SYMMETRY OF THE ETHENE MOLECULE

The effect of bringing two CH$_2$ units, each of C_{2v} symmetry, together to form an ethene molecule has the effect of generating additional symmetry elements. All of the symmetry elements of the ethene molecule are shown in Figure 4.1. As shown in this figure, each CH$_2$ fragment is turned into itself by the operations that were met when discussing the C_{2v} point group—that is, there are two perpendicular mirror planes, the intersection between them defining a twofold axis, common to the two CH$_2$ units. As Figure 4.1 shows, the union of the two CH$_2$ units to form ethene has the effect of generating two new C_2 rotation axes, the three twofold axes being mutually perpendicular. This immediately suggests their use as Cartesian coordinate axes, a suggestion which we shall follow. Note that each of the three C_2 axes is unique. This is rather important because later in this book sets of twofold axes which are not unique will be met. In ethene, each twofold axis is unique because there is no operation in the group which interchanges any pair of them. This assertion can be checked after reading the next few paragraphs and a complete list of the

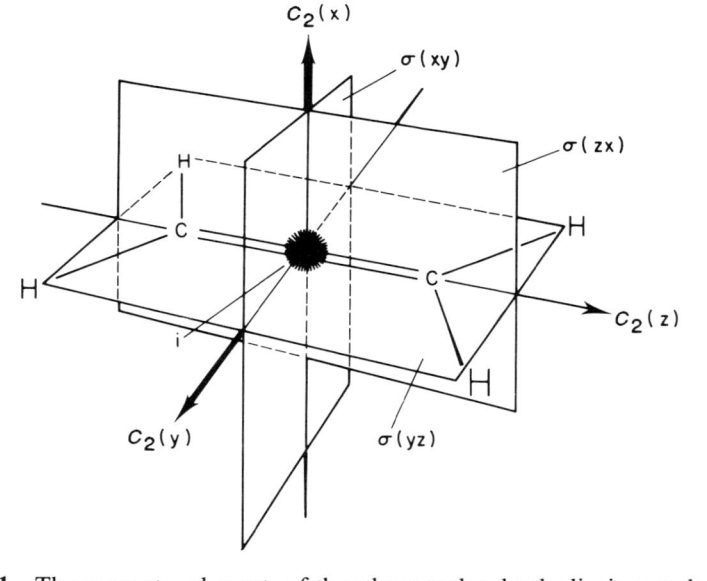

Figure 4.1   The symmetry elements of the ethene molecule; the list is complete except for the identity element, which has to be added.

ethene symmetry operations has been obtained. In high-symmetry molecules it is quite common for there to be a set of rotation axes (or mirror plane reflections or other operations) which are either interchanged or mixed† by other operations of the group. In such cases there is a corresponding complication in the character table and it is this complication which we seek to avoid here—it will be met in the next chapter—by working with another example of an Abelian‡ point group (the C_{2v} group of Chapters 2 and 3 is Abelian, something which makes it particularly easy to work with).

When, as in this case, there are several apparently equally good choices for the z axis, it is usual to choose that axis which contains the largest number of atoms and so we shall take as z the twofold axis which passes through the two carbon atoms. It follows that the (local) z axis of each CH_2 fragment (of C_{2v} symmetry) is coincident with the molecular z axis. Just as in the C_{2v} case, the labels of the other coordinate axes are determined by the convention that the planar molecule lies in the yz plane.

As is evident from Figure 4.1 the mirror planes of each CH_2 fragment persist as symmetry elements in the complete molecule. A third mirror plane exists in

---

† We include the word 'mixed' here to make the statement rigorous. Its meaning will only become clear after reading Chapters 5 and 6.
‡ It is not essential that the meaning of 'Abelian' is fully comprehended at the present point but it is given in Appendix 1 (Section A1.3); comments on pages 64, 102 and 104 are also relevant.

the ethene molecule. This passes through the mid-point of the carbon–carbon bond and is perpendicular to that bond. Figure 4.1 shows that each of the mirror planes is perpendicular to one of the coordinate axes; equally, it lies in the plane defined by the other two coordinate axes. Rather than use a σ_v notation for the mirror planes, each mirror plane is conventionally labelled by the molecular coordinate axes which it contains: thus $\sigma(xy)$, $\sigma(yz)$ and $\sigma(zx)$. There will be more to say about these labels shortly.

All of the symmetry elements listed so far are similar to those encountered in the C_{2v} group. Additionally, however, the ethene molecule contains a centre of symmetry, a point such that inversion of the whole molecule through it gives a molecule which is indistinguishable from the starting one. This centre of symmetry is indicated by the star-like point at the centre of Figure 4.1. More strictly, a centre of symmetry is such that inversion of any point of the molecule in it gives an equivalent point. Pictorially, if a straight line is drawn from any point (the starting point) in the molecule to the centre of symmetry and then extended an equal length beyond the centre of symmetry, the terminal point of the line is symmetry-equivalent to the starting point. This element, and the corresponding operation are conventionally denoted by the lower case symbol i. A centre of symmetry, if there is one, is always at the centre of gravity of a molecule. A molecule may possess several rotation axes, several mirror planes but it can never possess more than one centre of symmetry.

Problem 4.1 Draw a diagram of the ethene molecule and use it to compile a list of all the symmetry elements that it contains. Check your answer by reference to the discussion above and the list below.

The symmetry elements of the ethene molecule provide a better example of the use of the word 'point' when talking about a point group than do the symmetry elements of the water molecule. As is evident from Figure 4.1, all of the symmetry elements have one point in common, all pass through a common point, located at the centre of gravity of the molecule (in this context, the identity element is best thought of as corresponding to a C_1 rotation axis).

In summary, then, and talking now in terms of symmetry operations rather than symmetry elements, the symmetry operations which turn the ethene molecule into itself are:

$$E \quad C_2(z) \quad C_2(y) \quad C_2(x) \quad i \quad \sigma(xy) \quad \sigma(zx) \quad \sigma(yz)$$

This group of symmetry operations is commonly given the shorthand label D_{2h}. A detailed discussion of such shorthand labels will have to be deferred until Section 7.5 because it will not be until then that all of the symmetry operations on which the classification is based will have been met; up to that point they will have to be discussed individually. It is clear that the label D_{2h} requires some explanation. Point groups which contain a principal C_n axis and, perpendicular to this principal axis, n twofold axes are called *dihedral point groups* and hence carry the label D_n (*D* for dihedral). If one CH_2 group of the ethene molecule were to be slightly rotated about the *z* axis (so that the molecule becomes non-planar) then all of the mirror planes would be destroyed, as would the centre of symmetry. The resulting molecule would be of D_2 symmetry.

> **Problem 4.2**   Show that when the two CH_2 groups of ethene are rotated (twisted) by equal and opposite amounts about the *z* axis of Figure 4.1 the molecule obtained has D_2 symmetry (operations E, $C_2(z)$, $C_2(y)$, $C_2(x)$). Note that for the simpler rotation described in the text (rotate one CH_2 group by $\theta°$) it is necessary to rotate the *x* and *y* axes, and the C_2 axes associated with them, by $\theta/2$ about the *z* axis in the same sense to obtain an equivalent result.

If, perpendicular to the principal rotation axis—that of highest *n* value in C_n—in a molecule there is a mirror plane, that is, a plane horizontal with respect to the C_n axis, then this mirror plane is denoted σ_h (recall that in Chapter 2 a mirror plane which is vertical with respect to the principal rotation axes, was denoted σ_v). If a D_n group also contains a σ_h mirror plane then the point group is labelled D_{nh}. The present point group falls into this category once a difficulty has been overcome. This is that in the present group, D_{2h}, no mirror plane has been called σ_h (or σ_v, for that matter). The reason for this is that in D_{2h} there are three C_2 axes, any one of which might equally well be chosen as the principal axis. The mirror plane which should be labelled σ_h would depend upon which particular C_2 axis is nominated as principal axis. In this particular case, where all three mirror planes are equally good candidates for being labelled σ_h the egalitarian solution is to give none of them this label but, rather, designate them as has been done above. However, egality cannot alter the claim of the group to be recognized as one of the D_{nh} type; accordingly it is labelled D_{2h}.

4.2   THE CHARACTER AND MULTIPLICATION TABLES OF THE D_{2h} GROUP

In order to proceed further we must obtain the character table of the D_{2h} point group. The procedure which was adopted for the C_{2v} case—considering the transformations of a variety of basis functions—could be used to generate the D_{2h} character table; the procedure is entirely analogous. For this reason, space will not be devoted to it. Rather, the reader is invited to use this method himself or herself in the following problem.

> **Problem 4.3**   Derive the character table of the D_{2h} point group (Table 4.1). In Chapter 2 the irreducible representations of the C_{2v} character table were generated by considering the transformations of the orbitals of

Table 4.1 Character table of the D_{2h} group

D_{2h}	E	$C_2(z)$	$C_2(y)$	$C_2(x)$	i	$\sigma(xy)$	$\sigma(zx)$	$\sigma(yz)$	
A_g	1	1	1	1	1	1	1	1	$s, d_{z^2}, d_{x^2-y^2}$
B_{1g}	1	1	-1	-1	1	1	-1	-1	d_{xy}
B_{2g}	1	-1	1	-1	1	-1	1	-1	d_{zx}
B_{3g}	1	-1	-1	1	1	-1	-1	1	d_{yz}
A_u	1	1	1	1	-1	-1	-1	-1	f_{xyz}
B_{1u}	1	1	-1	-1	-1	-1	1	1	p_z
B_{2u}	1	-1	1	-1	-1	1	-1	1	p_y
B_{3u}	1	-1	-1	1	-1	1	1	-1	p_x

a unique atom (the oxygen in H_2O). In order to use this technique in the present problem it is necessary to first have a unique atom. This can be done by placing a hypothetical atom at the centre of gravity of the ethene molecule. Using just the familiar s, p and d orbitals it is not possible to generate the A_u irreducible representation of the D_{2h} point group. This irreducible representation can be generated using one of the f-orbitals, the f_{xyz} orbital, a diagram of which is shown in Figure 4.2. To enable a check on your answer to this problem, the atomic orbital(s) of the hypothetical atom which generates each irreducible representation are given at the right-hand side of Table 4.1.

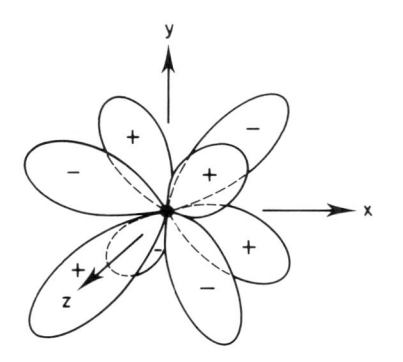

Figure 4.2 The f_{xyz} orbital of a hypothetical atom placed at the centre of gravity of the ethene molecule. Note that the phase of the orbital is positive in those regions of space in which the product xyz is positive.

For every group there exists a group multiplication table; that for the D_{2h} group is given in Table 4.2. Its derivation is analogous to the derivation of the C_{2v} group multiplication table. Just as in the C_{2v} case it will be found that when the appropriate substitution of characters for the corresponding symmetry operation is made in Table 4.2, any row of characters appearing in the D_{2h}

character table turns Table 4.2 into a multiplication table which is arithmetically correct.

Problem 4.4 By combining (multiplying) pairs of operations of the D_{2h} character table show that Table 4.2 is correct.

Some help with this problem is provided by Section 2.3. Indeed, if the hypothetical molecule considered there—$O(CH_2)_2$—is flattened

Table 4.2 Multiplication table for the D_{2h} group

D_{2h}	E	$C_2(z)$	$C_2(y)$	$C_2(x)$	i	$\sigma(xy)$	$\sigma(zx)$	$\sigma(yz)$
				First operation				
E	E	$C_2(z)$	$C_2(y)$	$C_2(x)$	i	$\sigma(xy)$	$\sigma(zx)$	$\sigma(yz)$
$C_2(z)$	$C_2(z)$	E	$C_2(x)$	$C_2(y)$	$\sigma(xy)$	i	$\sigma(yz)$	$\sigma(zx)$
$C_2(y)$	$C_2(y)$	$C_2(x)$	E	$C_2(z)$	$\sigma(zx)$	$\sigma(yz)$	i	$\sigma(xy)$
$C_2(x)$	$C_2(x)$	$C_2(y)$	$C_2(z)$	E	$\sigma(yz)$	$\sigma(zx)$	$\sigma(xy)$	i
i	i	$\sigma(yz)$	$\sigma(zx)$	$\sigma(xy)$	E	$C_2(x)$	$C_2(y)$	$C_2(z)$
$\sigma(xy)$	$\sigma(xy)$	i	$\sigma(yz)$	$\sigma(zx)$	$C_2(z)$	E	$C_2(x)$	$C_2(y)$
$\sigma(zx)$	$\sigma(zx)$	$\sigma(xy)$	i	$\sigma(yz)$	$C_2(y)$	$C_2(x)$	E	$C_2(z)$
$\sigma(yz)$	$\sigma(yz)$	$\sigma(zx)$	$\sigma(xy)$	i	$C_2(x)$	$C_2(y)$	$C_2(z)$	E

(Left margin label: Second operation, applying to the rows below $C_2(x)$)

symmetrically so that it becomes planar then there results a molecule of the same symmetry as ethene which also contains the unique atom required for Problem 4.3. A further hint on how to solve this problem is provided by Figure 4.3. Take a general point in space (indicated by the solid star). Perform the first operation (in Figure 4.3, $\sigma(zx)$) to give the cross-hatched star, follow it with the second operation (in Figure 4.2, i) to give the open star. Then ask 'what single operation turns the solid star into the open one' (in Figure 4.3, $C_2(y)$). One concludes that $\sigma(zx)$ followed by i is equivalent to $C_2(y)$.

Problem 4.5 Take any four of the irreducible representations of Table 4.1 and by substituting the appropriate character for each operation in Table 4.2 show that in each case an arithmetically correct multiplication table is obtained. If needed, Section 2.4 will provide guidance on this problem.

4.3 DIRECT PRODUCTS OF GROUPS

There are several interesting features of Table 4.2; for example, it is symmetric about either diagonal. Another is the way that it may be broken into four smaller blocks, pairs of which are identical. Similarly, the D_{2h} character table

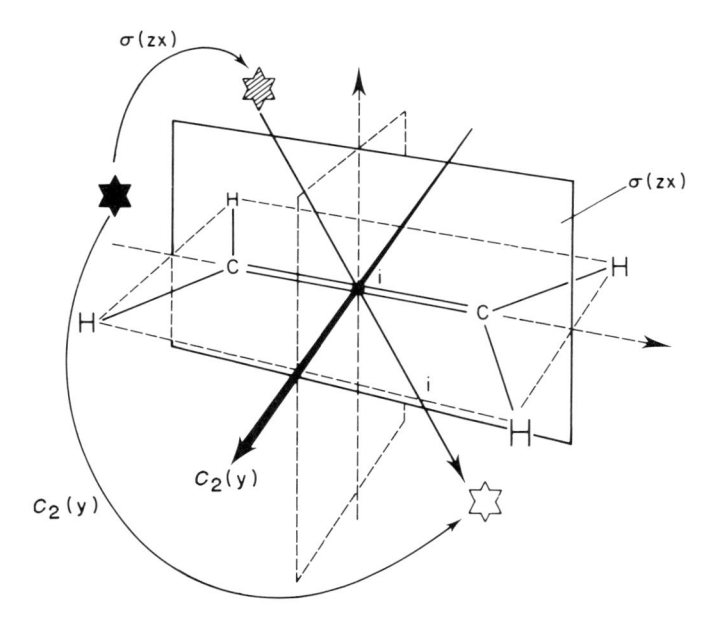

Figure 4.3 An illustration that $\sigma(zx)$ followed by i is equivalent to $C_2(y)$. In a similar way a C_2 can always be treated as σ combined with i; i can always be represented as C_2 combined with σ.

(Table 4.1) may also be broken into four blocks but now three of the blocks are identical and in the fourth the same set of characters appear, but with all signs reversed. There is a simple reason for these patterns. As is evident from Table 4.2, the operation $\sigma(xy)$ is equivalent to $C_2(z)$ followed by the inversion i. Similarly, $\sigma(yz)$ equals $C_2(x)$ followed by i and $\sigma(zx)$ equals $C_2(y)$ followed by i. It follows that the operation of the D_{2h} group may be rewritten as follows:

$$
\begin{array}{cccc}
E & C_2(z) & C_2(y) & C_2(x) \\
\end{array}
\qquad
\begin{array}{cccc}
i & \sigma(xy) & \sigma(zx) & \sigma(yz) \\
\end{array}
$$

$$
\text{are equivalent to} \qquad\qquad \text{are equivalent to}
$$

$$
\underbrace{
\begin{array}{cccc}
E & C_2(z) & C_2(y) & C_2(x) \\
\end{array}
}_{\text{Followed by } E}
\qquad
\underbrace{
\begin{array}{cccc}
E & C_2(z) & C_2(y) & C_2(x) \\
\end{array}
}_{\text{Followed by } i}
$$

That is, the operations of the D_{2h} group may be obtained by forming all possible products of members of the set E, $C_2(z)$, $C_2(y)$, $C_2(x)$ which, as has already been seen, form the D_2 group—with members of the set E, i—two operations which together form a group called the C_i group (pronounced 'cee eye'), to which we shall return shortly. Technically, one says that 'the D_{2h} group is the *direct product* of the D_2 and C_i groups'—where 'direct product' means 'form all possible products of one group of symmetry operations with all of the symmetry operations of the other'.† When the operations of a group may be expressed as direct products in this way so, too, may the corresponding character tables. That is, the character table of the D_{2h} group is the direct product of those of the D_2 and C_i groups (and the phrase 'direct product' now refers to characters but its meaning is otherwise unaltered). The character table of the D_2 group is given in Table 4.3 and that of the C_i group in Table 4.4. They should, together, be compared with Table 4.1. The four blocks in Table 4.1 are just the characters given in Table 4.3 with signs determined by Table 4.4. Thus in three of the blocks Table 4.3 reappears with unchanged sign and in the fourth all of the characters are multiplied by -1.

Table 4.3

D_2	E	$C_2(z)$	$C_2(y)$	$C_2(x)$
A	1	1	1	1
B_1	1	1	-1	-1
B_2	1	-1	1	-1
B_3	1	-1	-1	1

Table 4.4

C_i	E	i
A_g	1	1
A_u	1	-1

Problem 4.6 Multiply the character table 4.3 by the character table 4.4 and thus generate Table 4.1. Note that 'multiply' here means different things for operations and characters. For the latter it means simple arithmetic multiplication but for the former it means 'carry out the operations one after the other'. The way that the relevant operations multiply is indicated in this section; in order to generate Table 4.1 it is necessary to maintain the correct correspondence between products of characters and products of operations.

---

† In this section an *aufbau* approach is used, a larger group being built up from subgroups. An important notation appears if the inverse pattern is studied and the decomposition of a group into subgroups considered. Because both C_i and D_2 may be obtained from D_{2h} in one way and only one way they are both said to be *invariant subgroups* of D_{2h}. In contrast, some other subgroups of D_{2h}—one is C_{2v}; another is C_2—are not invariant subgroups because there is more than one way that they can be generated. This topic is dealt with more fully in Section 8.1; it is a topic which will become particularly important in Chapter 13.

An interesting thing about Table 4.3 is that the sets of characters that appear in it are the same as those of the C_{2v} point group (Table 2.4), although the operations and irreducible representation labels are not the same in the two groups. Groups which have character tables containing identical corresponding sets of characters are said to be *isomorphous* groups.† Isomorphous groups need have no operation in common— except, of course, the identity operation, which appears in all groups. However, isomorphism between character tables means that there is a close connection between the groups. Thus, something true in one group has a counterpart in an isomorphous group. An illustration of this is given in Section 9.4.

In the C_i character table (Table 4.4) the only distinction between the irreducible representations is the behaviour of the quantities they describe under the operation of inversion in the centre of symmetry, i. Both irreducible representations are denoted by A but something which is transformed into itself (i.e. is symmetric) under the inversion operation is distinguished from one which is turned into minus itself (i.e. is antisymmetric) by the subscripts g (from *gerade*, German for 'even') and u (from *ungerade*, German for 'odd') respectively; it is always true that a g suffix indicates an irreducible representation which describes something which is symmetric while the suffix u describes something which is antisymmetric with respect to inversion in a centre of symmetry. The reader will find it helpful to compare the use of g and u suffixes in the irreducible representation labels of Table 4.1 with the corresponding characters under the i operation.

Problem 4.7   For each irreducible representation in Table 4.1 which carries a g suffix (e.g. A_g) list the character under the i operation. Repeat this exercise for each irreducible representation carrying a u suffix (e.g. A_u). Compare your result with Table 4.4.

One final point. Lines have been included in Tables 4.1, 4.2 and 4.4 to clarify the discussion in the text. Normally they are omitted and, indeed, columns in these tables are sometimes permuted so that the pattern which is apparent from the way that these have been written here becomes less evident. In the compilation of character tables in Appendix 3 those that are direct products involving C_i have lines included (and, just for good measure, a few which are direct products involving a group which is isomorphous to C_i have lines included too).

4.4   THE SYMMETRIES OF THE CARBON ATOMIC ORBITALS IN ETHENE

The character table of the D_{2h} group given in Table 4.1 will now be used in a qualitative discussion of the electronic structure of the ethene molecule. At first

---

† Strictly, their multiplication tables must also show an analogous similarity but, in practice, the definition in the text is adequate.

sight one might expect that this discussion would be more complicated than that for the water molecule because we now have six atoms to consider. On the other hand, instead of working with a group with only four irreducible representations we now have eight, so that it might be hoped that the increase in symmetry will offset the greater molecular complexity. The first step is, as always, an investigation of the transformation properties of the various sets of atomic orbitals. Linear combinations of these orbitals will then be formed which transform as irreducible representations of the D_{2h} group. Finally, the interaction between orbitals of the same symmetry species will be included and a qualitative molecular orbital energy level diagram obtained.

The valence shell atomic orbitals that must be considered are the 2s and $2p_x$, $2p_y$ and $2p_z$ orbitals of the two carbon atoms and the four 1s orbitals of the terminal hydrogen atoms. Not one of these orbitals is unique—there is always at least one other, symmetry related, atom in the molecule with a similar orbital. This means that, in a sense, the present discussion must start at the point at which the discussion of the water molecule ended. Just as for the hydrogen 1s orbitals in the water molecule, the transformations of corresponding orbitals of symmetry-related atoms must be considered together. As a simple example, consider the 2s orbitals of the two carbon atoms (Figure 4.4).

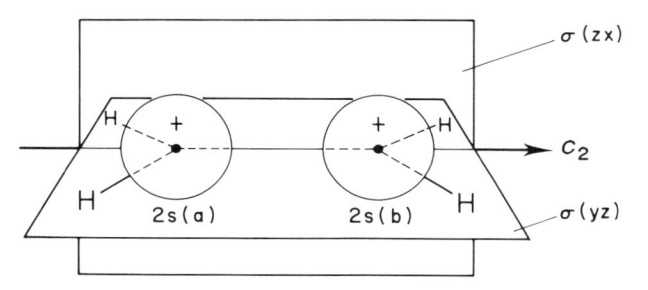

Figure 4.4 Those symmetry operations under which (together with the identity operation) the two carbon 2s orbitals of ethene are not interchanged.

Each of these orbitals remains itself under the $C_2(z)$ rotation, the $\sigma(zx)$ and $\sigma(yz)$ reflection operations and, of course, under the identity operation. For all the other symmetry operations of the group the two orbitals are interchanged. Now, if an orbital is unchanged by a symmetry operation it makes a contribution of unity to the resultant character, while if it goes into another member of the same set it contributes zero, so the characters describing the transformation of the carbon 2s orbitals are:

E	$C_2(z)$	$C_2(y)$	$C_2(x)$	i	$\sigma(xy)$	$\sigma(zx)$	$\sigma(yz)$
2	2	0	2	0	0	2	2

Either by trial and error, or by systematic use of the group orthonormal relationships which were met in Chapter 3 and which will be described in yet more detail in Chapter 5, it is concluded that this reducible representation has $A_g + B_{1u}$ components.

Problem 4.8 Use the orthonormality theorem in the way described in detail towards the end of Section 3.3 to show that the two 2s orbitals of the carbon atoms in ethene transform as $A_g + B_{1u}$. Your solutions to Problems 3.4 and 3.5 should give you any additional guidance you may need.

The 2s orbitals of the two carbon atoms in the ethene molecule, considered in isolation, resemble the two hydrogen 1s orbitals in the hydrogen molecule (or in the water molecule). It is reasonable, therefore, to anticipate that the linear combinations of these orbitals which transform as A_g and B_{1u} will be similar to those which were obtained when discussing the water molecule. That is, if we call the carbon 2s orbitals 2s(a) and 2s(b), as shown in Figure 4.4, then the correct linear combinations are of the form:

$$\frac{1}{\sqrt{2}}\left(2s(a) + 2s(b)\right) \quad \text{and} \quad \frac{1}{\sqrt{2}}\left(2s(a) - 2s(b)\right)$$

Later in this chapter a systematic way of deriving such linear combinations will be obtained and these functions can then be checked. Whenever there are only two symmetry-related orbitals to be considered, the correct combinations are sum and difference combinations—like those above—irrespective of the details of the symmetry. Of the two combinations given above it is easy to demonstrate that the first has A_g symmetry and the second B_{1u}.

Problem 4.9 By drawing diagrams of them and considering their transformations under the operations of the D_{2h} point group show that the two linear combinations given above transform as A_g and B_{1u}, respectively. If you get the answers the wrong way round read the next paragraph and compare the phases you have chosen with those in Figure 4.4.

There is a rather subtle aspect of this. Suppose that instead of choosing in Figure 4.4 to give the two carbon 2s orbitals the same phase they had been given opposite phases. The first combination above would, in this case, be an out-of-phase combination of the two orbitals, notwithstanding the + sign. The solution to this paradox is that in this case the first combination would have had B_{1u} symmetry and not A_g while the second would be the A_g combination. The systematic method of obtaining such functions takes account of our arbitrary choices of orbital phases and corrects for them. It is important to note that one cannot work with combination functions like those given above unless the phases chosen for the component atomic orbitals is known. The simplest way of giving this information is to state the symmetry of the combination, A_g or B_{1u}. One might think that the simple way would be to choose all orbitals to be of the same phase. Unfortunately such a simplification is not always possible. Thus, there are two alternative ways of drawing the $2p_z$ orbitals on the two carbon atoms; these are shown in Figure 4.5. In Figure 4.5(a) the $2p_z$ orbitals are chosen so that the phasing of the $2p_z$ orbitals coincides with that of the molecular coordinate axis system—the positive lobes point towards positive z and the negative lobes towards negative z. In Figure 4.5(b) the phase on one centre is reversed. This latter choice of phases has the advantage that under, say, the $C_2(x)$ rotation operation the $2p_z$ orbitals are simply interchanged. In the choice of Figure 4.5(a) they are not only interchanged by this operation but the phases of their lobes are also reversed. Here there is no simplification offered by convention; different people may, with equal validity, choose the phases differently. It follows that care has to be taken to check the basic choice of phases used by each person writing on the subject. If we choose the phases indicated in Figure 4.5(a) then the sum and difference combinations

$$\frac{1}{\sqrt{2}}\left(2p_z(a) + 2p_z(b)\right) \quad \text{and} \quad \frac{1}{\sqrt{2}}\left(2p_z(a) - 2p_z(b)\right)$$

are, respectively, the B_{1u} and A_g (C–C σ antibonding and bonding respectively) combinations of carbon $2p_z$ orbitals.

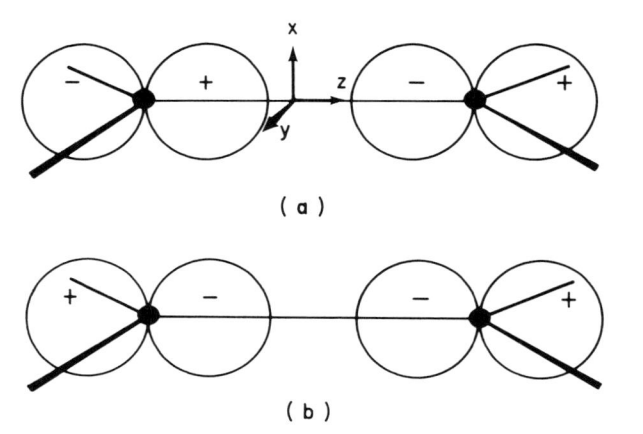

Figure 4.5 Alternative phase choices for the $2p_z$ orbitals of the carbon atoms in ethene. See the text for a discussion.

Problem 4.10 Because the transformation of the carbon $2p_z$ orbitals give rise to the same irreducible representations as do the transformations as the carbon 2s it follows that they transform in the same way (the jargon statement is 'they transform isomorphously'). This means that the $2p_z$ orbitals in both Figures 4.5(a) and 4.5(b) give the reducible representation given in the text just before Problem 4.8. Show that this is indeed so, irrespective of which set of phases is chosen for the $2p_z$ orbitals (Figures 4.5(a) and 4.5(b)).

The B_{1u} and A_g combinations of carbon $2p_z$ orbitals are shown schematically in Figure 4.6. Also, in this figure the A_g and B_{1u} combinations of carbon 2s orbitals are shown. In these diagrams the atomic orbitals on the two carbon atoms are shown as overlapping each other, although this overlap has been neglected in the expressions given above. This inconsistency is tolerated because it gives mathematical simplicity together with diagrammatic clarity.

Although detailed calculations did not fully justify it, in our discussion of the water molecule it was found to be convenient to mix together the oxygen 2s

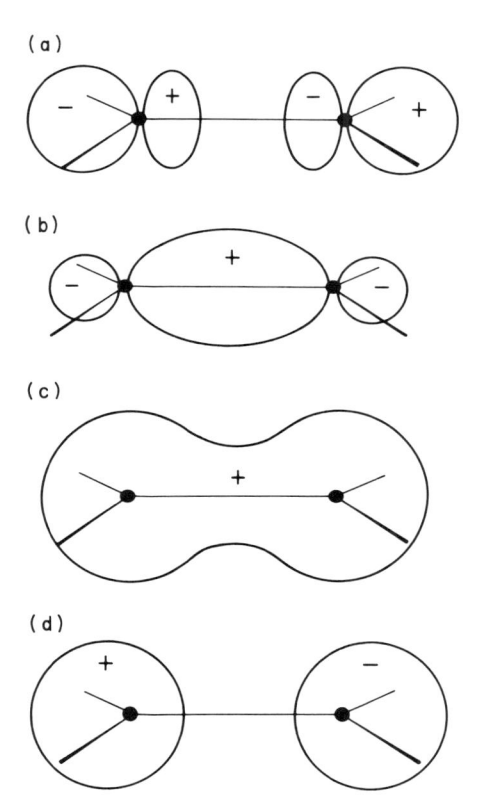

Figure 4.6 B_{1u} and A_g (bonding and antibonding) combinations of $2p_z$ and 2s orbitals in ethene.
(a) B_{1u} (antibonding) combination of $2p_z$ orbitals.
(b) A_g (bonding) combination of $2p_z$ orbitals.
(c) A_g (bonding) combination of 2s orbitals.
(d) B_{1u} (antibonding) combination of 2s orbitals.

and $2p_z$ orbitals. Since they had the same symmetry this mixing is allowed and the resultant picture that emerged was closely related to simple ideas on the bonding in the water molecule—and was an advantage when later looking at the latter in more detail. For the same reason the carbon 2s and 2p orbitals will be mixed in the present example forming, effectively, carbon sp hybrids. If these hybrids had been formed as a first step, a simpler discussion would have resulted. Unfortunately, this was not permissible because at that stage it had not been established that the carbon $2p_z$ and 2s orbitals transform isomorphously. Instead of going back to the start of the argument and working with sp hybrids it is easier to simply combine the A_g combinations of carbon 2s and $2p_z$ orbitals and similarly for the B_{1u}—the end result is the same. The result of mixing together—essentially, taking sum and difference combinations of—the two A_g orbitals of Figure 4.6 and (separately) the two B_{1u} orbitals is shown schematically in Figure 4.7. In each case both in-phase and out-of-phase combinations are shown. Two of these four orbitals will carry through, unmodified, into the final description of the ethene molecule. These are an A_g combination which is to be identified with the C–C σ bonding orbital (Figure 4.7a) and a B_{1u} combination which is the corresponding C–C σ antibonding orbital (Figure 4.7b). The other A_g and B_{1u} combinations (Figures 4.7c and d, respectively), which are largely directed away from the C–C bond, are involved in interactions with the terminal hydrogen atoms.

The other 2p orbitals of the carbon atoms are readily dealt with. For the pairs of $2p_x$ and $2p_y$ orbitals a similar phase ambiguity exists as for the $2p_z$ orbitals,

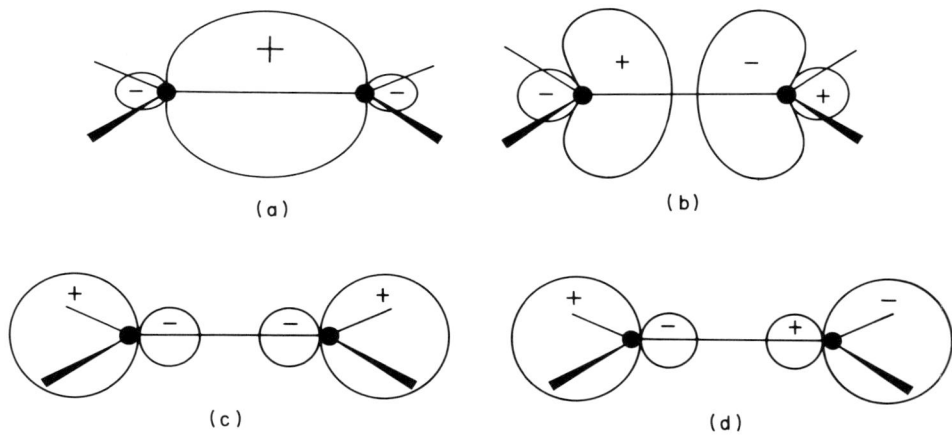

Figure 4.7   (a) The A_g C–C σ bonding orbital in ethene [this is, essentially, Figure 4.6(b) plus Figure 4.6(c)].
(b) The B_{1u} C–C σ antibonding orbital in ethene [this is, essentially, Figure 4.6(a) plus Figure 4.6(d)].
(c) The A_g carbon-based orbital involved in C–H bonding in ethene [essentially, Figure 4.6(c) – Figure 4.6(b) note the – sign].
(d) The B_{1u} carbon-based orbital involved in C–H bonding in ethene [essentially, Figure 4.6(d) – Figure 4.6(a)].

although it is usually found to be less troublesome. In this chapter the phases shown in Figures 4.8 and 4.9 have been chosen. These orbitals transform as follows:

$$2p_x: \quad B_{2g} + B_{3u}$$
$$2p_y: \quad B_{3g} + B_{2u}$$

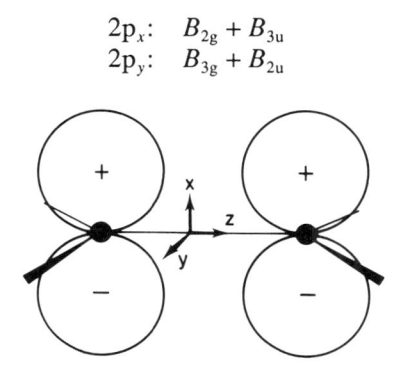

Figure 4.8 Carbon $2p_x$ orbitals in ethene. The phases of the orbitals have been chosen to be identical to those of the x axis.

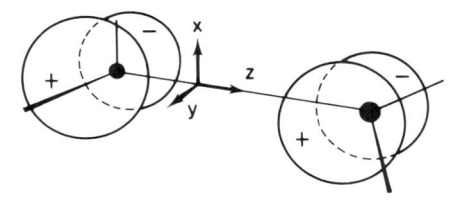

Figure 4.9 Carbon $2p_y$ orbitals in ethene. The phases of the orbitals have been chosen to be identical to those of the y axis.

Problem 4.11 Show that the transformations of the carbon $2p_x$ orbitals of Figure 4.8 and the $2p_y$ orbitals of Figure 4.9 give rise to the following reducible representations

	E	$C_2(z)$	$C_2(y)$	$C_2(x)$	i	$\sigma(xy)$	$\sigma(zx)$	$\sigma(yz)$
$2p_x$	2	-2	0	0	0	0	2	-2
$2p_y$	2	-2	0	0	0	0	-2	2

and then, using the orthnormality theorem method of Section 3.3, that these reduce to the irreducible components given in the text. Check that interchanging the phases of the lobes of one of the p orbitals in Figures 4.8 and 4.9 does not lead to any change in the above results.

Symmetry-correct linear combinations transforming as the above irreducible representations are sum and differences of the carbon $2p_x$ and $2p_y$ orbitals of

Figures 4.8 and 4.9 and are shown in Figures. 4.10 and 4.11. The B_{3u} combination of carbon $2p_x$ orbitals shown in Figure 4.10 is immediately identified as the carbon–carbon π bonding orbital and the B_{2g} combination as the carbon–carbon π antibonding orbital. Both of these will be carried through to the final energy level diagram.

This is a suitable point at which to define the labels σ and π. It is convenient to think of just two bonded atoms (which may be part of a larger molecule) and of a line which connects their nuclei. If an orbital—be it bonding or antibonding, localized or delocalized—has no nodal planes lying in the internuclear line then it involves a σ interaction in the region between the two nuclei. If there is a single nodal plane then the interaction is of π type; if two nodal planes then it is δ. If a molecule is planar then the σ/π distinction extends over the entire molecule and one can correctly distinguish σ molecular orbitals from π molecular orbitals (the former are symmetric and the latter antisymmetric with respect to reflection in the molecular plane).

There is an element of inconsistency in Figures 4.10 and 4.11. The only difference between the 2p orbitals shown in Figures 4.8 and 4.9 is that the former are rotated through 90° relative to the latter. One would therefore expect to find that Figure 4.11 is identical to Figure 4.10 except for this same rotation. In anticipation that the primary interaction involving the B_{2u} orbitals of Figure 4.11 is with the terminal hydrogen atoms, whereas there is no

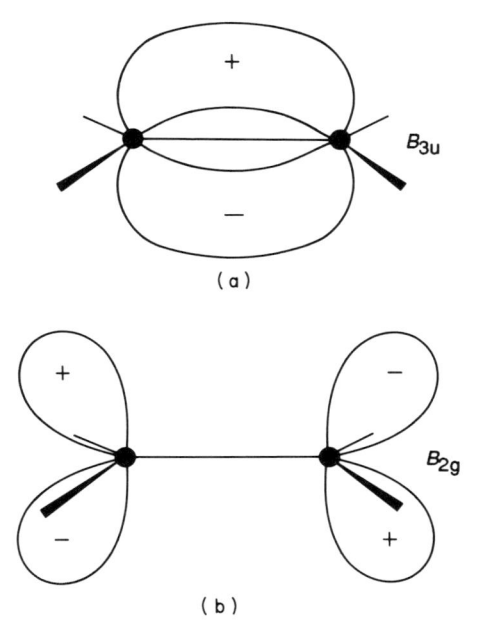

(a)

(b)

Figure 4.10 B_{3u} and B_{2g} combinations of carbon $2p_x$ orbitals in ethene, shown in perspective.

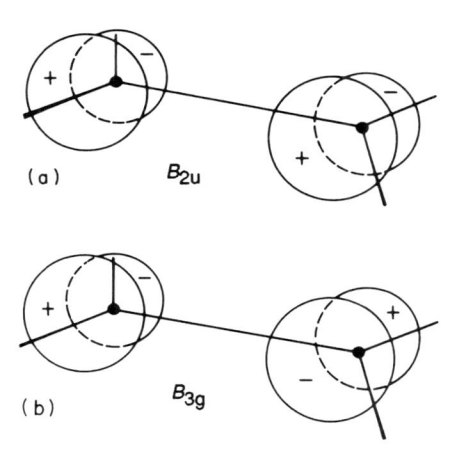

Figure 4.11 B_{2g} and B_{3g} combinations of carbon $2p_y$ orbitals in ethene, shown in perspective.

such interaction involving the orbitals of Figure 4.10, the carbon–carbon overlap has been ignored in Figure 4.11.

4.5 THE SYMMETRIES OF THE HYDROGEN 1s ORBITALS IN ETHENE

We now turn our attention to the four hydrogen atoms of the ethene molecule and consider the 1s orbital on each (which will be taken to each have the same phase). These orbitals are all equivalent one to another—they may be interconverted by the symmetry operations of the group—and so all four must be considered together. They are shown in Figure 4.12 together with the symmetry elements of the D_{2h} group. Of the entire set of corresponding symmetry operations only the identity operation and the $C_2(zx)$ operation leave any of the hydrogen 1s orbitals in their original position and each of these operations leave all four orbitals unmoved; all other operations interchange all of them. The transformations of the four hydrogen 1s orbitals therefore generate the reducible representation shown below. The reduction of this representation into its irreducible components provides a useful illustration of the use of the method described in Section 3.3—reducing it by trial and error could be a little tedious. First select an irreducible representation of the D_{2h}

E	$C_2(z)$	$C_2(y)$	$C_2(x)$	i	$\sigma(xy)$	$\sigma(zx)$	$\sigma(yz)$
4	0	0	0	0	0	0	4

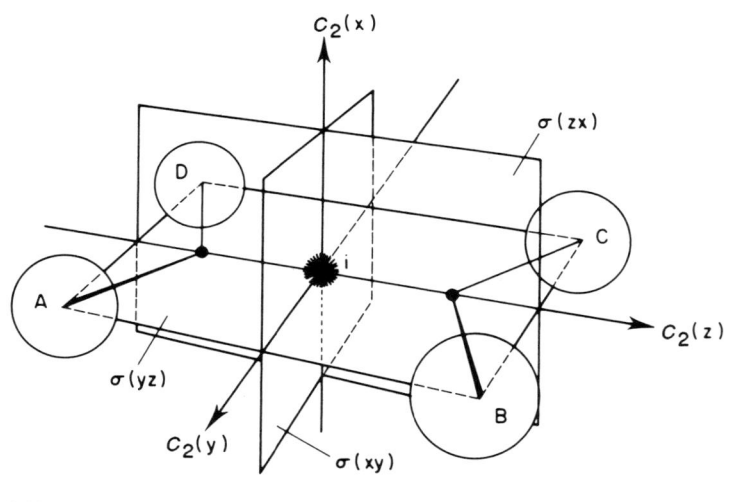

Figure 4.12 The four hydrogen 1s orbitals in ethene together with the symmetry elements of the D_{2h} group (excluding the identity).

group and multiply each character of the above reducible representation by the corresponding character of our selected irreducible representation. Add these products together and then divide by the order of the group (8 in the present case). The integer which results† is the number of times the selected irreducible representation appears in the reducible representation. Thus, if the A_g irreducible representation is selected the calculation proceeds as follows:

	E	$C_2(z)$	$C_2(y)$	$C_2(x)$	i	$\sigma(xy)$	$\sigma(zx)$	$\sigma(yz)$
	4	0	0	0	0	0	0	4
A_g	1	1	1	1	1	1	1	1
Products	4	0	0	0	0	0	0	4

The sum of products is therefore 8. Division by the order of the group yields the result that the A_g irreducible representation appears once. Proceeding in this way with each irreducible representation selected in turn from the character table, it is concluded that the reducible representation has components

$$A_g + B_{3g} + B_{1u} + B_{2u}$$

---

† If a nonsense answer is obtained (for example, a fraction) then either an arithmetical mistake has been made or the reducible representation has been wrongly generated—this is one way in which such mistakes are commonly discovered.

Problem 4.12 Show that the above reducible representation contains $B_{3g} + B_{1u} + B_{2u}$ components in addition to the A_g deduced in the text.

4.6 THE PROJECTION OPERATOR METHOD

Although not essential to a qualitative discussion of the bonding in the ethene molecule, it is very useful at this point to seek combinations of the four hydrogen 1s orbitals, which, separately, transform as A_g, as B_{3g}, as B_{1u} and as B_{2u}. This will provide a relatively simple introduction to the important *projection operator* method. As a bonus, some idea of the form of the C–H bonding molecular orbitals will be obtained.

We first consider the transformations of the individual hydrogen 1s orbitals in much greater detail. Previously, we have only been concerned with whether or not a hydrogen 1s orbital was turned into itself under a particular symmetry operation. If it did not do this the destiny of the hydrogen atom did not concern us. This is no longer the case. We shall now look in detail at one of the four hydrogen 1s orbitals and determine the precise effect of each symmetry operation on this chosen orbital. Label the hydrogen 1s orbitals as shown in Figure 4.12 and consider the transformation of the orbital which is labelled A. The following discussion will be made easier if an eye is kept on Figure 4.12 and another(!) on Table 4.1 and the individual characters that it contains. Under the identity operation, A remains itself; under the $C_2(z)$ rotation it becomes the orbital labelled D; under the $C_2(y)$ rotation it becomes B, and so on. A complete list of its transformations is given in Table 4.5; it is important that the reader checks that this table is correct.

Table 4.5

	E	$C_2(z)$	$C_2(y)$	$C(x)$	i	$\sigma(xy)$	$\sigma(zx)$	$\sigma(yz)$
Under the operation A becomes	A	D	B	C	C	B	D	A

Problem 4.13 Use Figure 4.12 to check that the above table of transformations is correct.

We are now in a position to generate symmetry-correct linear combinations of the hydrogen orbitals. We know that the set A, B, C and D gives rise to a B_{1u} combination and we shall now generate this combination. Consider orbital A and the effect of the $C_2(y)$ operation. Table 4.1 shows that under this operation a function transforming as B_{1u} changes sign. It follows, therefore that orbitals A and B must appear in the B_{1u} linear combination in the form

(A–B) since this expression changes sign under the $C_2(y)$ operation. Now consider the $C_2(x)$ and $C_2(z)$ operations—under which A interchanges with C and D, respectively. Because a B_{1u} function changes sign under $C_2(x)$ but retains its sign under $C_2(z)$ it is evident that C and D must appear as $-C$ and $+D$. It follows that the B_{1u} combination is of the (normalized) form:

$$\tfrac{1}{2}(A - B - C + D)$$

It is a simple matter to check that this combination does indeed transform correctly as B_{1u} under all of the operations of the group. The important thing to recognize is the way that the sign with which an individual orbital appears in the result is determined by the appropriate character of the irreducible representation.

The general method is at once evident. In order to generate a required linear combination we simply take the entries in Table 4.5 and multiply each entry by the corresponding character. The sum of the answers so obtained is the desired linear combination (although it will not be normalized). As an illustration of this method let us generate the B_{3g} linear combination of hydrogen 1s orbitals, by this, the *projection operator*, method:

	E	$C_2(z)$	$C_2(y)$	$C_2(x)$	i	$\sigma(xy)$	$\sigma(zx)$	$\sigma(yz)$
Under the operation								
A becomes	A	D	B	C	C	B	D	A
B_{3g}	1	−1	−1	1	1	−1	−1	1
Product	A	−D	−B	C	C	−B	−D	A

$$\text{Sum:} \quad 2A - 2B + 2C - 2D$$

The linear combination which is generated by this procedure is $(2A - 2B + 2C - 2D)$. This function is not normalized since the sum of squares of coefficients appearing is 16, not 1; to normalize we have to divide by $\sqrt{16} = 4$ and so obtain the normalized B_{3g} combination.

$$\tfrac{1}{2}(A - B + C - D)$$

The A_g and B_{2u} combinations are obtained in a precisely similar way. All four linear combinations are given in Table 4.6, and shown in Figure 4.13. Such combinations are often referred to as 'symmetry adapted combinations'.

Problem 4.14   Use the projection operator method to obtain the (normalized) A_g and B_{2u} combinations of hydrogen 1s orbitals.

It is to be emphasized that each of the four diagrams in Figure 4.13 pictures one orbital and *not* four. An instructive exercise at this point is to attempt to

Table 4.6

Symmetry species	Linear combination of 1s orbitals of hydrogen atoms in ethene
A_g	$\frac{1}{2}(A + B + C + D)$
B_{3g}	$\frac{1}{2}(A - B + C - D)$
B_{1u}	$\frac{1}{2}(A - B - C + D)$
B_{2u}	$\frac{1}{2}(A + B - C - D)$

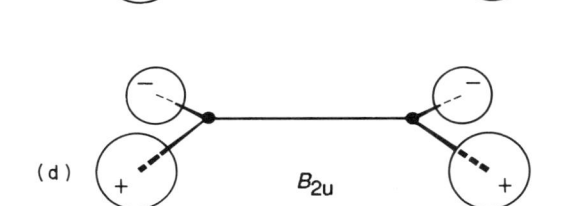

Figure 4.13 The symmetry-adapted combinations of hydrogen 1s orbitals in ethene. Note that the relative phases chosen for the individual hydrogen 1s orbitals is evident in the A_g combination. Here, all were chosen with identical phase but had one been chosen with phase opposite to all of the others then this one would appear with a $-$ phase in the A_g combination above.

generate from the data in Table 4.5 a combination transforming as an irreducible representation which is absent (and so is not listed in either Table 4.6 or Figure 4.13)—for example B_{1g}. It will be found that the projection operator method has the great advantage of being self-correcting!

> **Problem 4.15** Attempt to generate a combination of hydrogen 1s orbitals which does not, in fact, exist. Any of the irreducible representations B_{1g}, B_{2g}, A_u or B_{3u} may be chosen for this.

4.7 BONDING IN THE ETHENE MOLECULE

The symmetry-adapted linear combinations of hydrogen 1s orbitals which have been obtained are of the correct symmetries to interact with some of the carbon orbitals. Thus, the A_g and B_{1u} combinations interact with the carbon sp hybrids which were formed earlier and which are shown in Figures 4.7(c) and 4.7(d) respectively. The B_{3g} and B_{2u} are of the same symmetries as the carbon $2p_y$ combinations [Figures 4.10(b) and 4.10(a) respectively]. The resultant combinations are shown in Figure 4.14 where there is indicated, qualitatively, how they are derived from the earlier figures.

> **Problem 4.16** Check that the molecular orbitals shown in Figure 4.14 are correctly described by combining, qualitatively, the diagrams indicated below each molecular orbital.

There are just four primarily C–H bonding molecular orbitals and four corresponding C–H antibonding orbitals (these orbitals are also either weakly C–C bonding or weakly C–C antibonding). In order to obtain even a qualitative molecular orbital energy level diagram some idea of the relative energies of the various C–H and the C–C σ and π bonding molecular orbitals must be obtained. It is simplest first to look at those orbitals involved in C–H bonding; it will probably be found helpful to refer frequently to Figure 4.14 throughout the next few paragraphs.

There is no doubt about the most stable C–H bonding molecular orbital. This is the A_g orbital. It has two features which lead to its stability. First, just as the largely 2s(O) containing molecular orbital was the most stable in H_2O, so too here, an orbital containing an appreciable 2s(C) component is expected to be very stable. Second, the important interactions in which the A_g orbital is involved are bonding—it is both C–H and C–C σ bonding. Rather similar arguments hold for the B_{1u} largely C–H bonding molecular orbital. It contains a 2s(C) contribution and is C–H bonding but is C–C σ antibonding. It is fair to conclude that the B_{1u} orbital is next in stability after the A_g. The B_{2u} and B_{3g} C–H bonding molecular orbitals contain only carbon 2p orbitals so that they are expected to be at higher energy than the A_g and B_{1u}, which contain carbon 2s. Their relative energies can be related to the residual C–C bonding

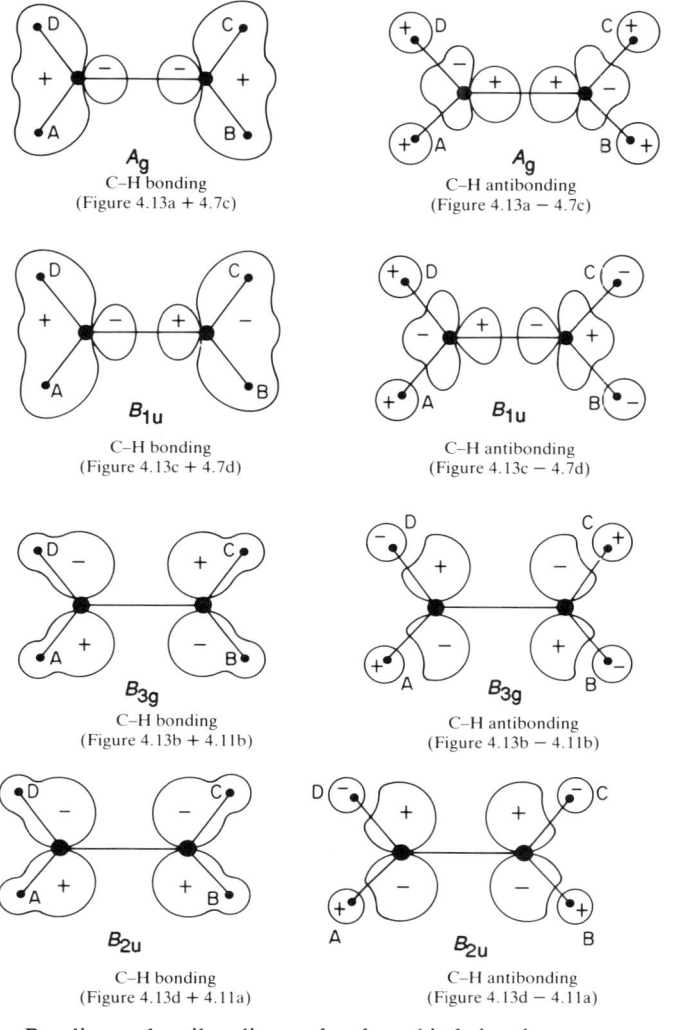

Figure 4.14 Bonding and antibonding molecular orbitals in ethene.

(which will be π in type) associated with each. The B_{2u} C–H bonding orbital is also C–C bonding but the B_{3g} is C–C antibonding. It seems clear that the B_{2u} orbital is the more stable. In summary, then, the C–H bonding molecular orbitals are expected to decrease in stability in the order:

$$A_g > B_{1u} > B_{2u} > B_{3g}$$

We now turn to the orbitals which are largely responsible for the carbon–carbon bonding. They are shown in Figures 4.7(a) and 4.9(a). There is no doubt that the A_g largely carbon–carbon σ bonding molecular orbital will be more stable than the B_{3u} carbon–carbon π bonding molecular orbital because one contains $2s(C)$ whereas the other contains $2p_z(C)$. However, it is not easy to unambiguously relate their energies to those of the C–H bonding molecular orbitals. The following argument is indicative. The bond energy of a single C–C σ bond is about 360 kJ mole^{-1}, although it is to be noted that this figure is appropriate to a bond length slightly longer than that found in ethene. In contrast, the energy of an average C–H bond is about 420 kJ mol^{-1}. It seems reasonable, then, to anticipate that the stabilization resulting from the C–H bonding interactions should be somewhat greater than that of the A_g C–C σ interaction. This means that it would be reasonable to expect the C–H bonding molecular orbitals which have a carbon 2s component (those of A_g and B_{1u} symmetries) to be lower in energy than the carbon–carbon bonding orbital with a 2s component (that of A_g symmetry). We have, then, the stability order:

$$A_g \text{ (C–H bonding)} > B_{1u} \text{ (C–H bonding)} > A_g \text{ (C–C bonding)}$$

The next lowest C–H bonding orbital is B_{2u} and the question is whether its stability is sufficient to make it lower in energy than the A_g (C–C bonding). If we interpret the bond energy data given above as 'the centre of gravity of the energies of the four C–H bonding interactions should be below the energy of the single C–C bonding interaction' then the order

$$B_{2u} \text{ (C–H bonding)} < A_g \text{ (C–C bonding)}$$

seems probable, although not certain. All that can be said is that it seems likely that the two will be of similar energies with perhaps the B_{2u} the more stable. In fact, this is the pattern experimentally observed.

The carbon–carbon π bonding molecular orbital, of B_{3u} symmetry, is also best placed by appeal to experiment. A great deal of spectroscopic and other information on carbon–carbon π bonded systems can be rationalized on the assumption that it is a carbon–carbon π orbital which is the highest occupied orbital. So, the B_{3u} (C–C bonding) is placed above the B_{3g} (C–H bonding). Together with the other arguments above, this leads to the molecular orbital energy level pattern shown in Figure 4.15. There are four valence electrons from each carbon and one from each hydrogen to be placed in these orbitals, a total of twelve. They occupy the six lowest orbitals in Figure 4.15; in this figure only one antibonding orbital, the lowest, C–C antibonding orbital of B_{2g} symmetry, is included.

Figure 4.15 can be checked in two ways. First, appeal can be made to detailed accurate calculations on this molecule. Second, the results of photoelectron spectroscopic measurements can be used. This theoretical and experimental work agree on the energy level sequence of ethene. The results are given below, the calculated values[1] being given in parentheses.

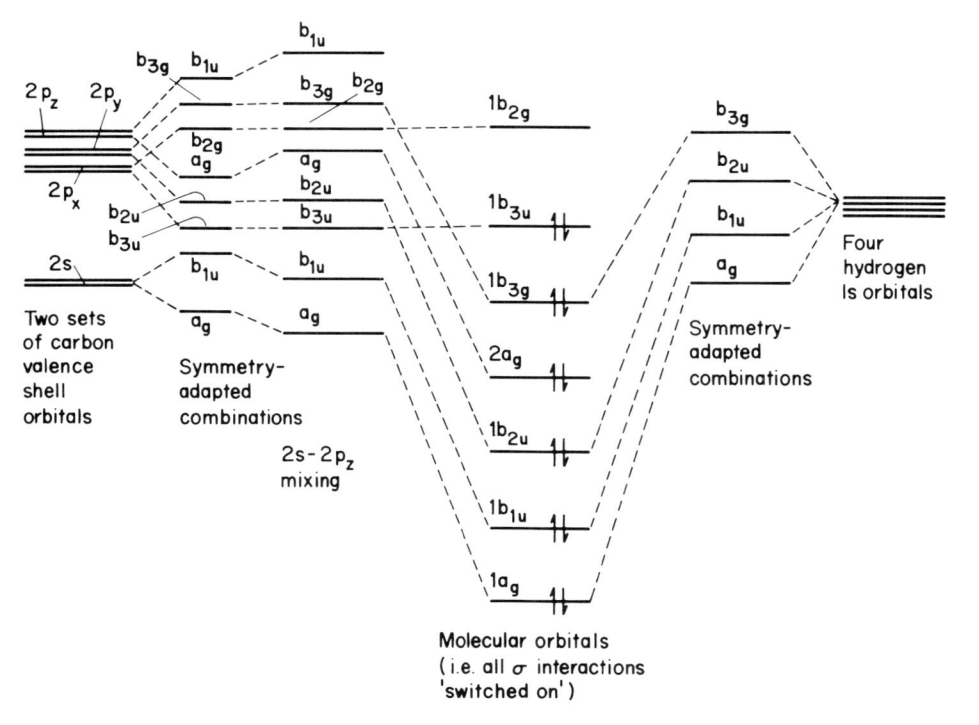

Figure 4.15 Schematic bonding molecular orbital energy level diagram for ethene.

$1b_{3u}$	(C–C bonding)	10.51	(10.44) eV
$1b_{3g}$	(C–H bonding)	12.85	(13.04) eV
$2a_g$	(C–C bonding)	14.66	(14.70) eV
$1b_{2u}$	(C–H bonding)	15.87	(16.07) eV
$1b_{1u}$	(C–H bonding)	19.1	(19.44) eV
$1a_g$	(C–H bonding)	23.5	(26) eV

The agreement with the qualitative picture developed above is excellent, giving some confidence in the arguments that have been used. In particular, the hope that increased molecular symmetry would offset the greater molecular symmetry would offset the greater molecular complexity compared with the water molecule seems to have been justified.

4.8 BONDING IN THE DIBORANE MOLECULE

Our discussion of the ethene molecule can be extended to another molecule, diborane. Diborane, B_2H_6, is of interest because it is the simplest of the boron hydrides (boranes). These, as a class, are often called 'electron deficient' because, whereas at least $(n-1)$ electron pairs are regarded as necessary to bond n atoms the boron hydrides all have fewer than $2(n-1)$ electrons. Thus, there are only twelve valence shell electrons available in diborane to bond eight atoms. However, as will be seen, the term 'electron deficient' is a misnomer because the molecular structure is such that all bonding molecular orbitals are filled with electrons. Whereas diborane posed such a problem for simple bonding models that it appeared necessary to give it a separate classification, a symmetry-based discussion shows that there is no need to invoke new concepts.

The structure of diborane is shown in Figure 4.16, from which it can be seen that it has four terminal hydrogen atoms and two borons which together have the same symmetry, D_{2h}, as ethene (although the bond lengths and angles are different, of course). In addition, diborane has two hydrogen atoms out of, what is for ethene, the molecular plane. These two hydrogens are usually called the 'bridge' hydrogen atoms. It is the presence of these bridging hydrogen atoms in place of the C–C π bond of ethene that plays a major part in leading diborane to have a rather different chemistry to ethene.

Figure 4.16 does not show all of the symmetry elements of diborane. Comparison with Figure 4.1 shows that the bridge hydrogens, located on the $C_2(x)$ axis of Figure 4.1, in no way diminish the D_{2h} symmetry of the ethene-like B_2H_4 unit. Diborane, like ethene, has D_{2h} symmetry. It follows that apart from that involving the bridge hydrogen atoms, the bonding in the diborane molecule must, qualitatively, be similar to that given in the previous section for ethene since boron, like carbon, has 2s and $2p_x$, $2p_y$ and $2p_z$ valence orbitals. It therefore seems reasonable to expect the retention of the same energy level sequence:

A_g (B–H$_t$ bonding) < B_{1u} (B–H$_t$ bonding) < B_{2u} (B–H$_t$ bonding) < B_{3g} (B–H$_t$ bonding)

where the suffix t has been added to distinguish terminally bonded hydrogens from the bridge hydrogens. There is little doubt that there is a substantial difference between the carbon–carbon bonding in ethene and the boron–boron bonding in diborane. This is shown by even a cursory study of the experimental data—the carbon–carbon bond length in ethene is 1.34 Å while the

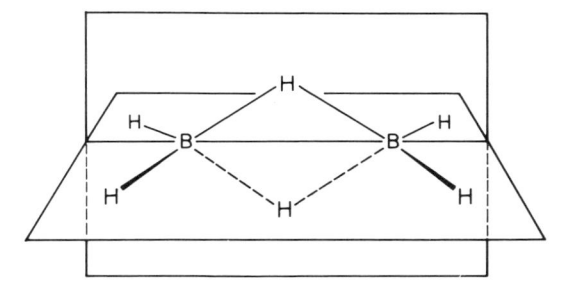

Figure 4.16   The structure of diborane, B_2H_6, shown in perspective.

boron–boron bond length is 1.77 Å. The details of the B–B bonding will also be different from the C–C bonding in ethene because only the former has bridge hydrogens. Clearly, our discussion of the B–B bonding must start with these bridge hydrogens.

The transformations of the 1s orbitals of the two bridge hydrogen atoms in diborane generate the following reducible representation:

E	$C_2(z)$	$C_2(y)$	$C_2(x)$	i	$\sigma(xy)$	$\sigma(zx)$	$\sigma(yz)$
2	0	0	2	0	2	2	0

a representation which has $A_g + B_{3u}$ components. As usual, the functions transforming as these irreducible representations are simply the sum and difference of the two 1s orbitals (which are labelled E and F, as shown in Figure 4.17 and taken to have the same phase). That is, they are as shown below.

Symmetry species	Linear combination of bridge hydrogen orbitals
A_g	$\dfrac{1}{\sqrt{2}} (E + F)$
B_{3u}	$\dfrac{1}{\sqrt{2}} (E - F)$

Problem 4.17 Check that the transformation of the two bridge hydrogen atoms in diborane are as given above. It is of particular importance to show that the two linear combinations of these orbitals transform as indicated.

The only orbitals shown in Figure 4.5 with which it is reasonable to expect any important interaction involving these bridge hydrogen orbitals are the

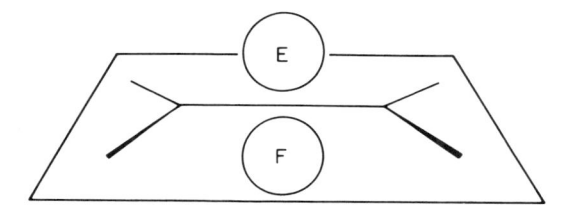

Figure 4.17 The 1s orbitals of the two bridging hydrogen atoms of diborane.

boron–boron π bonding orbital of A_g symmetry (which will be similar to that shown in Figure 4.7(a) but with boron atoms in place of carbon) and the boron–boron π bonding orbital of B_{3u} symmetry (which will resemble that shown in Figure 4.10(a)). The interactions between the bridge hydrogen orbitals and these boron–boron orbitals are shown qualitatively in Figures 4.18 and 4.19. Which of the bonding interactions shown in Figures 4.18 and 4.19 is the more important? For the B_{3u} (boron–boron π bonding) orbital the B_2H_4 plane is a nodal plane; its maximum amplitude must be out of this plane. In contrast, the maximum amplitude of the A_g (boron–boron σ bonding) orbital is in the B_2H_4 plane. Because the bridge hydrogens are above and below this plane it seems probable that the interaction will be greater with the B_{3u} boron combination than with the A_g. Whether this difference will lead to the orbital of B_{3u} symmetry being beneath that of A_g (in Figure 4.15 the B_{3u} is above the A_g) cannot be unambiguously predicted—in fact, it does. An additional reason for this pattern is the greater B–B bond length in diborane compared to the C–C bond in ethene. Because of this difference, it is likely that the A_g B–B σ bonding interaction is less than the C–C σ interaction in ethene.

In Figure 4.20 a schematic molecular orbital energy level diagram for the diborane molecule is given in which all of the above arguments are brought together. The left-hand side of this diagram shows schematically the ethene molecular orbital energy level pattern (Figure 4.15) which is then modified to take account of the bridge hydrogens. Qualitatively, the problem of the relative order of the B_{2u} (B–H$_t$ bonding) and A_g (B–H$_b$ bonding) orbitals, encountered for ethene, reappears here. The experimental and theoretical

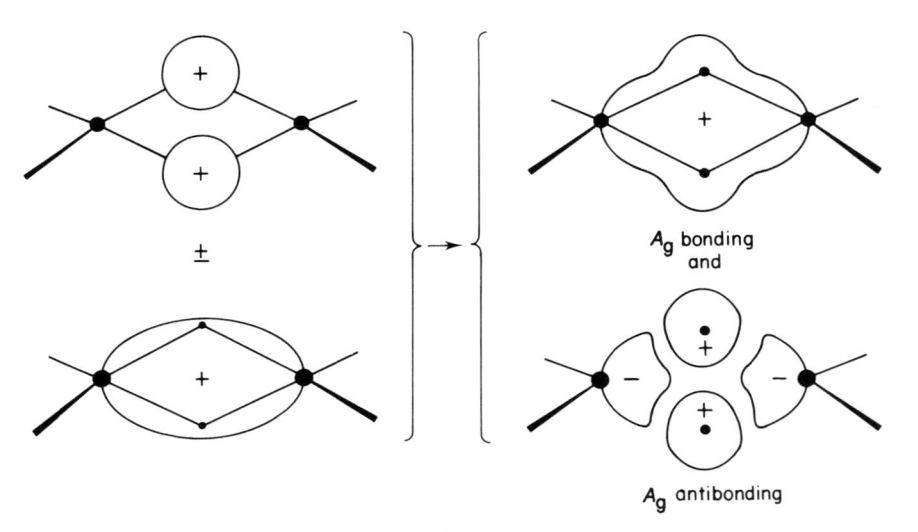

Figure 4.18 Schematic representation of the interactions of A_g symmetry involving the bridge hydrogens of diborane.

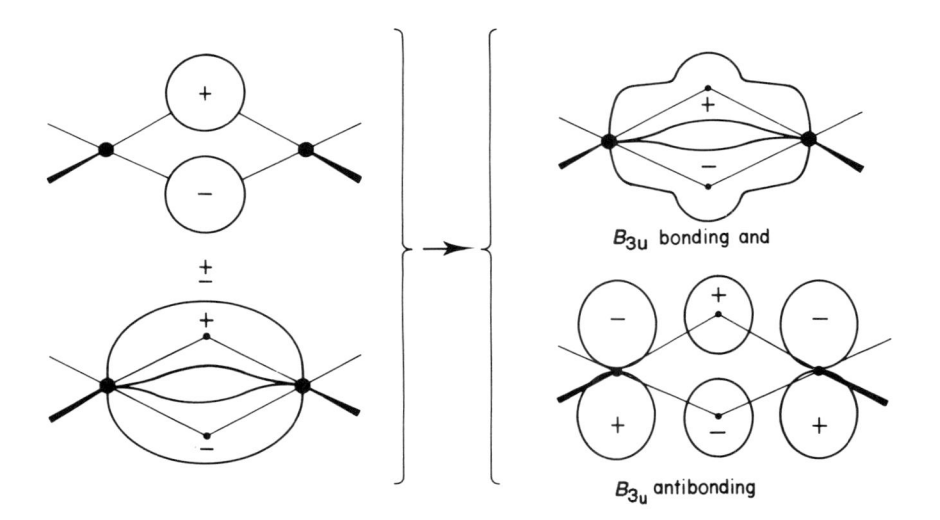

Figure 4.19 Schematic representation of the interactions of B_{3u} symmetry involving the bridge hydrogens of diborane.

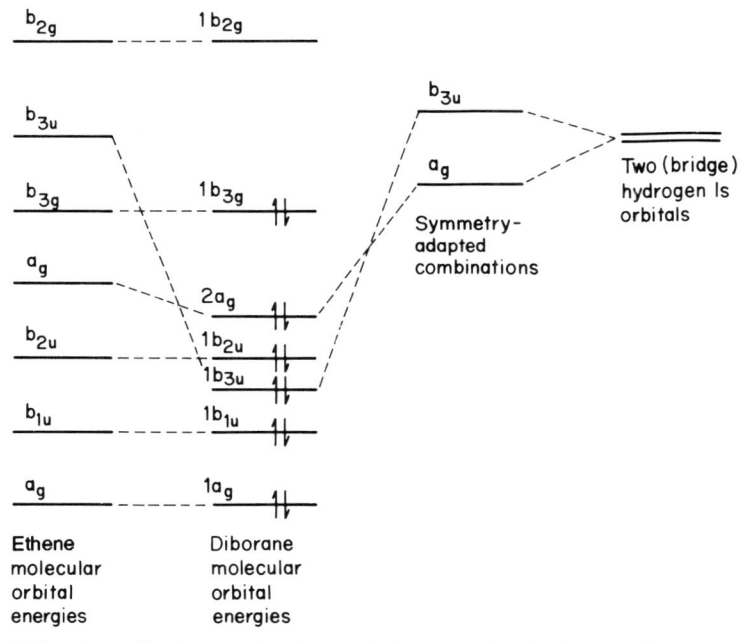

Figure 4.20 A qualitative molecular orbital energy level diagram for B_2H_6 and its relationship to that for C_2H_4.

data[2] (the latter in brackets) are given below (where the suffix b indicates bridge hydrogens):

$1b_{3g}$	(B–H$_t$ bonding)	11.81 (11.95) eV
$2a_g$	(B–H$_b$–B bonding)	13.3 (13.12) eV
$1b_{2u}$	(B–H$_t$ bonding)	13.9 (13.73) eV
$1b_{3u}$	(B–H$_b$–B bonding)	14.7 (14.04) eV
$1b_{1u}$	(B–H$_t$ bonding)	16.06 (16.34) eV
$1a_g$	(B–H$_t$ bonding)	21.4 (22.57) eV

Again, an excellent qualitative prediction of the orbital energies has been obtained using our simple symmetry-based model. It is interesting to note that with the sole exception of that of B_{3u} symmetry, every orbital in this list is at a higher energy than its counterpart in ethene, in accord with the higher chemical reactivity of diborane.

Problem 4.18 The molecule N_2H_4, unlike B_2H_6 and C_2H_4, does not have D_{2h} symmetry (it has a low-symmetry structure which may be regarded as similar to ethane with one hydrogen removed from each nitrogen atom). Use Figure 4.15 to explain why a D_{2h} structure is not stable for N_2H_4. The discussion in the text associated with Figure 4.15 hints at the answer to this problem.

4.9 COMPARISON WITH OTHER MODELS

Most discussions of the electronic structures of the ethene and diborane molecules concern themselves almost exclusively with the carbon–carbon double bond and the bridge bonding respectively. Some of these descriptions appear rather different to those which have been given in the present chapter and it is the purpose of this section to discuss the relationship between the various models.

Consider ethene. Two models are commonly presented for this molecule. In the first, each carbon atom is sp^2 hybridized, two of three sp^2 hybrids being involved in bonding with the terminal hydrogen atoms while the third is responsible for the carbon–carbon σ bonding. A π-bond is formed as a result of overlap between the 2p orbitals which were not hybridized. This model is pictured in Figure 4.21. The sp^2 hybrid orbitals on one carbon atom have been labelled a, d and e and those on the second carbon atom, b, c and f. The hybrids which are involved in carbon–hydrogen bonding are a, b, c and d. It is easy to show that the transformations of these orbitals under the operations of the D_{2h} point group follow (or, more precisely, are isomorphous to) those of the hydrogen 1s orbitals A, B, C and D which were considered earlier in this chapter (Section 4.5). It follows that this hybrid orbital model identifies

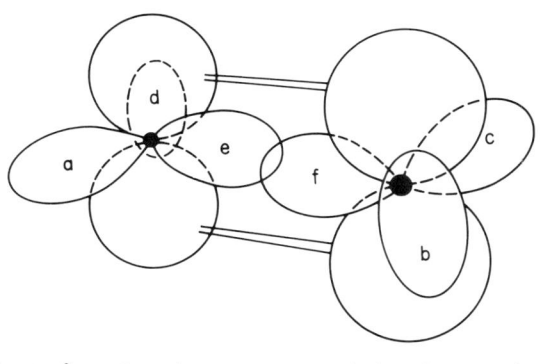

Figure 4.21 The 'sp^2 + p$_\pi$' carbon atom model for the bonding in ethene. For simplicity the hydrogen atoms are omitted. The sp^2 hybrids carry the labels a–f.

the C–H bonding molecular orbitals as being of A_g, B_{3g}, B_{1u} and B_{2u} symmetries, a conclusion identical to that reached above. It is also straightforward to show that the hybrid orbitals e and f form the basis for a reducible representation with A_g and B_{2u} components, which must correspond to the C–C σ bonding and antibonding orbitals. Again, the qualitative description of the C–C σ bonding is identical to that of the symmetry-based model. The main differences can be seen when the two orbitals e and f in Figure 4.21 are compared with their counterparts in the model used above. These orbitals were regarded as equal mixtures of the carbon 2s and 2p$_z$ orbitals whereas in the hybrid–orbital model the s-orbital contribution is only one third. However, it will be recalled that when the 2s and 2p$_z$ orbitals were mixed in equal amounts it was mentioned that this was an arbitrary mixing, made on grounds of simplicity. The 1:1 ratio could, accidentally, have been correct. Equally, the hybrid orbital model could be right in its ratio of 1:2. Detailed calculations show that both are wrong—there are two A_g orbitals contributing to C–C bonding, one largely involving 2s(C) the other 2p$_z$(C). The aggregate 2s:2p$_z$ ratio is 1:1.3 so, from this viewpoint, the sp model adopted in the text is not too bad. A point of apparent divergence between the two approaches is to be found in the carbon–hydrogen bonding orbitals of B_{3g} and B_{2u} symmetries. In the symmetry-based description these orbitals contain no contribution from the carbon 2s orbitals. In contrast, one might expect there to be such a contribution in the hybrid orbital description since each hybrid contains a 2s component. This is not the case. If the form of the hybrid orbitals is written out explicitly and the appropriate linear combinations of them obtained using the projection operator method (these combinations are those given in Table 4.6 but with capital letters replaced by lower case letters) then it will be found that the carbon 2s orbital contributions also vanish in the hybrid orbital description.

Problem 4.19   The explicit forms of the relevant carbon sp^2 hybrid orbitals are:

$$a = \frac{1}{\sqrt{3}}\, s(C_1) + \frac{1}{\sqrt{2}}\, p_y(C_1) - \frac{1}{\sqrt{6}}\, p_z(C_1)$$

$$d = \frac{1}{\sqrt{3}}\, s(C_1) - \frac{1}{\sqrt{2}}\, p_y(C_1) - \frac{1}{\sqrt{6}}\, p_z(C_1)$$

$$b = \frac{1}{\sqrt{3}}\, s(C_2) + \frac{1}{\sqrt{2}}\, p_y(C_2) + \frac{1}{\sqrt{6}}\, p_z(C_2)$$

$$c = \frac{1}{\sqrt{3}}\, s(C_2) - \frac{1}{\sqrt{2}}\, p_y(C_2) + \frac{1}{\sqrt{6}}\, p_z(C_2)$$

where C_1 and C_2 refer to the two carbon atoms. By substituting these in the explicit expressions for the B_{3g} and B_{2u} linear combinations given in Table 4.6 (but substituting the expression given above for a in place of A in Table 4.6 etc.) show that the carbon 2s orbital contributions vanish.

A model of the carbon–carbon double bond in ethene which is historically important and which is still encountered is that in which the carbon atoms are sp^3 hybridized and each bond of the double bond is equivalent, as shown in Figure 4.22. It is a simple matter to show that the two carbon–carbon bonding orbitals labelled a and b in Figure 4.22 provide a basis for a reducible representation with A_g and B_{3u} components. These are the symmetries which have already been deduced as those of the carbon–carbon bonding orbitals. Indeed, if the projection operator method is used to obtain A_g and B_{3u} combinations of a and b (they are the sum and difference of the two) then orbitals are obtained which are, essentially, identical to the carbon–carbon bonding orbitals shown in Figure 4.7 and Figure 4.10. That is, the use of sp^3 hybrids at each carbon atom is also consistent with the model of C–C bonding derived in this chapter, although such a description pictures the orbital on each carbon atom which is involved in this bonding as being one quarter composed of the carbon 2s orbital—the third value we have met! The use of sp^3 hybrids to explain the C–H bonding is also

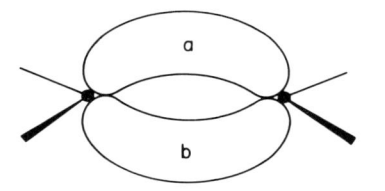

Figure 4.22   The sp^3 carbon atom model for the bonding in ethene. Hydrogens are omitted and the sp^3 hybrids to which they bond are represented by rods. Shown in the diagram are the bonding orbitals formed by the overlap of sp^3 hybrids on the two carbon atoms.

consistent with a symmetry-based discussion. Again C–H bonding molecular orbitals of A_g, B_{1u}, B_{2u}, and B_{3g} symmetries are obtained when the four C–H bonding sp^3 hybrids are used to generate a reducible representation of the group.

Perhaps the simplest description of the bonding of the bridging hydrogen atoms in diborane is the so-called banana bond picture. These bonds are shown in Figure 4.23; the close similarity with Figure 4.22 is immediately apparent. It is not at all difficult to show that the bridge bonds a and b in Figure 4.23 form the basis for two linear combinations, one of A_g symmetry and the other of B_{3u}. These symmetries are the same as those of the orbitals shown in Figure 4.20 as responsible for the bridge bonding. The similarity between the two descriptions follow at once.

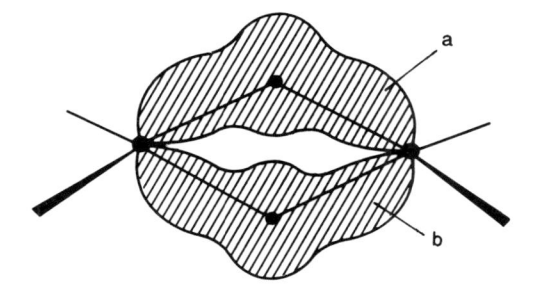

Figure 4.23 The 'banana bond' model for the bonding of the bridge hydrogens in diborane.

When a chemist speaks of a quantity such as 'the carbon–carbon bond' or 'the carbon–hydrogen bond' he or she is frequently referring to quantities which do not, themselves, transform as an irreducible representation of the point group of a molecule. In such cases several other symmetry-related bonds exist and these together provide a basis set from which appropriate irreducible representations can be generated. Thus, in the present context one can say that 'the C–H bonds in ethene (or B–H bonds in diborane) can be combined into combinations which transform as irreducible representations of the D_{2h} point group. Localized orbitals constructed so that they are equivalent to one another in this way and which can be used to derive symmetry adapted combinations are often referred to as *equivalent orbitals*. The C–H bonding molecular orbitals shown in Figure 4.14 are all different; in contrast most chemists prefer to think of equivalent orbitals, although they would probably prefer to call them localized orbitals. As was recognized in the case of the water molecule (Chapter 3), and again in the present chapter for ethene and diborane, these two types of pictures are usually equivalent to each other. Indeed, for all of the simple models which have been shown to be basically similar to the pictures obtained by the symmetry based approach used in this book, the orbitals which have been transformed to obtain reducible representations are all equivalent orbitals. That is, the approach developed in this chapter to the electronic structures of ethene and diborane is, fundamentally, no different from those with which the chemist is more familiar (the same is true of the discussion at the end of Chapter 3). On the other hand, the symmetry based approach has considerable advantages. Thus, the observation that the C–H bonds in ethene are equivalent does not imply that the removal of any one C–H bonding electron requires the same energy as the removal of any other. The fact that there are several ionization potentials—as shown by photoelectron spectroscopy—only becomes clear in a symmetry-based description of the bonding. Despite this emphasis on symmetry it must be recognized that symmetry arguments, by themselves, tell us nothing about energy levels. It is only when these arguments are elaborated by including additional concepts, such as nodality, orbital composition and relative magnitudes of interactions, that relative energies emerge.

One final point. In Section 4.2 it was seen that σ_v operations are equivalent to C_2 followed by i. This is a characteristic of all 'improper rotation' operations—they correspond to a proper rotation combined with i. Other examples will be met later in this book.

4.10 SUMMARY

In this chapter it has been seen that point groups may be related to each other. When a point group is the direct product of two smaller groups (the jargon is to refer to such smaller groups as 'invariant sub-groups' of the larger group) then the multiplication tables of the larger groups may be derived from those of the smaller groups (p. 68) as may its symmetry operations (p. 69), character table (p. 70) and (usually) labels for its irreducible representations (p. 70). The technique of using projection operators to obtain linear combinations of a particular symmetry is most important (p. 81). As in the previous chapter, symmetry based models led to qualitative predictions of electronic structure which were in accord with the results of theoretical calculations, photoelectron spectroscopic data (pp. 87, 91) and also consistent with more traditional bonding models (p. 92).

REFERENCES

1. W. Von Niessen, G. H. F. Diercksen, L. S. Cederbaum and W. Domcke, *Chem. Phys.*, **18** (1976), 469.
2. D. R. Lloyd, N. Lynaugh, P. J. Roberts and M. F. Guest, *J. Chem. Soc.* (*Faraday II*), **71** (1975), 1382.

5

The Electronic Structure of Bromine Pentafluoride, BrF$_5$

Although the object of this chapter is a discussion of the electronic structure of bromine pentafluoride, this topic represents only about a third of its contents. The group theoretical methods that have been developed in the previous chapters must be extended to enable a discussion of almost any molecule, irrespective of its symmetry. This generalization is the major purpose of this chapter and takes up most of it. However, it is simplest to work with an example in mind and bromine pentafluoride is a very convenient one. The structure of the bromine pentafluoride molecule is shown in Figure 5.1. The bromine is surrounded by four fluorines at the corners of a square and by a fifth, unique, apical, fluorine situated so that the five fluorines form a square-based pyramid around the bromine atom. Perhaps surprisingly, the bromine is slightly beneath the plane defined by the four co-planar fluorines. A valence electron count shows that there are two non-bonding electrons on the bromine atom. These are presumably in an orbital directed towards the obvious 'hole' around the bromine which if filled would mean that the bromine is surrounded by six groups at the corners of an octahedron (octahedral molecules will be the subject of Chapter 7). The electron pair repulsion (Sidgwick–Powell–Nyholm–Gillespie) model (Chapter 1) suggests that lone-pair bond-pair repulsion will have a greater effect on the four co-planar fluorine atoms than will the repulsion between these bromine–fluorine bonds and the apical one. The consequence of this inequality will be that the coplanar B–F bonds will be bent

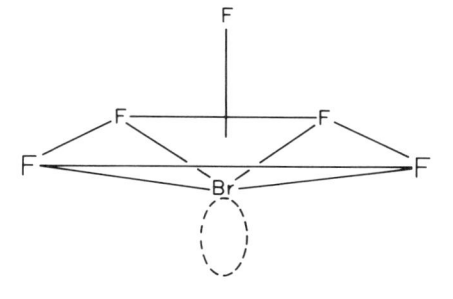

Figure 5.1 The actual structure of the BrF$_5$ molecule. For simplicity, in the text the bromine will be taken as coplanar with the four surrounding fluorines.

towards the apical fluorine atom, giving the observed geometry of the molecule. As a simplifying assumption, however, in this chapter it will be assumed that the central bromine is coplanar with the surrounding four fluorine atoms.

The chapter will start in the way that most group theory problems start in chemistry. One looks up the relevant character table in a compilation of these (such as that given in Appendix 3). So, the character table for the C_{4v} group will simply be presented—the bromine pentafluoride molecule has this symmetry. Hopefully, the reader will be unhappy with this procedure. The character table will, in fact, be derived later in the chapter, using a method rather different to those used in earlier chapters for the C_{2v} and D_{2h} character tables. There are many differences between the C_{4v} character table and these two. Much of this chapter will be occupied by an exploration of these differences—this study is important because it will lead to the generalization of group theoretical concepts and techniques referred to above. In the C_{4v} group there may be more than one symmetry operation corresponding to a single symmetry element and, correspondingly, the character table contains numbers other than 1 and −1. The most important generalizations will be of the orthonormality theorems. It is this generalization that will be used to generate the C_{4v} character table. This table is given in Table 5.1; it is helpful to see it at this point because the reader can then be made aware of the problems (and of their solutions!) before they are encountered.

Table 5.1

C_{4v}	E	$2C_4$	C_2	$2\sigma_v$	$2\sigma_v'$
A_1	1	1	1	1	1
A_2	1	1	1	−1	−1
B_1	1	−1	1	1	−1
B_2	1	−1	1	−1	1
E	2	0	−2	0	0

In the C_{4v} character table, confusingly, E appears in the list of irreducible representation labels as well as in the list of operations. In this new usage it labels an irreducible representation which describes the transformation of two things simultaneously (its character under the 'leave alone' operation is 2). Irreducible representations which describe the properties of two things simultaneously are often called 'doubly degenerate' irreducible representations. The reason for this will become evident later in this chapter. We shall start our discussion by looking in detail at the symmetry operations of the bromine pentafluoride molecule—of the C_{4v} group.

Problem 5.1 Both the D_{2h} and C_{4v} groups are of order eight—a total of eight operations is listed at the top of each table (compare Tables 4.1 and

5.1). However, their structures are rather different. Make a list of the qualitative differences between the two tables.

5.1 SYMMETRY OPERATIONS OF THE C_{4v} GROUP

The perspective shown in Figure 5.1 is not the best for seeing the symmetry of the BrF$_5$ molecule. This symmetry is most readily recognized by viewing the molecule along the bromine–axial fluorine bond as shown in Figure 5.2, from which it is clear that four fluorine atoms lie at the corners of a square. Evidently, the bromine–axial fluorine bond coincides with a fourfold rotation axis (i.e. a C_4) of the molecule. This brings with it something new. In all of the symmetries previously considered there has always been a single symmetry operation associated with each symmetry element of a molecule. As a result, the same symbol has been used for operation and for element, leaving it to the context to make clear which was the subject of discussion. Although we shall persist with the latter convention it must now be recognized that there is not always a 1:1 correspondence between symmetry elements and symmetry operations. In the present case, although there is just one fourfold rotation axis in the BrF$_5$ molecule there are two corresponding symmetry operations. The molecule is turned into itself by a rotation of 90° in either a clockwise or an anticlockwise direction about the fourfold axis. These two operations have the effect of interchanging the fluorine atoms of BrF$_5$ in different ways and so are distinct operations. The clockwise and anticlockwise C_4 rotations operations associated with the C_4 axis are inseparable—one cannot have one without the other. Usually such pairs of operations are grouped together and written as $2C_4$, thus recognizing both their distinction and similarity. It will be seen that they are written this way in the C_{4v} character table (Table 5.1). Operations paired and written in this way are said to be 'members of the same class'. Although this is an adequate definition of 'class' for most purposes, the concept of class is an important one in group theory and it is dealt with more formally and fully in Appendix 1. Because a fourfold axis exists it follows that a rotation of 180° (two steps of C_4 rotation) about this axis also turns the molecule into itself. However, this operation is *not* a C_4 rotation but a C_2, although, of course, you

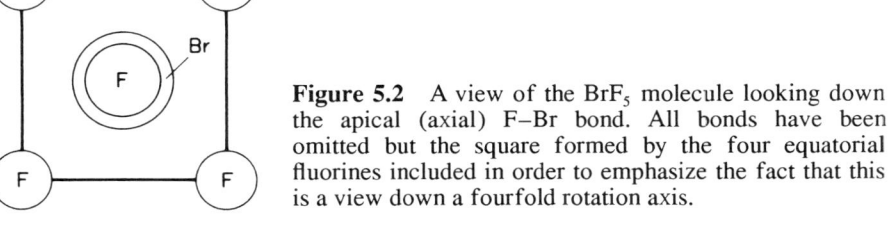

Figure 5.2 A view of the BrF$_5$ molecule looking down the apical (axial) F–Br bond. All bonds have been omitted but the square formed by the four equatorial fluorines included in order to emphasize the fact that this is a view down a fourfold rotation axis.

cannot have the former without also having the latter. Strictly, one should think of there being a C_2 axis coincident with the C_4 axis. The generality is clear—a high rotational symmetry may, automatically, imply the simultaneous existence of coincident axes of lower symmetry.

As has just been mentioned, the C_2 rotation operation may be regarded as a C_4 rotation operation carried out twice in succession (in the same clockwise or anticlockwise sense). Symbolically one can write

$$C_4 \times C_4 = C_4^2 \equiv C_2$$

where, following the discussion of Chapter 2, we have multiplied C_4 by C_4 to obtain C_2. In the same way it is easy to see that

$$C_4^3 \equiv C_4^{-1}$$

—carrying out three C_4 rotations in one sense, clockwise or anticlockwise, is equivalent to a single C_4 rotation in the opposite sense and that

$$C_4^4 = E$$

This is another point of difference with the groups met in previous chapters. For all of these groups it was found that any of their operations carried out twice in succession gave the identity, regenerated the original arrangement. For the C_4 operation it takes four steps (and for a general C_n rotation it takes n).

The other symmetry elements (and associated operations) of the BrF$_5$ are fairly evident. In addition to the identity operation, the two C_4 rotation operations and the associated C_2 rotation operation (which, it should be noted, comprises a class of its own) there are four mirror planes which are indicated in Figure 5.3. It can be seen from this figure that these mirror planes are of two types. First, those which we have labelled $\sigma_v(1)$ and $\sigma_v(2)$ in each of which lie the bromine and three fluorine atoms. It is impossible to have one of

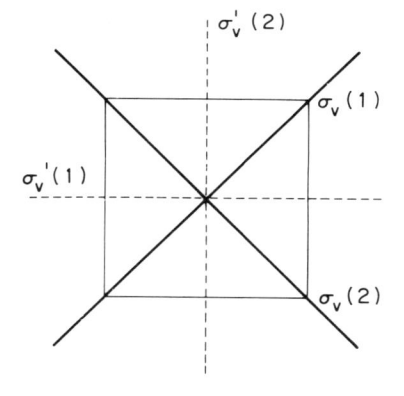

Figure 5.3 Mirror planes of symmetry in Figure 5.2 (bold). The square of Figure 5.2 is again shown.

these operations without the other because of the C_4 axis. A C_4 operation rotates one of these σ_v mirror planes into the other. They are therefore inextricably paired together. These operations therefore comprise a class which is written as $2\sigma_v$ and it appears in this form in the character table. Second, there are those mirror planes which have been labelled $\sigma_v'(1)$ and $\sigma_v'(2)$ in Figure 5.3. Each contains the bromine and the axial fluorine atom and again are interrelated by the C_4 axis. They comprise the class $2\sigma_v'$. Several comments are relevant at this point. First, all four of the mirror planes contain the C_4 axis and so are σ_v mirror planes, as they have been labelled. Second, many authors prefer to give the mirror planes which we have called $\sigma_v'(1)$ and $\sigma_v'(2)$ the labels $\sigma_d(1)$ and $\sigma_d(2)$, or, as a class, $2\sigma_d$. This is because in a closely related group— that of the symmetry operations of a square—they carry this label. Strictly, however, the loss in symmetry in going from this group to C_{4v} forbids the use of the σ_d symbol (as will be seen in Section 7.1, this symbol has a rather precise meaning which forbids its use here). Third, a comment on the fact that $\sigma_v(1)$ and $\sigma_v(2)$ are interconverted by a C_4 rotation, as also are $\sigma_v'(1)$ and $\sigma_v'(2)$. When symmetry elements are interconverted by another operation of the group, it is a sure sign that the corresponding operations fall into the same class. Finally, it is the presence of the C_4 axis, together with the vertical mirror planes that give rise to the shorthand symbol for the group, C_{4v}.

Collecting all of the symmetry operations of the C_{4v} group gives:

$$E \quad 2C_4 \quad C_2 \quad 2\sigma_v \quad 2\sigma_v'$$

and it is these operations that head the character table (Table 5.1). Although the next major task is to derive this character table, it is convenient first to consider a problem which will be encountered when using it.

Problem 5.2 Either draw a diagram or (better) make a model of the BrF_5 molecule and, by a study of this, make a list of the symmetry elements that it contains. Compare your list with that given above and explore the reason for any differences.

5.2 PROBLEMS IN USING THE C_{4v} GROUP

When considering the transformation of something—an orbital or set of orbitals, perhaps—what should be done when there are two operations in a class? How is character generated in such a case? Although the formal answer to this is unattractive, the practical answer is simple. Formally, the correct procedure is to consider the transformations under each of the individual operations in the class and to take the average of characters generated. However, it is invariably the case that each of the symmetry operations in a class always gives the same character. This means that, in practice, all that has to be done is to:

> Select a single symmetry operation from a class (and it quite often happens it is possible to set up the problem in such a way that there is one operation with which it is particularly easy to work) and take the character generated by this operation.

There is yet one more problem which it is as well to consider before turning to the C_{4v} character table. As has been seen, the axis of highest rotational symmetry is conventionally chosen as the z axis so that the C_4 axis of BrF$_5$ is clearly to be taken as the z axis. However, we are left with the problem of where to place the x and y axes. Perhaps the most obvious choice of direction is that shown in Figure 5.4, in which the four fluorines are taken to define the x and y axes, but what is wrong with the alternative choice given in Figure 5.5? The solution to this problem becomes clearer when it is noted that the x and y axes, just like the σ_v mirror planes in which they lie, are interchanged by the C_4 operations, irrespective of whether the orientation of Figure 5.4 or of Figure 5.5 is chosen for them. The orientations for x and y axes in these figures have an obvious attraction— they are choices which place the axes in mirror planes. A less attractive choice (but perfectly admissible one) such as that shown in Figure 5.6, still retains the property that x and y are interrelated by a C_4 rotation. Clearly, the x and y axes must be treated as a pair (if the choice for one is changed so too must that for the other), just as the two σ_v and the two σ_v' mirror planes have to be treated as a pair. This intimate pairing of x and y axes is basic to the difference between the C_{4v} character table and the Abelian character tables of earlier chapters in this book.†

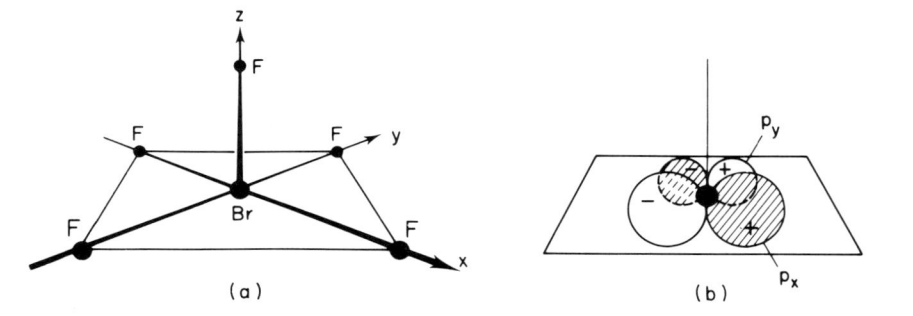

Figure 5.4   One choice of direction for x and y axes in BrF$_5$ (and consequent directions for the bromine $4p_x$ and $4p_y$ orbitals; the $4p_x$ orbital is shown cross-hatched).

---

† The argument presented here is not quite complete, as will be seen in Chapter 11, where the C_4 group will be explored. In that group the x and y axes behave as described above and yet the C_4 group is Abelian. As will be seen in Chapter 11, the dilemma is resolved in that the group contains an E irreducible representation, in which the two components are said to be 'separably degenerate'.

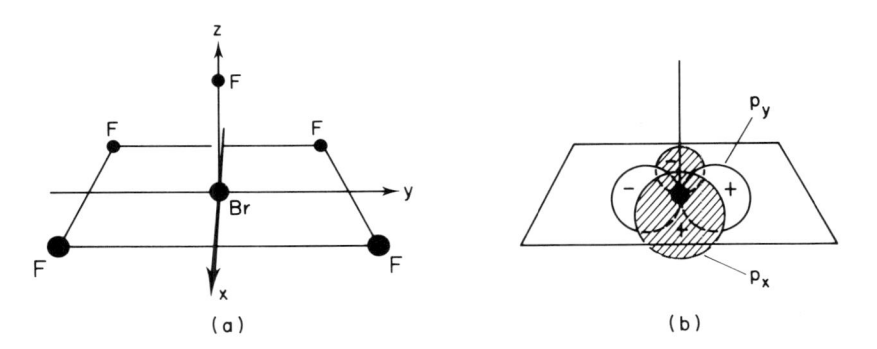

Figure 5.5 An alternative choice of direction for x and y axes in BrF_5 (and consequent directions for the bromine $4p_x$ and $4p_y$ orbitals).

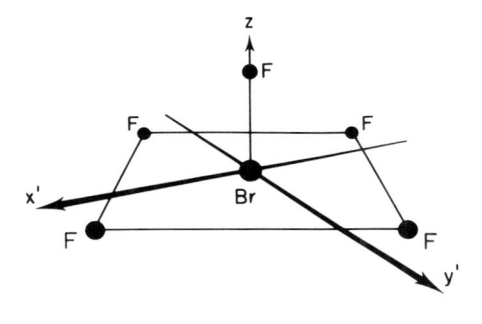

Figure 5.6 A third choice of direction for x and y axes in BrF_5.

The x and y axes are said to 'transform as a pair', a statement which will later be seen to be manifest in the fact that the x and y axes transform, together, as the E irreducible representation in Table 5.1. The choice of x and y axis directions—Figures 5.4, 5.5 and 5.6—is ultimately unimportant; after all, no physical property can in any way depend on the way we choose to place axes. Although this is an almost trivial statement, it is not so easy to elaborate it. This elaboration is discussed in more detail in Appendix 2.

5.3 ORTHONORMALITY RELATIONSHIPS

We now return to the problem of generating the character table of the C_{4v} point group. In principle, the procedure of Chapter 3 could be followed and the transformations of atomic orbitals of the bromine atom used to generate the irreducible representations of this group. Unfortunately, a complete set of irreducible representations could not be obtained even if f orbitals on the bromine atom were included—although if g orbitals were also included it would suffice!

If this is the only method available for the compilation of a character table, then for the more complicated groups one would be left wondering whether a study of yet higher orbitals might uncover further, previously unrecognized, irreducible representations. Fortunately there are systematic methods available for the generation of character tables. One of these methods will now be described, one which relies on the existence of the orthonormality theorems which have already been used in a simple form in earlier chapters. The form in which they are used here is a simple extension of their earlier form, adapted to take account of the fact that there can be more than one operation in a class (this was one of the simplifications involved in using Abelian groups—they have one operation in each class, never more). The general proof of these theorems is rather mathematical and is given in Appendix 2.

Of the theorems that follow, numbers 2, 3, 4 and 5 are those that are commonly called *the orthonormality theorems*.†

Theorem 1   In every character table there exists a totally symmetric irreducible representation.

Comment: The totally symmetric irreducible representation is the first given in any character table and has a character of 1 associated with each class of operation. It describes the symmetry properties of something which is turned into itself under every operation of the group. This really is a rather trivial theorem, introduced here for convenience. By definition, a molecule is turned into itself by every operation of the point group used to describe it. A totally symmetric irreducible representation must therefore exist for every point group.

Theorem 2   Take each element of any row of a character table (i.e. the characters of any irreducible representation), square each and multiply by the number of operations of the class to which the character belongs and add the answers together. The number that results is an integer which is equal to the order of the group (i.e. equal to the total number of symmetry operations in the group).

Comment: This theorem was first met in Chapter 3, when a systematic method of reducing a reducible representation into its irreducible components was obtained. Because an Abelian group was then involved, the number of operations in each class was one, so that there was no need to include the step of multiplying by the

---

† In this chapter and all others up to Chapter 10, all the characters that will be met are real. The theorems that follow apply only to character tables with such real characters. When complex characters are encountered, changes to the theorems have to be made to deal with this. The changes that are needed are covered in Chapter 11.

number of operations in a class. For a non-Abelian group, for which there is invariably at least one class containing more than one operation, care must be taken to include this additional step, otherwise the number obtained at the end of the summation will not be that of the order of the group.

Problem 5.3   Apply Theorem 2 to the irreducible representations of the C_{4v} point group (Table 5.1). The order of this group is eight.

Theorem 3	Take any two different rows of a character table (i.e. any two irreducible representations) and multiply together the two characters associated with each class. Then, in each case multiply the product by the number of operations in the class. Finally, add the answers together. The result is always zero.

Comment: This theorem again, is one already met—when reducing reducible representations. On that occasion, because there was only one operation in each class there was no need to explicitly include multiplication by the number of elements in the class. In general, however, this step must be included.

Problem 5.4   Apply Theorem 3 to at least five pairs of irreducible representations of the C_{4v} point group (Table 5.1). The E irreducible representation should be included in at least two cases.

As will be seen, Theorems 2 and 3 are at the heart of the method used to reduce reducible representations into their irreducible components. They are sometimes referred to as if the others did not exist and called, 'the orthonormality relationships'.

Problem 5.5   Look back at Section 3.1 and read the discussion on orthonormality given there. Why are Theorems 2 and 3 referred to as 'orthonormality relationships'?

The fourth and fifth theorems are similar to the Theorems 2 and 3 but relate to the columns of a character table instead of the rows. They are new to the reader but it can readily be checked that they are correct when applied to all of the character tables met so far.

Theorem 4	Consider any class (column) of a character table and square each of the elements in it; sum the squares and multiply the answer by the number of operations in the class. The answer is always equal to the order of the group.

Problem 5.6 Apply Theorem 4 to the columns of the C_{4v} character table.

Theorem 5   Consider any two different classes (columns) of the character table. This selects two characters of each irreducible representation. Multiply these pairs of characters of the same irreducible representation together, and sum the results. The answer is always zero.

Comment: In this case no explicit allowance has been made for the number of elements in a class. This is because multiplying by any factor which is common to all contributions to the sum would not change the final answer—it would still be zero.

Problem 5.7 Apply Theorem 5 to at least five pairs of columns of the C_{4v} character table.

Theorem 6   This states that a character table is always square—it has the same number of columns as it has rows; there are as many irreducible representations as there are classes of symmetry operations.

Comment: Yet again, it is easy to see that this theorem holds for all the character tables that have been encountered so far in this book.

Problem 5.8 As was indicated in a footnote at the end of Section 2.4. some character tables contain complex numbers. Sometimes, authors of introductory texts attempt to protect their readers from such horrors by manipulation of the character table. The 'character table' for the group C_4 taken from one such text is given below.

C_4	E	$2C_4$	C_2
A	1	1	1
B	1	-1	1
E	2	0	-2

Show that this 'character table' does not fully obey Theorems 2, 4 and 5. (The correct character table will be discussed in detail in Chapter 11 and is given in Table 11.1.)

5.4 THE DERIVATION OF THE C_{4v} CHARACTER TABLE USING THE ORTHONORMALITY THEOREMS

In this section the C_{4v} character table is derived systematically, in contrast to the hit-or-miss method of previous chapters (where the examples were chosen to give hits, of course). There are several methods of deriving character tables; that in this section is the easiest to understand, follow and use. The derivation of the C_{4v} character table starts by using Theorem 6. The total number of symmetry operations in the C_{4v} group is eight and it has already been seen that they fall into five classes. Because the character table must be square (Theorem 6) it follows that there are just five irreducible representations. Theorem 4 requires that the sum of squares of characters lying in the column corresponding to the identity operation is eight (the order of the group). Further, because of the nature of the identity operation (it counts a number of objects), none of these integers can be negative or zero. The only set of integers which satisfies these conditions is the set 1, 1, 1, 1 and 2 $(1^2 + 1^2 + 1^1 + 1^2 + 2^2 = 8)$. Including the totally symmetric irreducible representation (Theorem 1) we can write down the skeleton character table shown in Table 5.2, where the quantities a → p have yet to be determined.

Table 5.2

E	$2C_4$	C_2	$2\sigma_v$	$2\sigma_v'$
1	1	1	1	1
1	a	b	c	d
1	e	f	g	h
1	i	j	k	ℓ
2	m	n	o	p

Because an irreducible representation which describes the behaviour of a single object has characters which can only be 1 or −1, more strictly, are always of modulus unity (the object always goes into itself or minus itself, never into a different object under a symmetry operation) the entries a to ℓ in Table 5.2 all have values of either 1 or −1.

Consider now the column corresponding to the C_2 rotation operation. Again, by Theorem 4, the sum of characters in this column has to equal eight and since from the last paragraph the squares of b, f and j are each +1 it follows that n^2 must be 4 so that n is either +2 or −2. In fact, it follows from the previous paragraph that each of the entries a to ℓ, when squared, gives the number 1. From Theorem 4, and because each of the $2C_4$, $2\sigma_v$ and $2\sigma_v'$ classes have two operations in them, the elements m, o and p must each be equal to zero. If they had any other value, the sum of squares of elements in each column when multiplied by two, the order of the class, would give a number greater than eight. We are thus led to Table 5.3, in which all ± signs are to be regarded as independent of each other.

Table 5.3

E	$2C_4$	C_2	$2\sigma_v$	$2\sigma_v'$
1	1	1	1	1
1	±1	±1	±1	±1
1	±1	±1	±1	±1
1	±1	±1	±1	±1
2	0	±2	0	0

Consider the identity column in Table 5.3 together with one of the columns corresponding to any class of order two. For Theorem 5 to be satisfied (i.e. zero obtained when the products of corresponding characters are summed), the three characters listed for each class as ±1 must in fact, contain one +1 and two −1's. Since at this point in the argument the middle three rows of Table 5.3 are identical, two −1's may be arbitrarily selected for any one class of order two. This we shall do for the $2C_4$ class. The new form of Table 5.3 could be written down but first it is convenient to apply Theorem 3 (the sum of the products of characters multiplied by class orders must be zero) using the double degenerate irreducible representation given in Table 5.3 together with the first (totally symmetric) irreducible representation. Theorem 3 is only satisfied if the ±2 entry under the C_2 class of the doubly degenerate irreducible representation is actually −2. The characters of the doubly degenerate irreducible representation have therefore all been obtained. These results are brought together in Table 5.4.

There are many ways of completing the generation of the character table. For example, apply Theorem 5 (the sum of products of elements of the two classes must be zero) to the columns headed by the E and C_2 operations (the E and C_2 classes) in Table 5.4. The theorem can only be satisfied if all of the characters in the C_2 class are +1. Remembering this result, consider the first two rows (irreducible representations) of Table 5.4 and apply Theorem 3 (the sum of the products of characters multiplied by class orders must be zero). The only way in which a sum of zero can be obtained is if the two ±1 characters in

Table 5.4

E	$2C_4$	C_2	$2\sigma_v$	$2\sigma_v'$
1	1	1	1	1
1	1	±1	±1	±1
1	−1	±1	±1	±1
1	−1	±1	±1	±1
2	0	−2	0	0

the second irreducible representation are actually -1. These results are summarized in Table 5.5.

Table 5.5

E	$2C_4$	C_2	$2\sigma_v$	$2\sigma_v'$
1	1	1	1	1
1	1	1	-1	-1
1	-1	1	±1	±1
1	-1	1	±1	±1
2	0	2	0	0

Perhaps the most evident thing about the residual unknowns in Table 5.5 is that the characters associated with the third and fourth irreducible representations are the same. The application of either Theorem 3 or 5 readily shows that the four ±1 characters in Table 5.5 must be either

$$
\begin{array}{rr}
1 & -1 \\
-1 & 1
\end{array}
\quad or \quad
\begin{array}{rr}
-1 & 1 \\
1 & -1
\end{array}
$$

Substitution of these sets of numbers alternately into Table 5.5 shows that they generate the same two irreducible representations; the alternatives merely differ in the order in which the irreducible representations are listed. The generation of the C_{4v} character table is complete!

Problem 5.9 The derivation of the C_{4v} character table has been explained in some detail. It is important that each step is followed closely because this will give valuable practice in the use of the orthonormality theorems. If it has not already been done in reading this section, carefully check each step in the derivation of the C_{4v} character table.

The final character table is given in Table 5.6 where the commonly adopted symbols for the irreducible representations have also been included.

Table 5.6

C_{4v}	E	$2C_4$	C_2	$2\sigma_v$	$2\sigma_v'$	
A_1	1	1	1	1	1	$z, z^2, x^2 + y^2$
A_2	1	1	1	-1	-1	
B_1	1	-1	1	1	-1	$x^2 - y^2$
B_2	1	-1	1	-1	1	xy
E	2	0	-2	0	0	$(x, y), (zx, yz)$

Note the difference between irreducible representations labelled B and those labelled A. Both are singly degenerate but the B's are antisymmetric with respect to a rotation about the axis of highest symmetry (C_4) whereas the A's are symmetric. This particular distinction may be compared with that discussed in Section 2.4, where the A's and B's in the C_{2v} point group were distinguished by their behaviour under a C_2 rotation operation. The generalization is clear—for a group for which the highest rotational axis is C_n, A's are symmetric with respect to this operation whereas B's are antisymmetric.

Problem 5.10 A fragment of the C_{8v} character table is shown below. Complete this fragment.

C_{8v}	E	$2C_8$	$2C_4$	$2C_8^3$	C_2	$\ldots$
A_1	1					$\ldots$
A_2	1					$\ldots$
B_1	1					$\ldots$
B_2	1					$\ldots$

Hint: Within this fragment there is no distinction apparent between A_1 and A_2 or between B_1 and B_2. Consider first behaviour under C_8; the other entries follow because C_4 and C_8^3 are multiples of C_8 and the characters under these operations must be consistent with that for C_8. This problem illustrates yet another approach to the compilation of character tables and the sort of relationships that exist within them. A discussion which parallels that required to answer this problem is to be found at the end of Section 11.5.

Problem 5.11 Use the second part of the hint in Problem 5.10 to explain why there are no B irreducible representations in the character table of the C_{7v} group (or, indeed, any group containing a C_n axis when n is odd).

We are now in a better position to discuss Table 5.6 than when it was first met as Table 5.1. There are five aspects of it on which it is appropriate to comment, all associated with the E irreducible representation. The first has already been mentioned, the label E itself. This is identical to the label used to describe the identity operation. Although this appears confusing, in practice it is not. This is because the contexts in which the two labels are used are always quite different; the context tells which is intended. In some texts, however, the ambiguity is avoided by a difference in typeface or, more simply but less frequently, by the use of the label I for Identity operation. Second, the occurrence of the characters 2 and 0 in this irreducible representation is something new and requires comment. The appearance of the character 2 for the identity, leave alone, operation means that two things are being left alone, that the E irreducible representation describes the behaviour of a pair of objects simultaneously. The x and y axes of BrF_5, no matter which choice is made for them, is such a pair. Another, closely related, example which will be considered in detail shortly are the valence shell p_x and p_y orbitals of the bromine atom in BrF_5. The character 2 here, then, means the same as when it was met in Chapter 3; that two objects remain themselves. However, the way that the 0 appears is something new. Previously, this character was obtained because every object under consideration moved as a result of a symmetry operation. There is another way in which 0 can appear. This is when each object which remains unchanged is matched by one which changes its sign. The sum of 1 and -1 is, of course, 0. An example of this will be met when the p_x and p_y orbitals of the bromine atom in BrF_5 are considered. Third, note the way in which the members of a basis set for the E representation are written (in the extreme right-hand column of the character table). The x and y axes are such a pair and are written (x, y) in contrast to the listing of functions which, separately and independently, provide a basis for a representation. The functions z and z^2 each, separately, forms a basis for the A_1 irreducible representation. The way that either can do this independently of the other is indicated by the way they are written: z, z^2. Fourth, this is a convenient point at which to formally introduce a piece of useful jargon (which we have already used!). Irreducible representations which describe the transformation of two objects simultaneously are said to be *doubly degenerate* whereas those describing the transformation of one are said to be *singly degenerate*. There are fundamental reasons for this usage but simplest is to note that if the objects the irreducible representations describe are orbitals, then for E irreducible representations there *must* be two orbitals with exactly the same energy. Were they not the same, the act of carrying out, for example, a C_4 rotation would have the effect of changing energies (because it interchanges the orbitals, p_x and p_y, for instance). The energy of an orbital would depend on whether or not we chose to do a C_4 operation and this clearly is ridiculous. This dilemma is only avoided by the orbitals having the same energy, being degenerate.

Fifth, if we were to construct a group multiplication table for the C_{4v} group we would find that only the characters of the various A and B irreducible representations could be substituted for their corresponding operations to give an arithmetically correct multiplication table. The substitution fails for the E irreducible representation. The reason is that we should really use 2×2 matrices to describe the E irreducible representation, not a simple number. When these *matrices* are substituted for the corresponding characters, and multiplied by the laws of matrix multiplication then a correct multiplication table is obtained. This is explained in more detail in Appendix 2. At this point it is appropriate only to comment that ordinary numbers may be regarded as 1×1 matrices (whereupon the laws of matrix multiplication reduce to the ordinary laws of numerical multiplication) so that those irreducible representations containing only characters value $+1$ or -1 may also be regarded as involving matrices. The connection between matrices and character tables is profound and important. Indeed, the name 'character' is the name given to the sum of the elements along the leading diagonal (top left to bottom right) of a matrix. This is no accident, as Appendix 2 makes clear. However, in almost all of the applications of group theory to chemistry there is no need to make explicit use of matrix algebra. Hence this book, in which there is no use of matrix algebra in the body of the text.

Two quite different methods of generating character tables have now been encountered, that of using the transformations of suitable basis functions and that of the use of the character table theorems. As has been indicated, there are also other methods too. They will not be discussed. From this point on in the text, character tables will not be systematically derived. The procedure will be that almost invariably used—a character table is taken from a compilation such as that in Appendix 3. When appropriate, however, comments on various aspects of those that are met will be included in the text.

5.5   THE BONDING IN THE BrF$_5$ MOLECULE

We now return to the problem with which this chapter started, that of the bonding in the BrF$_5$ molecule. The discussion will be simplified by considering only σ-interactions between the fluorine and bromine atoms. Further, the possibility that d orbitals on the bromine may be involved in the bonding will be ignored. These are reasonable simplifications but it is as well to anticipate their consequences. First, each fluorine atom will have six valence-shell nonbonding electrons, a total of thirty in the molecule. There will be peaks arising from these electrons in the photoelectron spectrum of the molecule which may make it difficult to test the final model. Second, the neglect of bromine d-orbitals will mean that we will find, at most, three bonding molecular orbitals responsible for the σ bonding of the four coplanar fluorines to the bromine atom. Of the bromine's four valence shell orbitals one, $4p_z$, has a node in the plane containing the four fluorines and so cannot be involved in this bonding, leaving only three orbitals potentially available to bond to the four fluorines.

As usual, the first thing to do is to consider the transformation properties of the bromine valence shell orbitals, 4s, $4p_x$, $4p_y$ and $4p_z$ (for simplicity, the prefix 4 will not be used in the following discussion). It is a simple matter to show that the bromine s and p_z orbitals separately transform as A_1. The only likely point of any difficulty arises from the fact that there are three classes containing two operations. What has to be done? As indicated above, the answer is 'consider either operation' (or, in the more general case in which there are more than two operations in the same class, consider any one—often one will be a particularly convenient and easy choice). Whichever of the alternative operations is chosen, the same set of characters will result.

Problem 5.12 Show that the bromine s and p$_z$ orbitals do indeed transform as A_1.

Hint: The z coordinate axis is shown in Figure 5.4. Viewing the orbitals from the direction in Figure 5.2 should prove helpful.

Problem 5.13 Repeat Problem 5.12 but using the other choice of symmetry operation for the $2C_4$, $2\sigma_v$ and $2\sigma_v'$ classes to that used in Problem 5.12.

As has been indicated earlier, p$_x$ and p$_y$ transform together as E. However, this has to be shown, as does the fact that the result is independent of the choice of orientation of x and y axes (although the demonstration of this latter point will be incomplete because only the alternative axis sets of Figures 5.4 and 5.5 will be considered). The transformation of the p$_x$ and p$_y$ orbitals of the bromine atom under the eight symmetry operations of the C_{4v} group are detailed in Table 5.7 for the two choices of x and y axes. It is most important that this table should be worked through carefully. Note, in particular, that the behaviour of the two sets of p orbitals under the mirror plane reflections depends on the choice of x and y axis directions. Despite these differences, the character resulting from the transformations is the same for either choice, a most important result. Similarly, the sum of characters generated by p$_x$ and p$_y$

Table 5.7 The transformations of the bromine p$_x$ and p$_y$ orbitals in BrF$_5$. The table shows the orbital obtained when each operation operates on p$_x$ and p$_y$. Its contribution to the aggregate character is given in parentheses after each orbital

	E	C_4	C_4^3	C_2	$\sigma_v(1)$	$\sigma_v(2)$	$\sigma_v'(1)$	$\sigma_v'(2)$
p$_x$ (Figure 5.4) becomes	p$_x$(1)	$-$p$_y$(0)	p$_y$(0)	$-$p$_x$($-$1)	$-$p$_x$($-$1)	p$_x$(1)	p$_y$(0)	$-$p$_y$(0)
p$_y$ (Figure 5.4) becomes	p$_y$(1)	p$_x$(0)	$-$p$_x$(0)	$-$p$_y$($-$1)	p$_y$(1)	$-$p$_y$($-$1)	p$_x$(0)	$-$p$_x$(0)
p$_x$, p$_y$ together	2	0	0	-2	0	0	0	0
p$_x$ (Figure 5.5) becomes	p$_x$(1)	$-$p$_y$(0)	p$_y$(0)	$-$p$_x$($-$1)	$-$p$_y$(0)	p$_y$(0)	$-$p$_x$($-$1)	p$_x$(1)
p$_y$ (Figure 5.5) becomes	p$_y$(1)	p$_x$(0)	$-$p$_x$(0)	$-$p$_y$($-$1)	$-$p$_x$(0)	p$_x$(0)	p$_y$(1)	$-$p$_y$($-$1)
p$_x$, p$_y$ together	2	0	0	-2	0	0	0	0

C_{4v}	E	$2C_4$	C_2	$2\sigma_v$	$2\sigma_v'$
Representation generated by p$_x$ and p$_y$ together	2	0	-2	0	0

is the same for all operations in any one class. The − consensus-characters generated by the p$_x$ and p$_y$ orbitals under the operations of the C_{4v} point group are given at the bottom of Table 5.7. The choice of x and y axes of Figure 5.6 (or any other arbitrary choice) would also lead to the same set of characters as those given in Table 5.7, although the truth of this is not self-evident. For the proof of the statement the use of matrix algebra is unavoidable. The reader who wishes to check out this particular aspect will have to turn to Appendix 2, where the proof is given. Comparison with Table 5.6 shows that the representation which has been generated using p$_x$ and p$_y$ as bases is the E irreducible representation of the C_{4v} group. Because the x and y axes transform similarly to p$_x$ and p$_y$ (just drop the p's in Table 5.7 to obtain the transformation of the axes) it follows that these too transform as E, as asserted earlier in this chapter.

The next task is to determine the irreducible representations spanned by the fluorine σ orbitals involved in bonding with the bromine. No attempt will be made to specify in detail the composition of the fluorine σ orbitals. Each will be a mixture of s and p orbitals but the participation of each of these components is not symmetry determined and, in any case, the choice does not affect the qualitative conclusions that will be reached. For simplicity, in the diagrams in this chapter these hybrid orbitals will be drawn as spheres (in contrast, it is more convenient that they be drawn as pure p orbitals in Appendix 4). The next step is the usual one, a consideration of the transformation properties of these fluorine hybrid orbitals. That of the axial fluorine lies on all of the symmetry elements of the C_{4v} group. All of the corresponding operations turn the orbital into itself. It therefore transforms as the totally symmetric irreducible representation of the C_{4v} point group (A_1). The hybrid σ orbitals of the four symmetry-related fluorine atoms transform as a set and form a basis for a reducible representation which must be decomposed into its irreducible components. The generation of the reducible representation is straightforward but two comments are relevant. First, care has to be taken in the definition of σ_v and σ_v' mirror planes. In this chapter the choice shown in Figure 5.3 will be followed. This choice is arbitrary but it is important to be consistent, otherwise meaningless results will be obtained. Second, as before, for the classes containing two symmetry operations either operation may be chosen to obtain the character. So, use that which is the easier (most people find that C_4 is easier than C_4^3, for example). Following this procedure it should readily be found that the reducible representation generated by the fluorine σ orbitals is:

E	$2C_4$	C_2	$2\sigma_v$	$2\sigma_v'$
4	0	0	2	0

Now, this reducible representation has to be reduced to its irreducible components. Again, recognition has to be made of the fact that three classes contain two operations. The reduction of a reducible representation depends on the group theory orthogonality relationships given earlier in this chapter. In particular, Theorems 2 and 3 are relevant. So, the above representation has to be multiplied by the characters of each irreducible representation in turn. These products are then multiplied by the number of operations in the class and the results summed. If no mistake has been made, the sum is a multiple of 8 (the order of the C_{4v} group), the multiplication factor giving the number of times that the chosen irreducible representation appears in the reducible representation. This is worked out for the case of the B_1 irreducible representation below.

	E	$2C_4$	C_2	$2\sigma_v$	$2\sigma_v'$
Reducible representation					
	4	0	0	2	0
B_1	1	−1	1	1	−1
Multiply	4	0	0	2	0
Number of operations in class					
	1	2	1	2	2
Multiply last two rows					
	4	0	0	4	0

Add the entries in the last row; the sum $= 8$. We conclude that the B_1 irreducible representation occurs once in the reducible representation. Repetition of this process shows that the reducible representation has $A_1 + B_1 + E$ components.

Problem 5.14 Show that the reducible representation generated above has $A_1 + E$ components in addition to the B_1.

Labelling the fluorine orbitals as indicated in Figure 5.7, and proceeding as in Section 4.6, the normalized form of the linear combinations of fluorine hybrid orbitals which transform as the A_1 and B_1 irreducible representations—the symmetry adapted combinations—are obtained. The discussion in Section 4.6 is so close to that needed at this point that it will not be repeated. The only

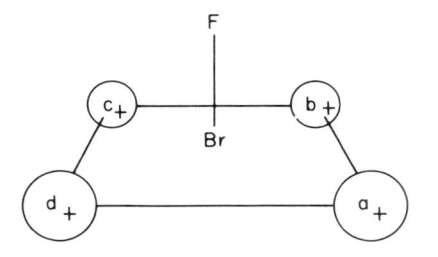

Figure 5.7 The labelling and phases of the σ hybrid orbitals of the four coplanar fluorines. For simplicity these hybrid orbitals are drawn as circles; the square of Figure 5.2 is shown in perspective in this and following diagrams in order to locate the fluorine atoms without including all of them.

thing new is to point out that each and every operation of the group must be considered (so, *both* C_4 and C_4^3). The character that is used to generate a particular symmetry-adapted combination is that in the character table, applied to *each* operation in a class. So, for the B_1 irreducible representation, *both* C_4 and C_4^3 are associated with a character of -1.

Symmetry species	Symmetry-adapted combinations of fluorine orbitals
A_1	$\frac{1}{2}(a + b + c + d)$
B_1	$\frac{1}{2}(a - b + c - d)$

Problem 5.15   Working with the labels of Figure 5.7, use the projection operator method to generate the A_1 and B_1 functions given above.

The generation of the two combinations which transform as E is a more difficult problem, and it will be considered in some detail. Those who experienced difficulty with Problem 5.15 should find the following discussion helpful. As for the A_1 and B_1 combinations the projection operator method described in the last chapter will be used. Using the fluorine hybrid orbital labelled a in Figure 5.7 as generating element and mirror plane operations as labelled in Figure 5.3 the following transformations are found (they were probably generated in tackling Problem 5.15):

Operation	E	C_4	C_4^3	C_2	$\sigma_v(1)$	$\sigma_v(2)$	$\sigma_v'(1)$	$\sigma_v'(2)$
Under the operation orbital a becomes								
	a	d	b	c	c	a	b	d
The E irreducible representation								
	2	0	0	-2	0	0	0	0
Multiply	2a	0	0	-2c	0	0	0	0

(C_4 and C_4^3 are clockwise and anticlockwise rotations of 90° respectively.)

The sum of products is $2a - 2c$ which gives, on normalization:

$$\frac{1}{\sqrt{2}}(a - c)$$

as one of the E functions. Note, as emphasized several times already, that in the above derivation each operation of the group is listed separately. So, when there is a class comprising two symmetry operations, below *each* operation the corresponding character of the E irreducible representation is given.

The wavefunction obtained, $(1/\sqrt{2})(a - c)$, is one member of the pair of functions transforming as E. How may we obtain its partner? In this function there is no contribution from the orbitals b and d; we might reasonably expect them to contribute to the orbital we are seeking. If we consider the transformations of either of these orbitals and follow the projection operator technique used above it is a simple task to show that the function, $(1/\sqrt{2})(b-d)$ is generated. This is the second function for which we have been looking.

Problem 5.16   Generate the second E function, $(1/\sqrt{2})(b-d)$.

The functions $(1/\sqrt{2})(a-c)$ and $(1/\sqrt{2})(b-d)$ transform as a pair under the E irreducible representation of the C_{4v} group and are shown in Figure 5.8. The method used to obtain the second member of the degenerate pair was based on an enlightened guess. In the next chapter a more systematic method of generating such functions will be presented. One final word on these combinations. Whereas in the A_1 combination adjacent fluorine σ orbitals have the same phase—and so any interaction between them is bonding—in the B_1 they are always of opposite phase. Any interaction between them is antibonding. In each of the E combinations there is no interaction between adjacent σ orbitals (only *trans* orbitals appear in any one combination). This argument leads us to expect a relative energy order $A_1 < E < B_1$, a sequence which will be reflected in the presentation of the molecular orbital energy level diagram for BrF$_5$ (Figure 5.12).

We are now almost ready to consider the interaction between bromine and fluorine σ orbitals. First, however, recall that the s and p$_z$ bromine orbitals separately transform as the A_1 irreducible representation. Analogous situations have been encountered in earlier chapters, when the corresponding orbitals were combined to obtain two mixed, hybrid, orbitals of the form $(1/\sqrt{2})(s \pm p_z)$. This simplifying procedure will also be followed in the present case. One of the mixed orbitals is orientated in a way that should give good overlap with the σ orbital of the apical fluorine atom, an orbital which, as has been seen, is also of A_1 symmetry; these steps are shown schematically in Figures 5.9 and 5.10. The remaining s – p$_z$ mixed orbital on the bromine atom points in the direction

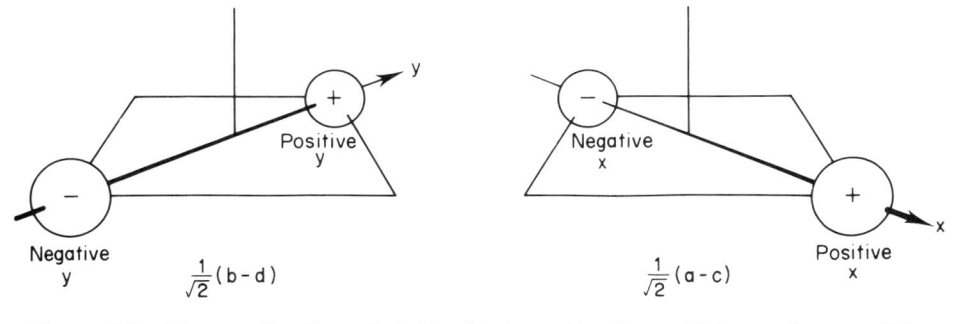

Figure 5.8   The two fluorine σ hybrid orbital combinations which transform as E in the C_{4v} point group and which are derived in the text.

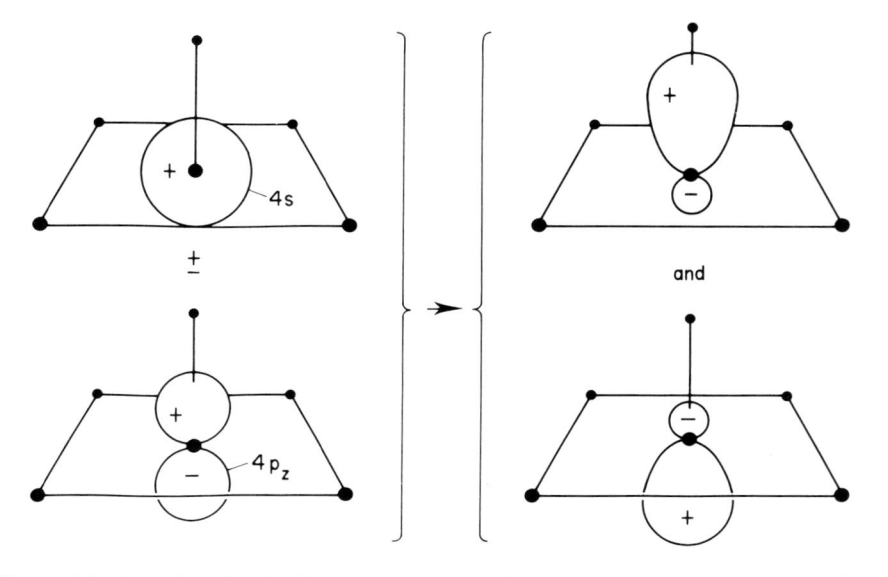

Figure 5.9 Hybrid orbitals (right) derived from bromine atomic orbitals (left) of A_1 symmetry in BrF$_5$.

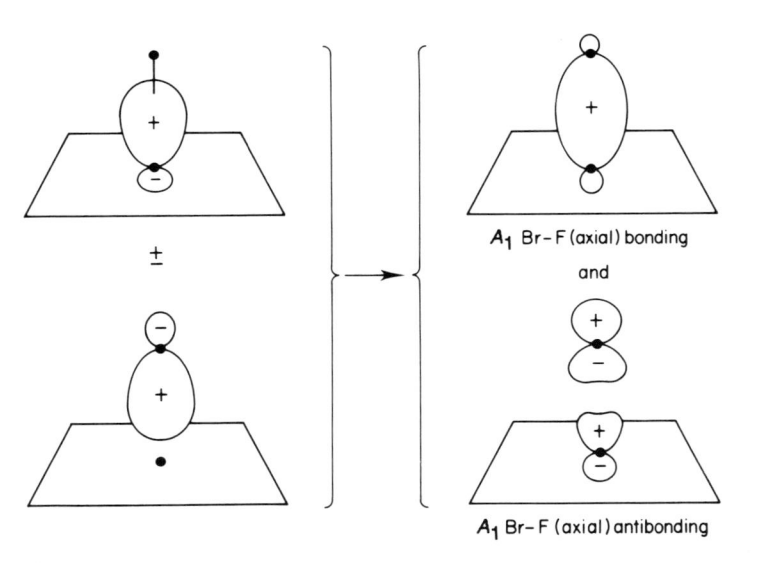

Figure 5.10 Bonding of the axial fluorine to the bromine atom in BrF$_5$; some form of sp hybrid is envisaged as involved on each atom.

indicated by dotted lines in Figure 5.1 and therefore might be regarded as the orbital which (in the electron-pair repulsion model) causes the distortion of the molecule that was noted at the beginning of this chapter. Unfortunately, as will be seen, reality is perhaps more complicated than this. The complication arises from the fact that there is also a combination of σ orbitals from the planar fluorines which has A_1 symmetry. Clearly, it can interact with A_1 orbitals of the bromine. However, one of these latter A_1 orbitals is p_z and this has a nodal plane in which the fluorines lie (in our simplified geometry of coplanar bromine and fluorines). The basal plane fluorine A_1 combination interaction will therefore be almost entirely with the bromine s orbital. Correspondingly, it seems probable that the bromine s orbital involvement with the axial fluorine and in the basal lone pair will be rather less than assumed above.

The only other valence orbitals on the bromine atom are p_x and p_y which, together, are of E symmetry. They are shown in Figure 5.4(b). They interact with the two fluorine σ orbital combinations of E symmetry which were generated earlier in this section and which are shown in Figure 5.8. Provided the p orbitals and the fluorine orbital combinations are properly chosen—and this means that the same set of coordinate axes is used for each—then each p orbital only interacts with one σ orbital combination. The orbitals in Figures 5.4(b) and 5.8 are properly chosen and the results of their interactions to give what have been represented as sum (bonding) and difference (antibonding) combinations are shown in Figure 5.11.

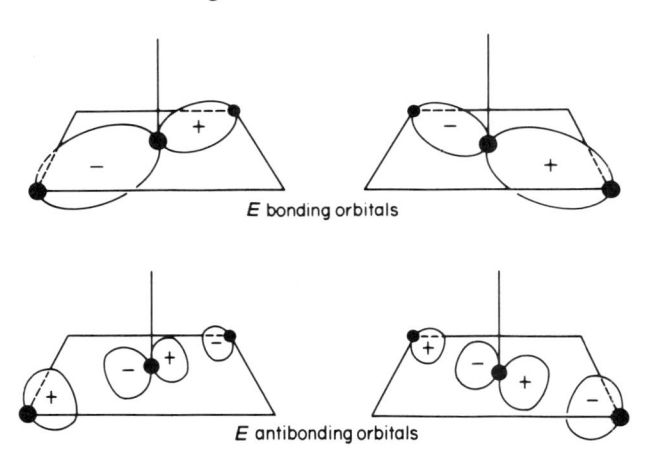

Figure 5.11   The two bonding orbitals and the two antibonding orbitals of E symmetry in BrF₅ arising from interactions of the four coplanar fluorines with the central bromine atom.

Problem 5.17   Repeat the above discussion of the interactions of orbitals of E symmetry using the choice of coordinate axes shown in Figure 5.5 (use the bromine p orbitals shown in Figure 5.5(b)).

Hint: The projection operator method used in the text automatically selected the coordinate axis choice shown in Figure 5.8 because of an (implicit) choice to consider the transformation of an individual fluorine σ orbital (a). The method can be forced to give combinations appropriate to the axes of Figure 5.5 by considering, instead, the transformation of a pair of neighbouring σ orbitals. Thus, the pairs (a + b) and (a + d) are suitable pairs to use in tackling this problem.

The above discussion is summarized in Figure 5.12, where, as has been recognized, there must be some uncertainty about the details of the positions of the orbitals of A_1 symmetry. Into the orbital pattern shown in this figure a total of twelve electrons (seven from the bromine and one from each fluorine orbital) have to be placed. It will be remembered that this diagram does not include the fluorine non-bonding electrons.

There are some interesting consequences of Figure 5.12 and of our discussion of the bonding of the BrF$_5$ molecule. We have suggested that one molecular orbital (of A_1 symmetry) is primarily involved in the bonding of the axial fluorine to the bromine. If this view is correct then this bromine–fluorine bond involves two electrons. In contrast, in the picture developed above, the strongly bonding molecular orbitals involving the planar fluorine atoms are of E symmetry (although there will be a smaller contribution from an A_1 orbital). That is, the four coplanar fluorine atoms are bonded to the central bromine atom by little more than two molecular orbitals. If this conclusion is correct, it suggests that the bonding between the bromine and each of the four coplanar fluorines is rather weak, a view supported by the fact that bromine

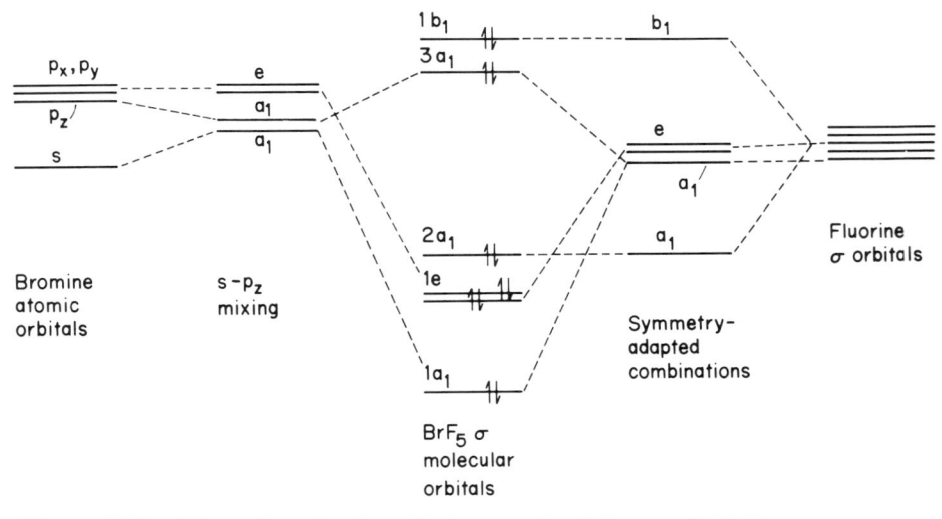

Figure 5.12   Schematic molecular orbital energy level diagram for BrF$_5$.

pentafluoride is an extremely powerful fluorinating agent. Some further support for this difference between axial and planar fluorines is to be found in molecular structure determinations which show Br–F bond lengths of 1.68 Å (axial) and 1.78 Å (equatorial).

Both theoretical calculations and photoelectron spectroscopic data are available for BrF₅. For us they are complicated by the fact that in the above discussion all the electrons on the bromine and on the fluorine atoms which are not involved in σ bonding have been omitted. These electrons, then, have to be regarded as non-bonding and so are expected to be relatively easy to ionize. Fortunately, some of the symmetries they span are not included in the σ bonding set (A_2, for example) and this helps to identify them. The highest lying, most easily ionized, electrons (at 13.5 eV in the photoelectron spectrum[1]) are believed to be the lone pair of electrons on the bromine atom. Then come at least three peaks (between 15 and 17 eV) corresponding to ionization of the fluorine 2p non-bonding electrons, included amongst these are the $1b_1$, $2a_1$ and $3a_1$ electrons of Figure 5.12. Between 18 and 22.5 eV are two peaks which are almost certainly composites but which have been reported as including ionization from the $1e$ and $1a_1$ Br–F σ-bonding molecular orbitals of Figure 5.12. Theoretical calculations are available for both ClF₅[2] (which has a structure similar to that of BrF₅) and BrF₅[3] itself. The two sets of calculations, which used somewhat different theoretical models, are in good qualitative agreement with each other and with the experimental data, although they suggest that perhaps the $1a_1$ orbital of Figure 5.12 is just a little too low in energy to be seen in the photoelectron spectrum. For us the most important general conclusion is the promising result that, once again, our relatively simple symmetry-based arguments lead to an energy level pattern which is in good qualitative agreement both with experiment and with detailed theoretical calculations.

Problem 5.18   In our discussion of the bonding in BrF₅ we have ignored the presence of 4d orbitals on the bromine. The justification for this is that the 4d orbitals of the isolated bromine atom are so large and diffuse that they cannot overlap effectively with a valence shell atomic orbital of any other atom unless there is something which causes them to contract. Something may exist in BrF₅ because the polarity of each of the Br–F bonds will be such that there will presumably be a significant build-up of positive charge on the bromine atom. One effect of this would be to lower the energy and decrease the size of the bromine 4d orbitals and thus perhaps make them available for chemical bonding. If this occurs we should have included the d orbitals in our discussion. This is an attractive hypothesis but one that is extremely difficult to test, even by detailed calculations.

Show (a) that the bromine d_{z^2} orbital has A_1 symmetry and its $d_{x^2-y^2}$ orbital B_1 symmetry. This latter orbital is shown (contracted!) in Figure 5.13 together with the B_1 combination of fluorine σ orbitals with which it

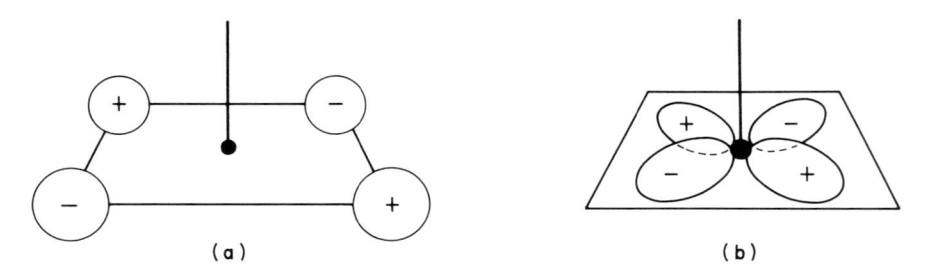

Figure 5.13   (a) The B_1 combination of fluorine σ orbitals. (b) The $d_{x^2-y^2}$ (B_1) orbital of bromine.

potentially interacts. (b) That both of the labels B_1 and $d_{x^2-y^2}$ would have to be changed if the coordinate axis set of Figure 5.5 were used in the discussion.

5.6   SUMMARY

In this chapter it has been found that operations may be divided into classes (p. 99) and that when some classes contain more than one operation the character table contains at least one degenerate representation (pp. 98, 111). The presence of a degenerate representation in the C_{4v} group enabled the orthonormality relationships to be presented in a more general form (p. 103). The procedures previously used to reduce a reducible representation has to be modified in the more general case although the projection operator technique is basically unchanged (p. 113). Application of these techniques to the problem of the bonding in BrF_5 suggested both a reason for its existence—polar Br–F bonds possibly enabling participation of bromine d orbitals in the bonding (Problem 5.18)—and for its reactivity—the four coplanar fluorines are not strongly bonded (p. 119).

REFERENCES

1. R. L. De Kock. B. R. Higginson and D. R. Lloyd, *Chem. Soc. Faraday Disc.*, **54** (1972), 84.
2. M. B. Hall, *Chem. Soc. Faraday Disc.*, **54** (1972), 97.
3. G. L. Gutzev and A. E. Smolyar, *Chem. Phys. Letts.*, **71** (1980), 296.

6

The Electronic Structure of the Ammonia Molecule

In the first chapter of this book four different qualitative descriptions of the bonding in the ammonia molecule were discussed in outline. The symmetry-based approach has now been developed to a point at which this problem may be reconsidered in more detail. At the same time a problem encountered in the last chapter will reappear—that of the choice of directions of x and y axes. The form in which this problem appears is one which will lead to a general solution, a solution which will enable molecules of high symmetry, such as those which will be subject of Chapter 7, to be tackled.

6.1 THE SYMMETRY OF THE AMMONIA MOLECULE

The structure of the ammonia molecule is given in Figure 6.1 which also shows the symmetry elements possessed by this molecule. The axis of highest rotational symmetry (which will therefore be taken as the z axis) is a C_3 rotation axis and has associated with it clockwise and anticlockwise rotation operations (which, to help the reader remember this distinction, will be called C_3^+ and C_3^-; in a more general notation they would be called C_3 and C_3^2, following that used in the last chapter). In addition, there are three mirror planes each of which contains the threefold axis (and are vertical with respect to it—that is, they are σ_v mirror planes), with one hydrogen atom lying in each mirror plane. The symmetry operations which turn the ammonia molecule into itself are therefore,

$$E \quad C_3^+ \quad C_3^- \quad \sigma_v(1) \quad \sigma_v(2) \quad \sigma_v(3)$$

This group is called the C_{3v} point group, the shorthand symbol C_{3v} indicating the coexistence of the C_3 axis and the vertical mirror planes.

Problem 6.1 Show that this set of operations comprises a group.
Hint: It will be found helpful to refer back to Problem 4.4. The group multiplication table for the C_{3v} group is given in Table 8.2.

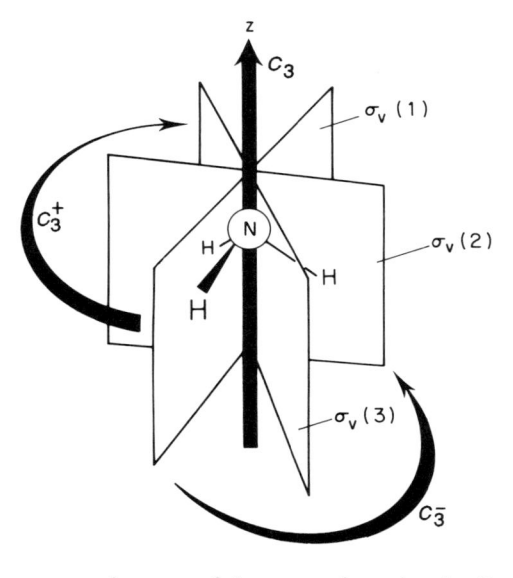

Figure 6.1 The symmetry elements of the ammonia molecule. One hydrogen atom is located in each mirror plane.

The class structure of the symmetry operations of the C_{3v} group is suggested from the similarities between the various operations and is

$$E \quad 2C_3 \quad 3\sigma_v$$

Alternatively, the formal methods described in Appendix 1 may be used to deduce this class structure (the $2C_3$ class is given as a worked example in Appendix 1). The character table of the C_{3v} point group is given in Table 6.1.

In this table we have followed Table 5.6 and given the usual presentation of character tables. On the right hand side of the table are shown functions which are a basis for a particular representation. Thus the z axis, chosen following the convention which locates it along the C_3 axis, transforms as A_1 and the x and y axes, together, are a basis for the E irreducible representation.

Table 6.1

C_{3v}	E	$2C_3$	$3\sigma_v$	
A_1	1	1	1	$z, z^2, x^2 + y^2$
A_2	1	1	-1	
E	2	-1	0	$(x, y), (zx, yz), (xy, x^2 - y^2)$

Problem 6.2   Use the theorems of Section 5.3 to derive Table 6.1.
Note: this is a relatively short problem but one that gives excellent practice in the use of the orthonormality theorems.

There is one particular point about the C_{3v} character table which has to be discussed in detail. This concerns the axis pair (x, y) which, as shown in Table 6.1, transform as the doubly degenerate irreducible representation E. Because they must be perpendicular to the z axis, the x and y axes lie in a plane perpendicular to the C_3 axis. But where in this plane do they lie? This problem is similar to one which was discussed in Chapter 5 where, in the C_{4v} point group, it was found that a variety of directions could be chosen for the x and y axes. So too, in the present problem there is no unique choice for the x and y axis directions. However, the present problem is more difficult than that encountered in the C_{4v} case and so it will be examined in some detail. Suppose the x axis is chosen so that it lies in one of the σ_v mirror planes as is shown in Figure 6.2, which gives a view looking down the threefold axis. Two, related, problems at once arise. First, there is no evident reason why a particular mirror plane should be selected rather than one of the others. Second, the choice which has been made for the x axis means that the y axis is forced to be quite differently orientated in space. However, having made a choice we will stay with it and move on to the next problem, that of the effect of a C_3 rotation operation on these x and y axes, shown in Figure 6.3. It is seen from this figure that the x axis is rotated so that it lies along one of the directions which could have originally been taken as the x axis but was not. Similarly, the y axis is rotated into a direction appropriate to this second choice of x axis. In Figure 6.3 the alternative x and y axes are indicated by primes (so that x is rotated into x' and y into y'). This is a quite new situation. So far in this book symmetry operations have turned objects into themselves or interchanged them. Here, a symmetry operation has generated something which did not previously exist, or so it seems. Well, the truth is that the x' and y' axes did previously exist—it is

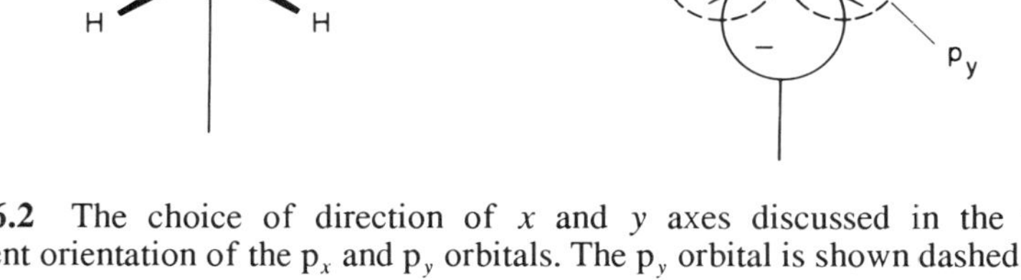

Figure 6.2   The choice of direction of x and y axes discussed in the text and consequent orientation of the p_x and p_y orbitals. The p_y orbital is shown dashed.

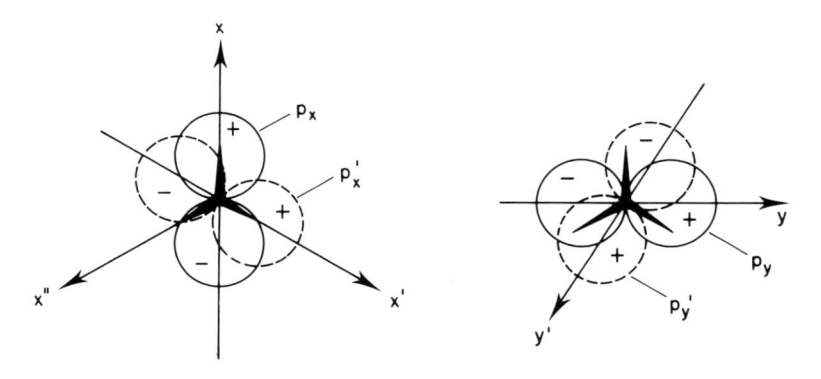

Figure 6.3 The x and y axes of Figure 6.2 together with an alternative set (x' and y') produced by a C_3 rotation of x and y. In both cases the corresponding p orbitals are also shown. The x'' axis will be referred to later in the text (following Problem 6.6).

just that they were not revealed. However, a little work is involved in showing that this must be the case.

As is clear from Table 6.1 the C_3 rotation acting on the x, y axis pair which converts them into the x', y' axis pair is associated with a character of -1 (this is the character of the E irreducible representation under C_3 rotations). In some way or other the x', y' set is -1 times the x, y. How? This problem is tackled by investigating the relationship between two axis sets (x, y) and (x', y') related by a rotation by an angle α (later α will be taken as equal to $120°$, as appropriate to the C_{3v} point group). If an object were to start at the origin of coordinates in Figure 6.4 and be displaced along the x' axis it is evident that this displacement could, alternatively, be represented as a sum of displacements along the original x

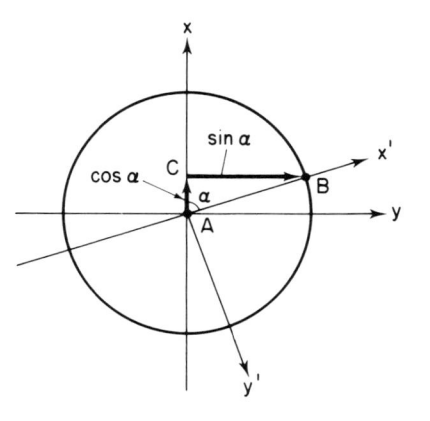

Figure 6.4 In this figure the circle is taken to be of unit radius. It follows that the unit displacement AB is the sum of the displacements AC and CB which, respectively, have magnitudes of cos α and sin α.

and y axes. As shown in this figure, for an angle α relating the x and x' axes, a unit displacement along the x' axis is equivalent to a displacement of $\cos \alpha$ along x combined with a displacement of $\sin \alpha$ along y. The rotated x axis, x', is a mixture of the original x and y, as too is the rotated y, y'.

When determining the contribution to a character made by the transformation of something such as an x axis, so far in this book we have asked the question 'is the x axis turned into itself, into minus itself or into something different', and have associated the characters of 1, -1 and 0 with these three situations. We have now encountered a situation in which the x axis is rotated into an axis which may be described as in part containing the original x axis. Accordingly, our question must be modified to the simpler, but more general, form, 'to what extent is the old axis contained in the new?'. As is evident from Figure 6.4, and the discussion above, the numerical answer to this question is $\cos \alpha$, where α is the angle of rotation. An axis which is left unchanged by a rotation corresponds to $\alpha = 0$, so $\cos \alpha = 1$, the character that we have associated with this situation. Similarly, for a rotation of 180°, $\cos \alpha = -1$, again the character that the rotation of a coordinate axis by 180° gives. For $\alpha = 90°$ (when the x axis is rotated so that it becomes the y axis) $\cos \alpha = 0$, again the expected answer. The general rule is clear:

> When an axis is rotated by an angle α by a symmetry operation its contribution to the character for that operation is $\cos \alpha$.

Comment: This statement holds for axes; for products of axes it has to be modified. Thus, because

$$x \text{ contributes } \cos \alpha$$

it follows that $\quad x^2$ contributes $\cos^2 \alpha$
and that $\quad x^3$ contributes $\cos^3 \alpha$
and so on.

Note that this rule applies to products of axes which are perpendicular to the axis of rotation. Thus, if the rotation axis is the z axis then the function xz will vary as $\cos \alpha$ because only the x axis is perpendicular to the rotation axis—the z axis is left unchanged. However, the function xy will vary as $\cos^2 \alpha$ because both x and y separately vary as $\cos \alpha$.

Problem 6.3 Show that the transformation of x^2 under a rotation of α about the z axis is given by the factor $\cos^2 \alpha$.
Hint: it is sufficient to check that this relationship holds for particular values of α; $\alpha = 0°$, 90°, 180°, 270° and 360° are particularly convenient.

Problem 6.4 (a) In the C_{5v} point group a pair of functions transforming as the doubly degenerate irreducible representation E_1 have a character of $2\cos 72°$ under a C_5 rotation. Suggest a pair of functions which might form a basis for this irreducible representation.

(b) Repeat this problem for the E_2 irreducible representation, for which the character is $2\cos 144°$. Solution of this problem requires a small extension of the argument developed above. (Solutions to both these problems will be found in the character table for the C_{5v} group in Appendix 3.)

Returning to the case of the C_{3v} point group, it is concluded that the x and y axes each make a contribution of $\cos 120° = -\frac{1}{2}$ to the character under the C_3, 120°, rotation operations. The sum of these two, -1, is, indeed, the character of the E irreducible representation under this operation. In Section 3.2 it was first mentioned that two quantities, such as axes or orbitals, can be mixed by the operations of a group. We are now able to understand just what this means. The effect of a C_3 rotation on the original x and y axes is to rotate them to give new axes, each of which is a mixture of the original axes. In such cases the contribution that each axis makes to the character is always fractional. Everything that has been said about the x and y axes also holds for the $2p_x$ and $2p_y$ orbitals of the nitrogen atom in ammonia because the transformations of $2p_x(N)$ is isomorphous to that of x, as is that of $2p_y(N)$ to y (Figure 6.3). This parallel has already been anticipated by taking the molecular x and y axes to pass through the nitrogen atom—although they could be chosen to pass through any point along the C_3 axis—so that the above argument could be used as a basis for a discussion of the bonding in the ammonia molecule without need to redefine axes.

6.2 THE BONDING IN THE AMMONIA MOLECULE

We now complete this chapter by a discussion of the bonding in the ammonia molecule. As is evident from the discussions above, the transformation properties of the nitrogen valence shell 2p orbitals follow those of the coordinate axes given in Table 6.1. The nitrogen $2p_z$ has A_1 symmetry and $2p_x$ and $2p_y$, as a pair, have E symmetry; it is a trivial exercise to show that the nitrogen 2s orbital is totally symmetric (this orbital is spherical and lies on all symmetry elements; it therefore transforms as A_1). The transformation of the three hydrogen 1s orbitals under the operations of the group gives rise to the reducible representation

E	$2C_3$	$3\sigma_v$
3	0	1

which is a linear sum of the irreducible representations A_1 and E.

Problem 6.5 Reduce the above reducible representation into its irreducible components.
Hint: If this problem is found difficult, an explicit solution to a similar problem has been given in the previous chapter.

Much of the discussion so far in this chapter has developed from the fact that the operation of rotation by 120° has the effect of mixing functions which provide a basis for the E irreducible representation. This same problem reappears again when we try to determine the symmetry adapted combinations of hydrogen 1s orbitals in the ammonia molecule which transform as the E irreducible representation, a problem which will now be considered in detail. Labelling the hydrogen 1s orbitals as indicated in Figure 6.5 and considering the transformation of the orbital labelled a, the six symmetry operations of the group are found to lead to the following transformations, where the σ_v mirror planes are labelled as in Figure 6.1:

E	C_3^+	C_3^-	$\sigma_v(1)$	$\sigma_v(2)$	$\sigma_v(3)$
a	b	c	a	c	b

Application of the projection operator technique described in Section 4.6 shows the A_1 function to be:

$$\frac{1}{\sqrt{3}}(a + b + c)$$

There is no difficulty in obtaining one of the E functions. The steps involved are shown in Table 6.2 and lead to the function

$$\frac{1}{\sqrt{6}}(2a - b - c)$$

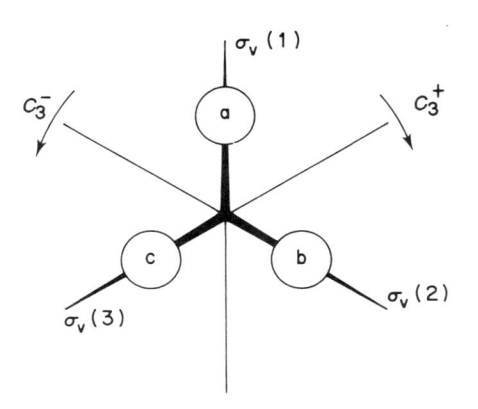

Figure 6.5 The labels used in the text for the hydrogen 1s orbitals of the ammonia molecule.

Table 6.2

Operation	E	C_3^+	C_3^-	$\sigma_v(1)$	$\sigma_v(2)$	$\sigma_v(3)$
a is turned into	a	b	c	a	c	b
Characters of the E irreducible representation	2	−1	−1	0	0	0
Multiply	2a	−b	−c	0	0	0
Sum		$2a - b - c$				
Normalize		$\dfrac{1}{\sqrt{6}}(2a - b - c)$				

A problem arises when we try to obtain the second E function. A similar problem was met in Section 5.5 when discussing the four orbitals of the coplanar fluorine atoms in BrF_5. In that case the problem was relatively simple because the first E function contained contributions from only two of the σ orbitals; the projection operator technique applied to one of the other σ orbitals immediately gave the second E function. There is no such simple solution to the present problem; all three hydrogen 1s orbitals appear in the E function that has been generated, although not all with the same weights. Following the procedure described in Chapter 5 the transformations of either the hydrogen 1s orbital b or c could be used as a basis for the projection operation method—but which? If b is used then the combination

$$\frac{1}{\sqrt{6}}(2b - c - a)$$

is obtained while if c is used the function

$$\frac{1}{\sqrt{6}}(2c - a - b)$$

is obtained.

Problem 6.6 Show, by constructing tables analogous to Table 6.2, that the transformation of the hydrogen 1s orbitals b and c lead to the E functions

$$\frac{1}{\sqrt{6}}(2b - c - a) \quad \text{and} \quad \frac{1}{\sqrt{6}}(2c - a - b), \text{respectively.}$$

We have, apparently, obtained three quite different functions transforming as E—yet we know that, for a doubly degenerate irreducible representation, there can only be two. As indicated above, this problem is closely related to the three possible choices for the x axis that were discussed earlier and the solution of the problem is also similar. What we have done in using a, b and c separately to generate an E function is to generate functions appropriate to the x, x' and x'' axes, respectively, of Figure 6.3 as indicated in Figure 6.6(a). The functions corresponding to the x' and x'' axes, like these axes themselves, are mixtures of the functions appropriate to the original x and y axes. It is the latter pair that we are seeking. The first member of the pair we have is pure but the second only as part of a mixture.

There are many ways of obtaining the second E function from the mixture. Perhaps the simplest is to exploit the fact that if the first function corresponds to the x axis then the second corresponds to the y axis. This vector (axis-like)

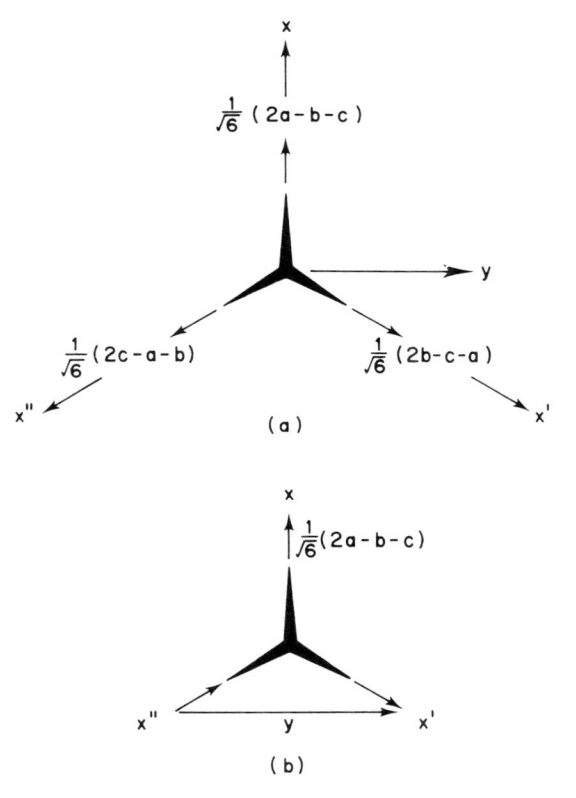

(a)

(b)

Figure 6.6 (a) Alternative symmetry-adapted combinations of hydrogen 1s orbitals in NH$_3$ corresponding to the axes x, x' and x'' of Figure 6.3. Just as one of these axes has to be selected so, too, does one of the three symmetry-adapted combinations. (b) The (vector) sum of displacements along $-x''$ and x' is a displacement along y.

property of the functions is indicated by the arrows in Figure 6.6(a). If, as shown in Figure 6.6(b), the direction of the vector pointing in the direction x'' is reversed and added to that pointing in the x' direction a vector pointing in the y direction is obtained. These steps now have to be repeated using functions rather than vectors.

The negative of the function associated with x'' is

$$-\frac{1}{\sqrt{6}} (2c - a - b)$$

and adding it to the function associated with x'

$$\frac{1}{\sqrt{6}} (2b - c - a)$$

gives

$$\frac{1}{\sqrt{6}} (-2c + a + b + 2b - c - a) = \frac{1}{\sqrt{6}} (3b - 3c)$$

That is, the second E function is of the form

$$(b - c)$$

or, normalized,

$$\frac{1}{\sqrt{2}} (b - c)$$

Problem 6.7 The sum of vectors pointing along x' and x'' of Figure 6.6(a) is the negative of a vector pointing along x. Show that an analogous statement is true for the corresponding E functions.

Problem 6.8 The fact that the two E functions which have just been obtained have quite different mathematical forms tends to be received with suspicion. Show that their forms are such that the orbitals a, b and c make equal total contributions to the E functions.
Hint: Sum the squares of coefficients in the normalized E functions.

The symmetry-adapted combinations of hydrogen 1s orbitals which have just been generated are shown in Figure 6.7 together with the nitrogen orbitals of the same symmetry with which they interact. Note that in Figure 6.7 the unequal contribution of a, b and c to each of the two symmetry-adapted combinations of E symmetry is reflected in the diagrammatic representation of the orbitals. It will also be noted that in Figure 6.7 the approximate procedure of taking a combination of nitrogen 2s and $2p_z$ orbitals as the nitrogen orbital which interacts with the A_1 combination of hydrogen 1s orbitals has been

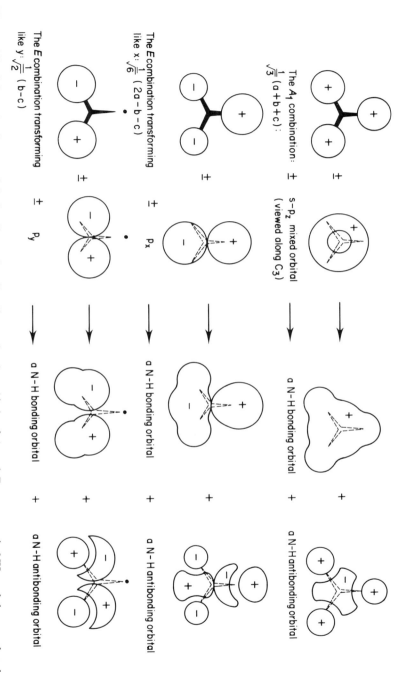

The A_1 combination:
$$\frac{1}{\sqrt{3}}(a+b+c):$$

s–p_z mixed orbital
(viewed along C_3)

a N–H bonding orbital

a N–H antibonding orbital

The E combination transforming like x: $\frac{1}{\sqrt{6}}(2a-b-c)$

p_x

a N–H bonding orbital

a N–H antibonding orbital

The E combination transforming like y: $\frac{1}{\sqrt{2}}(b-c)$

p_y

a N–H bonding orbital

a N–H antibonding orbital

Figure 6.7 Schematic pictures of the bonding and antibonding molecular orbitals of A_1 and E symmetry in NH$_3$ and the way that they are derived from atomic and group orbitals.

followed. The resulting schematic molecular energy level diagram of ammonia is shown in Figure 6.8. There are eight valence electrons which have to be allocated to these orbitals (five from the nitrogen and one from each of the three hydrogens) and they are accommodated in the lowest molecular orbitals of A_1 and E symmetry, all of which are M–H bonding and the second A_1 orbital, which is, essentially, the nitrogen lone-pair orbital. These qualitative conclusions are to be compared with the results of detailed calculations and with the results of photoelectron spectroscopy.

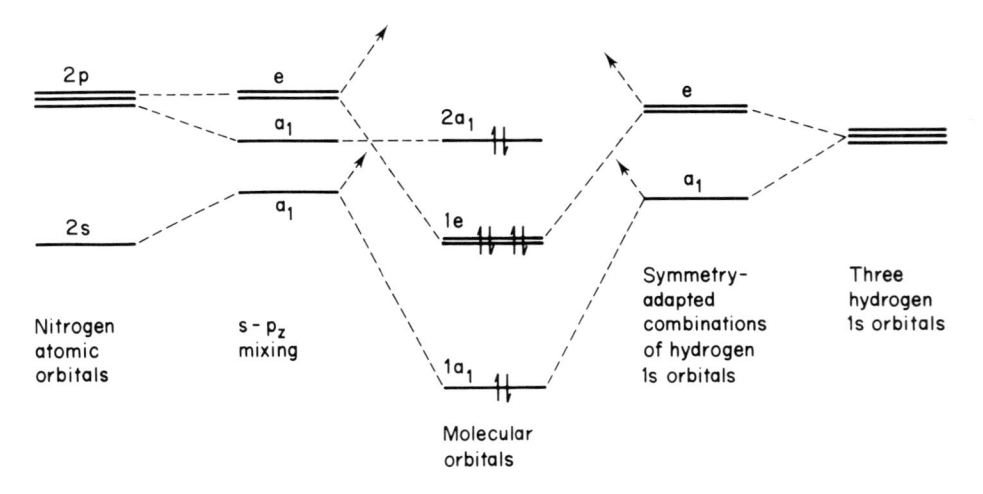

Figure 6.8   A schematic molecular orbital energy level diagram for NH_3.

Calculations[1] show that the A_1 orbitals of the ammonia molecule have energies of *ca* −11.6 and *ca* −31.3 eV and that of E symmetry *ca* −17.1 eV. Of these, the more stable of the A_1 orbitals has, as experience leads us to expect, a major contribution from the nitrogen 2s atomic orbital. These data are in general agreement with the photoelectron spectroscopic results[2] which give energies of about 10.2 eV, 27.0 and 15.0 eV for these levels, respectively. Again, the encouraging result that a symmetry-based model is in good qualitative agreement both with detailed calculations and with experiment.

Ammonia is a molecule for which, like the water molecule, it is a simple matter to describe the angles at which the various contributions to the molecular bonding maximize. Using arguments entirely similar to those of Chapter 3 for the water molecule, it is concluded that the bonding interactions of A_1 symmetry involving the nitrogen $2p_z$ orbital (Figure 6.9(a)) maximizes at small bond angles, whereas the interactions between orbitals of E symmetry maximize for the planar molecule (Figure 6.9(b)). These angular variations are conveniently summarized in a Walsh diagram, just as for the water molecule in Chapter 3. This diagram is given, qualitatively, in Figure 6.10.

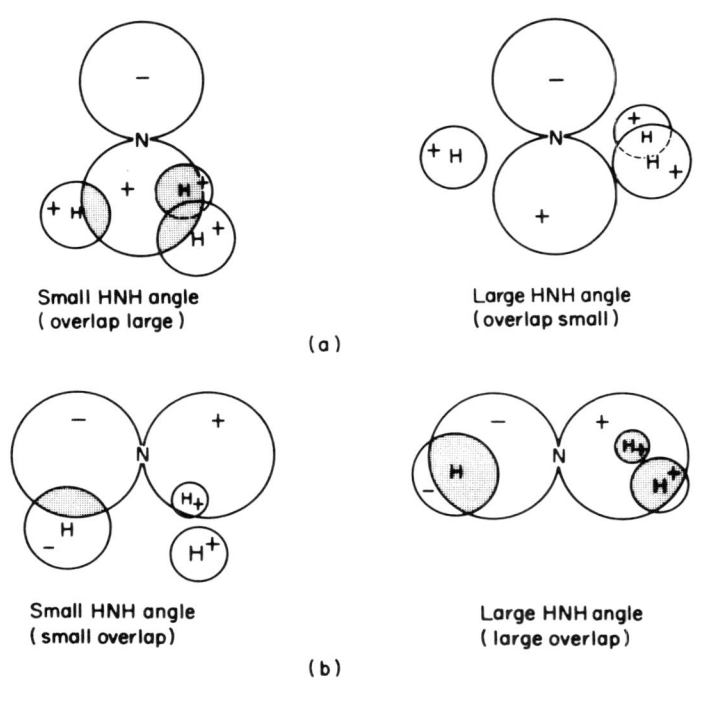

Figure 6.9 (a) The overlap between the A_1 symmetry-adapted hydrogen 1s combination and the nitrogen $2p_z$ orbital decreases as the HNH bond angle increases (this decrease is related to the fact that the hydrogen combination and the $2p_z$ orbitals have different symmetries in the planar molecule).
(b) The overlap between an E symmetry-adapted hydrogen 1s combination and a nitrogen $2p_x$ orbital increases with the HNH bond angle.

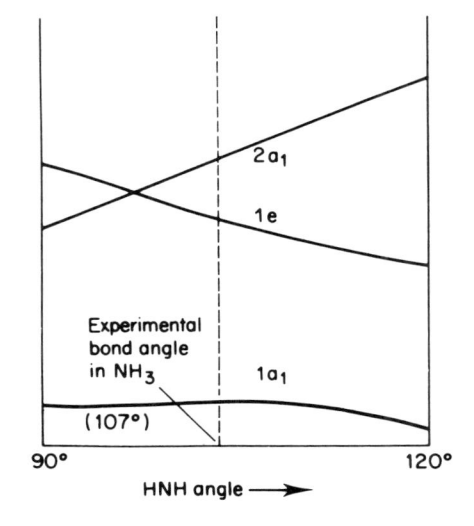

Figure 6.10 A Walsh diagram for NH_3.

Problem 6.9 Check that Figure 6.10 does, indeed, summarize the discussion of the above paragraph. What can be concluded about the non-bonding nature of the highest A_1 orbital from this diagram?

As indicated in Chapter 1, calculations show that the total bonding in the ammonia molecule is a maximum when the molecule is planar so it can be concluded that E interactions dominate. However, this argument neglects the effects of repulsive forces on the molecular geometry and, as stated in Chapter 1, the same calculations show that it is these that—just—lead to the molecule adopting a pyramidal shape. At the observed bond angle there are both A_1 and E contributions to the bonding. Were we to remove an electron from the highest A_1 molecular orbital, which contains a large nitrogen $2p_z$ contribution, and which, despite our simplified discussion, makes a contribution to the molecular bonding, it would be reasonable to expect that a more nearly planar molecule would result. Experiment, indeed, indicates that in its ground state NH_3^+ is a planar molecule.

Although the bonding in planar NH_3 or NH_3^+ has not been discussed in this text, it is, none the less, of interest to consider a related planar species. This is the molecule trisilylamine, $(SiH_3)_3N$. In this molecule the Si_3–N framework is planar (unlike the C_3–N, skeleton in trimethylamine $(CH_3)_3N$, which is pyramidal like ammonia). The question of why trisilylamine should be planar has been widely discussed in the past and frequently associated with Si–N π bonding involving the empty 3d orbitals of silicon accepting electrons from the lone pair on nitrogen (which in this geometry occupy a pure 2p orbital, Figure 6.11). Our discussion has indicated that the planarity of this molecule could arise if the delicate balance between bonding and repulsive forces found for ammonia—and which appears to occur for many such molecules—is such as to favour the planar form of trisilylamine. This argument, of course, does not

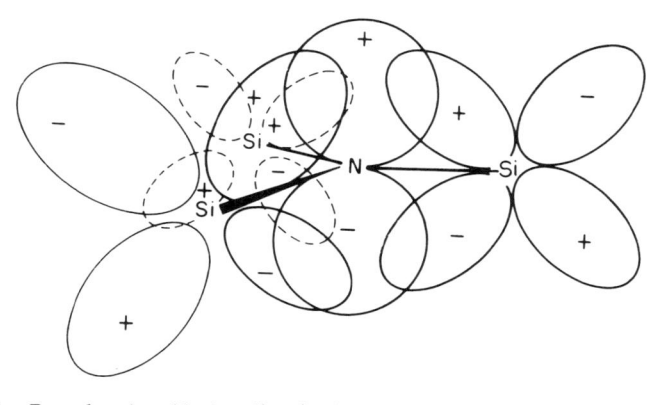

Figure 6.11 Postulated p_π/d_π bonding in the planar molecule trisilylamine, $N(SiH_3)_3$. In fact, the silicon $3d_\pi$ orbitals are much larger and diffuse than pictured here.

require the existence of any Si–N π bonding. It could be, of course, that the presence of a small amount of π bonding is decisive in tipping a delicate balance. Equally, such an interaction might be important not for any— small—π bonding stabilization which results but because the resulting more diffuse electron distribution leads to a reduction in the destabilization resulting from electron repulsion. However, it is important to recognize that the observed planar geometry of trisilylamine does not of itself prove the existence of significant d–p π bonding in this molecule. Indeed, there are now (fairly complicated) organic species known which have a similar planar structure around a nitrogen atom and for which it would be difficult to advance a π bonding argument.

6.3 SUMMARY

In this chapter the problem of the transformation of functions has been discussed which form the basis for a degenerate reducible representation but which appear to be differently oriented with respect to the symmetry elements and may, indeed, have different mathematical forms (pp. 125, 127). Despite these superficial differences, the fact that they are mixed or interchanged by some operations of the group is sufficient to ensure their ultimate equivalence (p. 128).

REFERENCES

1. C. D. Ritchie and H. F. King, *J. Chem. Phys.*, **47** (1967), 564.
2. A. W. Potts and W. C. Price, *Proc. Roy. Soc.*, **326** (1972), 181.

7

The Electronic Structures of some Cubic Molecules

The methods developed so far in this book will be exploited to the full in this chapter, where the electronic structure of the octahedral molecule SF_6 will be considered in detail. This is quite a large molecule but its symmetry is also considerable; enough to enable us to consider not only the bonding between sulfur and fluorine but also the non-bonding electrons on the fluorines. There are short-cuts which can be used in symmetry discussions and the present discussion will enable several of them to be introduced. Throughout this book new symmetry operations have been introduced in each chapter. This is also true for the present chapter; the operations complete the types encountered in point groups and so a general review of point group classification will be included. This will prepare the way for the following chapter, in which the relationships between point groups will be covered. Although the molecule SF_6 is the major subject of this chapter the content also includes an important class—transition metal complexes of octahedral symmetry. Both tetrahedral and octahedral molecules have x, y and z axes equivalent to each other and the chapter also contains a discussion of both methane and of tetrahedral transition metal complexes.

A cube is shown in Figure 7.1, together with an octahedron and a tetra-hedron. An octahedron is closely related to a cube. If the mid-points of faces of a cube are joined together the figure that is generated is an octahedron. The octahedron has eight faces, but what is of more importance is the fact that it has six apices because when these apices are occupied by six atoms around an atom at the centre of the figure an octahedral molecule results (Figure 7.2). In

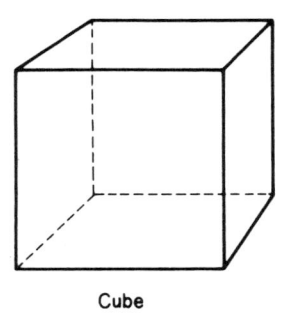
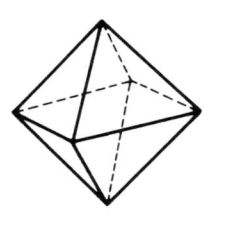
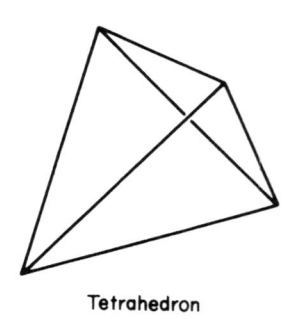

Cube   Octahedron   Tetrahedron

Figure 7.1

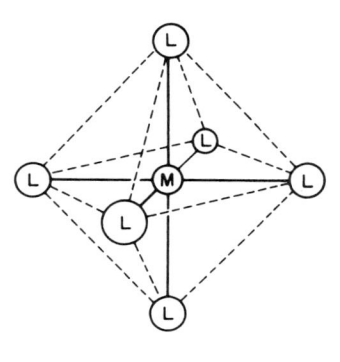

Figure 7.2 An octahedral molecule ML$_6$.

the majority of octahedral ML$_6$ compounds the central atom is a metal ion, while the surrounding atoms or ions are usually those of an electronegative element and are called ligands. Such species are referred to as 'octahedral complexes'. Although it is not convenient to start the discussion with such molecules they will be looked at in more detail later in the chapter.

A tetrahedron (Figure 7.1) is also derived from a cube, as was recognized in Chapter 1 (see Figure 1.4 and the discussion in Section 1.2.4). The fact that both the octahedron and tetrahedron are related to the cube means that it is possible to give a common discussion of the electronic structure of octahedral and tetrahedral transition metal complexes. In this book we shall not embark on this discussion although the starting point will be indicated. First, however, we look at the symmetry of the octahedron in more detail.

7.1 THE SYMMETRY OPERATIONS OF THE OCTAHEDRON

Figure 7.3(a) shows those pure rotational symmetry operations which turn an octahedral ML$_6$ molecule into itself. The octahedron contains three fourfold rotation axes and, of necessity, three coincident twofold rotation axes. There are also six twofold axes which are quite distinct from those that are coincident with the fourfold axes. Figure 7.3(a) shows a rather bewildering array of symmetry axes but there is a simple way of reducing the complexity. This is by associating symmetry elements with geometrical features. Thus, each C_3 axis passes through the mid-points of a pair of equilateral triangular faces on opposite sides of the octahedron. There are eight faces and so four pairs of opposite faces. It follows that there are four different C_3 axes. Similarly, the C_4 and coincident C_2 axes pass through opposite pairs of apices; there are six apices and so just three C_4's and C_2's. The other, C_2', axes pass through the mid-points of pairs of opposite edges. Because the octahedron has twelve edges there are six C_2' axes. The fact that the operations associated with each

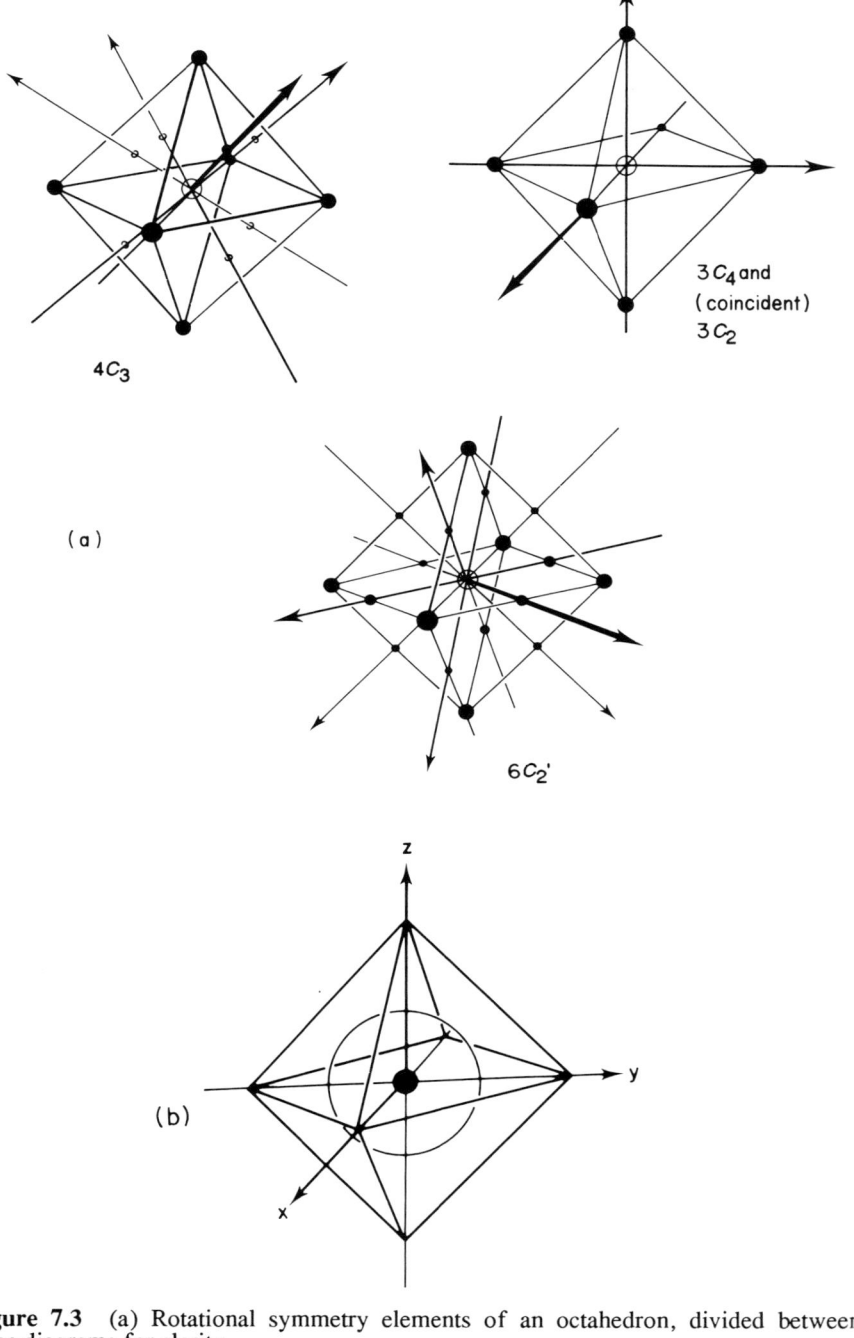

$4C_3$

$3C_4$ and
(coincident)
$3C_2$

(a)

$6C_2'$

(b)

Figure 7.3 (a) Rotational symmetry elements of an octahedron, divided between three diagrams for clarity.
(b) The conventional choice of coordinate axes for an octahedron.

set of axes form separate classes is actually evident from the way that rotation axes are interchanged by other operations of the group (for instance, a C_3 operation interchanges the C_4 axes).

Problem 7.1 Use Figure 7.1 to obtain the rotational axes of an octahedron (i.e. work through the above argument). Your answers can be checked by reference to Figure 7.3.

Problem 7.2 Use Figure 7.1 to obtain the rotational axes of a cube. Compare your answer with that found for an octahedron.

For each of the C_3 and C_4 rotation axes there are two distinct symmetry operations—those of rotation clockwise and rotation anticlockwise. They have been met in the previous two chapters, although it was not always convenient there to refer to them as clockwise and anticlockwise rotations. It follows that the rotational symmetry operations which turn an ML_6 molecule into itself are

$$E, \quad 8C_3, \quad 6C_4, \quad 3C_2 \quad \text{and} \quad 6C_2'$$

where the identity operation has been included and the $6C_2'$ refer to those twofold axes which pass through pairs of opposite edges of the octahedron. This group of 24 operations comprise the point group O. The fact that it is a complete group may be shown by constructing the group multiplication table.

Problem 7.3 Construct the multiplication table for the group O.
Note: This means constructing a 24×24 table and so will take some time. A good model (perhaps made of cardboard) is almost essential. Follow the transformations of a general point (i.e. one not lying on a symmetry axis).
Hint: As is invariably true, each operation appears once, and once only, in each row and each column of the multiplication table. The fact that this is so demonstrates that the set of 24 operations form a group.

The character table for the group O may be derived using the theorems met in Chapter 5 (although the task is not a trivial one) and is given in Table 7.1.c

Problem 7.4 Derive the character table for the group O using the theorems of Section 5.3.
Hint: It may help to look again at the solution to Problem 6.2.

There are several aspects of Table 7.1 which call for comment. For the first time triply degenerate irreducible representations are encountered; they are labelled T (with various suffixes). Their existence was implied earlier in this chapter when it was stated that 'octahedral molecules have x, y and z axes equivalent to each other'. Either these axes provide the basis for a reducible or

Table 7.1

O	E	$8C_3$	$6C_4$	$3C_2$	$6C_2'$	
A_1	1	1	1	1	1	$(x^2 + y^2 + z^2)$
A_2	1	1	-1	1	-1	
E	2	-1	0	2	0	$[1/\sqrt{3}(2z^2 - x^2 - y^2), (x^2 - y^2)]$
T_1	3	0	1	-1	-1	(x, y, z)
T_2	3	0	-1	-1	1	(xy, yz, zx)

an irreducible representation. In the event, it is irreducible and, as indicated by the basis functions given at the right-hand side of Table 7.1, they actually form a basis for the T_1 irreducible representation.

Problem 7.5 Show that the x, y and z axes, as a set, form a basis for the T_1 irreducible representation of the point group O.
Hint: Take the Cartesian axes to coincide with the C_4 axes (Figure 7.3(b)). For each class of operation select that individual operation which makes the transformation simplest to follow. The answer to this problem will be detailed—in an equivalent form—at the beginning of Section 7.2.

On the right-hand side of Table 7.1 are shown more basis functions than have previously been met. The reason is that the discussion of transition metal complexes later in this chapter will require a knowledge of how the d orbitals of the transition metal at the centre of the octahedron transform. Table 7.1 shows that the d orbitals d_{xy}, d_{yz} and d_{zx} are degenerate and transform as T_2 while d_{z^2} (or, more accurately, $d_{(1/\sqrt{3})(2z^2-x^2+y^2)}$) and $d_{(x^2-y^2)}$ are degenerate and transform as E. The function $x^2 + y^2 + z^2$—which, like an s orbital, has spherical symmetry, transforms as A_1.

It is evident from Figures 7.1 and 7.2 that an octahedron—and a cube—contain symmetry elements in addition to the rotations that have so far been listed. It contains a centre of symmetry, i, σ_h and σ_d mirror planes, and some rotation–reflection axes which are denoted S_n. The octahedron contains S_4 and S_6 rotation axes. This type of element is not an easy one to fully appreciate and they will be looked at in detail shortly. All are shown in Figure 7.4. Of these, the i and σ_h (a mirror plane horizontal with respect to an axis of highest symmetry, here C_4) have been met in Chapter 4. The σ_d mirror plane is something new. Mirror planes that bisect the angle between a pair of twofold axes are called σ_d mirror planes, the suffix d being the first letter of the word dihedral (the same word which gives its initial letters to groups such as D_2, D_{2h} and D_{3h}, groups which have, respectively, two, two and three twofold axes perpendicular to the axis of highest symmetry). In the octahedron there are six σ_d mirror planes. Although they, indeed, bisect the angles between the C_2 axes

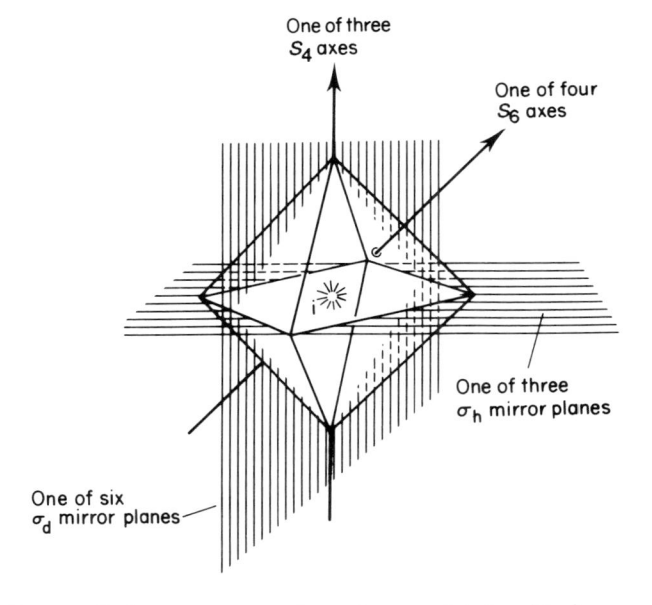

Figure 7.4 Some of the symmetry elements associated with improper rotation operations of an octahedron.

it is easier to count them by nothing that each σ_d mirror plane cuts opposite edges of the octahedron, just like the C_2' axes. There are six such pairs of edges and so six σ_d mirror planes.

Problem 7.6 Start with the definition of σ_d mirror planes as those that bisect the angle between pairs of C_2 axes and thus show that there are six σ_d mirror planes.
Hint: How many pairs of C_2 axes are there?

Note that the mirror planes which have been labelled σ_h bisect the angles between pairs of C_2' axes. These mirror planes could have been labelled as σ_d'. However, convention dictates that the label σ_h takes precedence over σ_d whenever both are applicable.

Operations such as S_6 and S_4 are interesting because, as will be seen, they are two-part operations, conventionally taken as a rotation part and a reflection part. Hence they are called rotation–reflection operations. It has been seen that the cube and octahedron have the same rotational symmetry (Problem 7.2) and it will transpire that they also have the same additional operations also. If this is so, it follows that both have S_6 and S_4 axes. The S_4 axes are easier to see for the cube and are illustrated in Figure 7.5(a). As this figure shows, the operation consists of a rotation by 90° (clockwise and

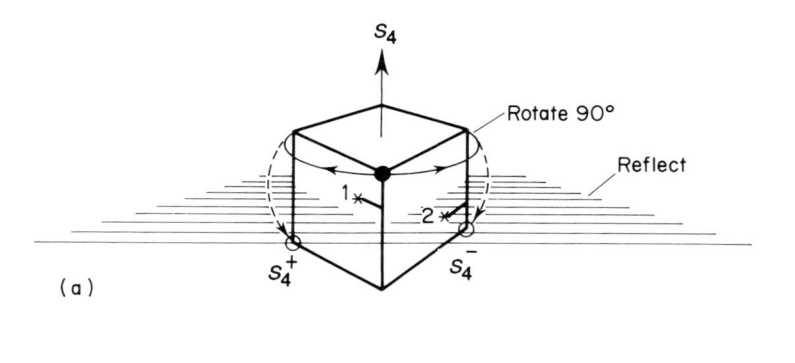

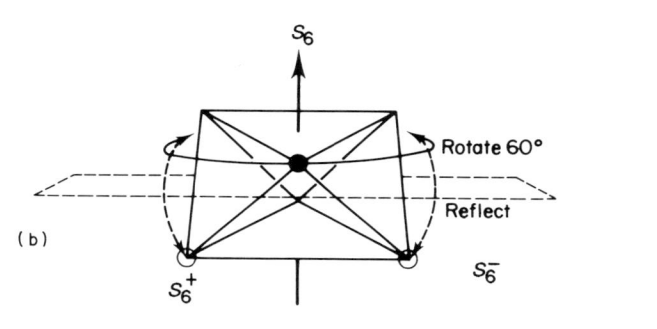

Figure 7.5   (a) An S_4 symmetry operation. (b) An S_6 symmetry operation.

anticlockwise rotations being associated with different S_4 operations) followed by reflection in a mirror plane perpendicular to the axis about which the 90° rotation was made. It is clear that this operation interconnects corners of the cube, what is not so clear is that it is necessary—for the pairs of corners connected by the S_4 operations in Figure 7.5(a) are also connected by C_2 operations (the C_2 axes emerging through mid-points of the cube faces on the right- and left-hand sides of Figure 7.5(a)). The difference between the S_4 and C_2 operations is shown by the stars in Figure 7.5(a). The star labelled 1 moves to the position occupied by star 2 under the S_4 operation but these two points are not interconnected by a C_2 rotation. The S_6 operation (rotate by 60° and then reflect in a perpendicular mirror plane) is most readily seen for an octahedron standing on a face and is illustrated in detail in Figure 7.5(b). In the case of the S_4 operations both the 90° rotation and reflection have an independent existence as C_4 and σ_h operations, respectively. In the case of the S_6 operations the rotation and reflection do not exist in their own right as symmetry operations of the octahedron and cube.

The S_n operations seem rather strange because the operation involves two operations, C_n and σ_h which may or may not have an independent existence. This apparently paradoxical situation may be made more acceptable by returning for the moment to the C_{2v} point group, discussed in Chapter 2. Figure 7.6 shows the water molecule and the C_2 operation which interrelates the two

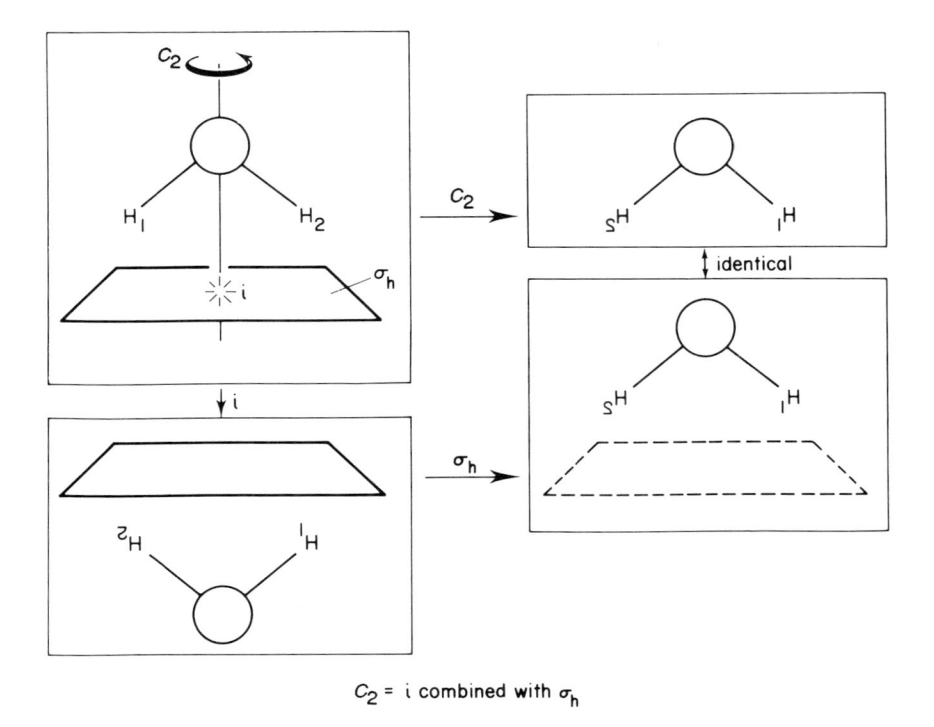

C_2 = i combined with σ_h

Figure 7.6 The two water molecules at the right-hand side of this diagram are the same, showing that the C_2 operation of the H_2O molecule (shown at the top) is equivalent to an inversion at some point along this axis followed by reflection in a σ_h mirror plane containing this inversion centre (shown at the left and bottom).

hydrogen atoms. As is seen from this figure, completely equivalent to this single C_2 operation is the combined operation of inversion through *any* point along the C_2 axis followed by reflection in a mirror plane perpendicular to the C_2 axis followed by reflection in a mirror plane perpendicular to the C_2 axis and containing the inversion centre. Neither the operations i or σ_h (or the infinity of counterparts which arise from the freedom of the pair to be located anywhere along the C_2 axis) are operations of the C_{2v} point group, yet their combination is. In the C_{2v} point group the combination of i and σ_h is not used because there is a much simpler alternative, the C_2. In the case of S_4 and S_6 operations, no such simpler form exists and there is no alternative but to use a composite. It is perhaps helpful to comment that *all* improper rotation operations may be regarded as a (correctly chosen) proper rotation operation combined with inversion in a centre of symmetry.

Problem 7.7 Determine what combinations of two independently non-existent operations are equivalent to the (real) σ_v and σ_v' operations of the C_{2v} point group.

Hint: The three operations C_2, σ and i form a complementary trio. Any one can be represented in terms of the other two (provided that the other two are correctly oriented).

Problem 7.8 Although the S_2 operation exists it is seldom mentioned as such. This is because a simpler form—and different label—exists for it. What is its alternative name?

It is necessary to count the S_4 and S_6 operations. The number of each follows from their correspondence with C_4 and C_3 operations; there are six S_4 and eight S_6. A deeper reason for this numerical connection will emerge shortly.

Problem 7.9 Show that the operation S_4 carried out twice is equivalent to C_2

$$S_4^2 = C_2$$

that S_6 carried out twice is equivalent to C_3

$$S_6^2 = C_3$$

and that S_6 carried out thrice is equivalent to i

$$S^3 = i$$

It is concluded that the complete list of symmetry operations of the octahedron (or cube) is:

$$E \quad 8C_3 \quad 6C_4 \quad 3C_2 \quad 6C_2' \quad i \quad 8S_6 \quad 6S_4 \quad 3\sigma_h \quad 6\sigma_d$$

The shorthand symbol for this set of operations is O_h (pronounced 'ooh aiche'). The character table of the O_h group is given in Table 7.2. With some considerable effort a group multiplication table for the O_h group may be constructed (it is a 48×48 table); the character table may be derived using the methods of Section

Table 7.2

O_h	E	$8C_3$	$6C_4$	$3C_2$	$6C_2'$	i	$8S_6$	$6S_4$	$3\sigma_h$	$6\sigma_d$	
A_{1g}	1	1	1	1	1	1	1	1	1	1	$x^2 + y^2 + z^2$
A_{2g}	1	1	-1	1	-1	1	1	-1	1	-1	
E_g	2	-1	0	2	-1	2	-1	0	2	-1	$\left[\frac{1}{\sqrt{3}}(2z^2 - x^2 - y^2), \frac{1}{\sqrt{3}}(x^2 - y^2)\right]$
T_{1g}	3	0	1	-1	-1	3	0	1	-1	-1	(R_x, R_y, R_z)
T_{2g}	3	0	-1	-1	1	3	0	-1	-1	1	(xy, yz, zx)
A_{1u}	1	1	1	1	1	-1	-1	-1	-1	-1	
A_{2u}	1	1	-1	1	-1	-1	-1	1	-1	1	
E_u	2	-1	0	2	-1	-2	1	0	-2	1	
T_{1u}	3	0	1	-1	-1	-3	0	-1	1	1	(T_x, T_y, T_z) (x, y, z)
T_{2u}	3	0	-1	-1	1	-3	0	1	1	-1	

5.3 but fortunately an easier method exists. This arises from the fact that the O_h group is the direct product of the groups O and C_i (the group containing E and i). The concept of a group being the direct product of two other groups was met in Chapter 4, where the D_{2h} character table was seen to be the direct product of those of C_{2v} and C_i. In the same way, the character table for O_h, Table 7.2, is the direct product of Table 7.1 (the character table for O) and Table 4.4 (the character table for C_i). That this is so is evident from the way Table 7.2 is set out; it consists of four blocks containing the characters of Table 7.1 modulated by the signs of the four characters of Table 4.4 (which, for convenience, is repeated again as Table 7.3).

Table 7.3

C_i	E	i
A_g	1	1
A_u	1	-1

Problem 7.10   Show that the labels used for the irreducible representations in Table 7.2 may be derived immediately from those of the character tables Table 7.1 and Table 7.3.

Direct product relationships, of course, apply both to operations and to characters. To the operations listed at the head of a character table as well as to the characters within the table itself. Because of the relationship between the groups O and O_h it is often possible to pretend that the symmetry of an octahedral molecule is O and then determine the g or u nature of the irreducible representations obtained by simply considering the effect of the inversion operation, i. Thus, a p orbital is *ungerade*—undergoes a change of phase— under the i operation.This, together with the knowledge that a set of p orbitals transform as T_1 in the group O (Table 7.1), is sufficient to establish that they transform as T_{1u} in O_h. Incidentally, but relevant to the set of p orbitals, when in this chapter the effects of a C_3 rotation operation are illustrated the particular C_3 operation shown in Figure 7.7 will always be the one chosen. The effect of this choice is that coordinate axes, and thus labels, always permute as follows:

These permutations apply equally to produces of axes. So, when Figure 7.24 is reached it will be seen that this C_3 operation turns $d_{x^2-y^2}$ into $d_{y^2-z^2}$—which is just what the permutation gives. The relationship between the groups O_h, O and C_i means that those operations possessed by O_h which are not present in O may be written in such a way that each is equivalent to some operation of O together with the operation i. The operations of O are *proper* (or *pure*) *rotation*

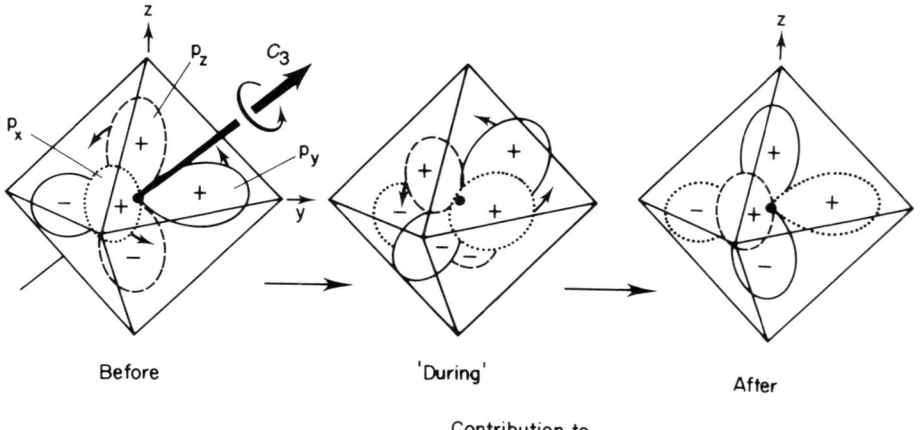

Before   'During'   After

Contribution to
to the character

p_x becomes $p_y \longrightarrow 0$
p_y becomes $p_z \longrightarrow 0$
p_z becomes $p_x \longrightarrow \underline{0}$
                              0

Figure 7.7   The transformation of a set of p orbitals of a central atom of an octahedral molecule under a C_3 rotation operation.

operations, the additional operations are *improper rotations*. As was seen at the end of Chapter 4, this latter name is used to denote any point group operation which is not a pure rotation. The precise correspondence between proper and improper rotations in the O_h group is evident from the way that Table 7.2 is written and is

E combined with i gives i
C_3 combined with i gives S_6
C_4 combined with i gives S_4
C_2 combined with i gives σ_h
C_2' combined with i gives σ_d

The existence of these relationships immediately explains why there is the same number of S_6 operations as C_3 and the same number of S_4 as C_4—and so on—in O_h.

Problem 7.11   The above discussion indicates a different definition of S_n operations from that used earlier in this chapter. They may be defined as 'rotation–inversion' operations and, indeed, this is the way that they are described by crystallographers. This definition is 'Rotate by $(180 + \theta)°$, where $\theta = (360/n)°$ and follow by inversion in a centre of symmetry'. C_n (rotate by $(360/n)°$) and i may, or may not, exist in their own right as operations in a group containing S_n. The two definitions of

S_n operations, rotate–reflect and rotate–invert are entirely equivalent. A little thought will show that the duality exists because of the connection between i, C_2 and σ, detailed above for the C_{2v} point group—but the connection is general.†

Problem 7.12 Show that the operation S_4^- (defined as a rotation–reflection operation) is the same as the operation S_4^+ (defined as a rotation–inversion operation). Here, the superscripts denote the direction of 90° rotation (thus avoiding having to add on 180°). This equivalence is the one normally made.
Hint: Use Figure 7.5(a).

7.2 THE BONDING IN THE SF₆ MOLECULE

As is so often the case, the O_h point group is easier to use than to talk (or write!) about. To illustrate its use (or, more correctly, how its use may often be avoided), we now turn to a discussion of the bonding in the SF₆ molecule. The valence shell atomic orbitals of the sulfur atom will be taken as 3s and 3p, ignoring the 3d. The behaviour of d orbitals in octahedral molecules is of major importance in transition metal chemistry and this is the context in which they will be discussed later in this chapter. Throughout our discussion of SF₆ it will be convenient to work in the point group O rather than the correct group O_h. The reason for this lies in the structure of the O_h group. It is twice as large as O (48 operations compared with 24) and so is more cumbersome to handle. But, as has been seen, O_h is the direct product of O with C_i so that the only additional information that O_h has compared with O is that of behaviour under the additional operation introduced by the group C_i—that is, behaviour under the operation i. It is easier to ask of a basis function 'how does it transform under i' and to add either g (*gerade* = symmetric with respect to inversion in a centre of symmetry) or u (*ungerade* = antisymmetric respect to inversion in a centre of symmetry) as a suffix to the irreducible representation of O than to plough through the whole set of O_h operations.

It is easy to show that the sulfur 3s orbital, shown in Figure 7.3(b), transforms as the totally symmetric, A_1, irreducible representation of the point group O. In O_h, of course, it has A_{1g} symmetry. In Figure 7.3(b) is shown the axis system that will be used in the discussion, although in many of the following figures, the octahedron is drawn from a different viewpoint from that shown in Figure 7.3. Like the coordinate axes of Problem 7.5, the sulfur 3p orbitals transform together as the T_1 irreducible representation (T_{1u} in O_h). This particular problem will be looked at in some detail because it illustrates how to handle the sometimes bewildering task of working with several

---

† So, the 180° in $(180 + \theta)°$ arises from the C_2.

equivalent objects in a high symmetry environment. Our discussion is largely diagrammatic because good diagrams—perhaps more than a good model—are very important. It is therefore essential that each figure is studied carefully and the transformations that it shows followed in detail. Figures 7.7–7.10 illustrate the transformations of the set of 3p orbitals under a representative operation of each class of the O point group. For clarity of presentation the lobes of the p orbitals are shown more as ellipses than the circles used so far in this book; the different p orbitals are distinguished by the way they are outlined. As an aid to visualizing the interconversions these figures not only show the starting arrangement but also show them at some point while the operation is in progress as well as in the final arrangement. The actual transformations brought about by these operations are listed in the figures and the compilation of the corresponding characters detailed. Figures 7.7–7.10 should be very carefully studied; the characters to which they give rise are:

$$
\begin{array}{ccccc}
E & 8C_3 & 6C_4 & 3C_2 & 6C_2' \\
3 & 0 & 1 & -1 & -1
\end{array}
$$

Comparison with Table 7.1 confirms that this set of characters is the T_1 irreducible representation.

Problem 7.13 By a detailed study of Figures 7.7–7.10 derive the above set of characters.

The s and p orbitals of the sulfur atom in SF_6 bond with those fluorine orbitals that point towards the sulfur atoms; without defining their composition

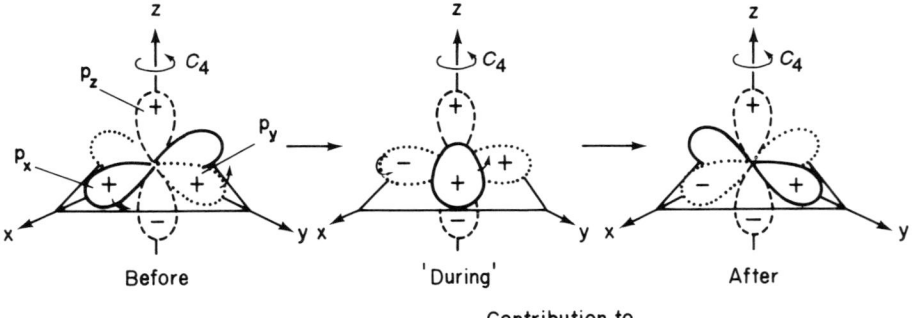

Contribution to
the character

p_x becomes p_y → 0
p_y becomes $-p_x$ → 0
p_z becomes p_z → $\dfrac{1}{1}$

Figure 7.8 The transformation of a set of p orbitals of a central atom of an octahedral molecule under a C_4 rotation operation.

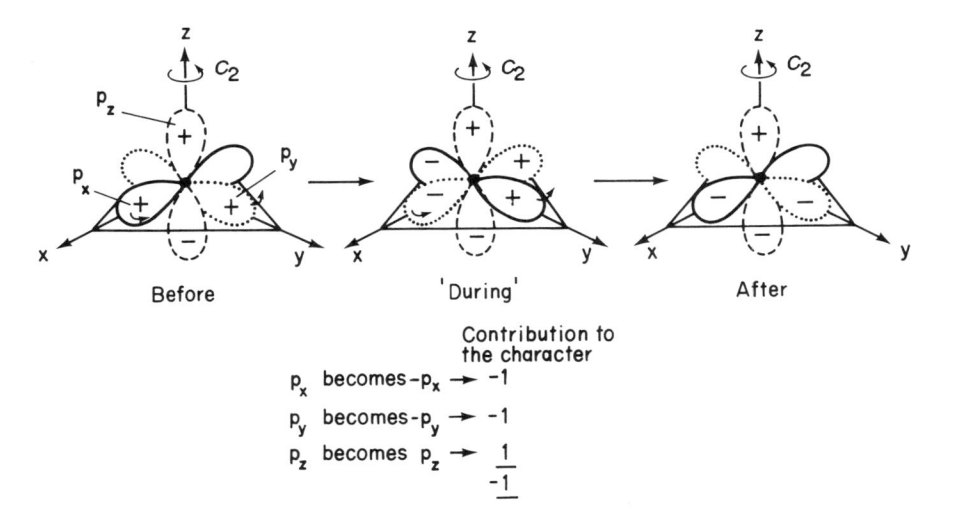

Figure 7.9 The transformation of a set of p orbitals of a central atom of an octahedral molecule under a C_2 rotation operation.

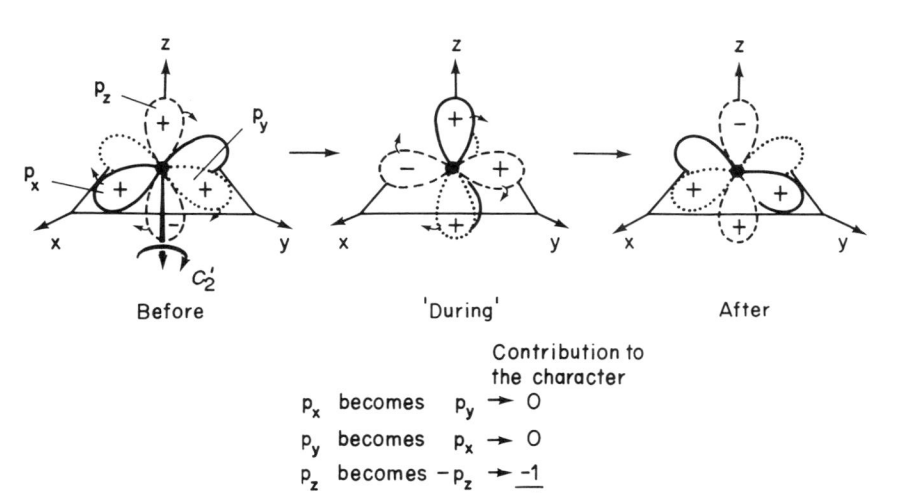

Figure 7.10 The transformation of a set of p orbitals of a central atom of an octahedral molecule under a C_2' rotation operation.

further they will simply be called the fluorine σ orbitals. This set of orbitals is shown schematically in Figure 7.11 and, because all are symmetry-related, the transformations of the six must be considered as a set. This is not a difficult task but some mental gymnastics can be avoided by remembering a general principle; only if an object lies on a symmetry element can it give a non-zero

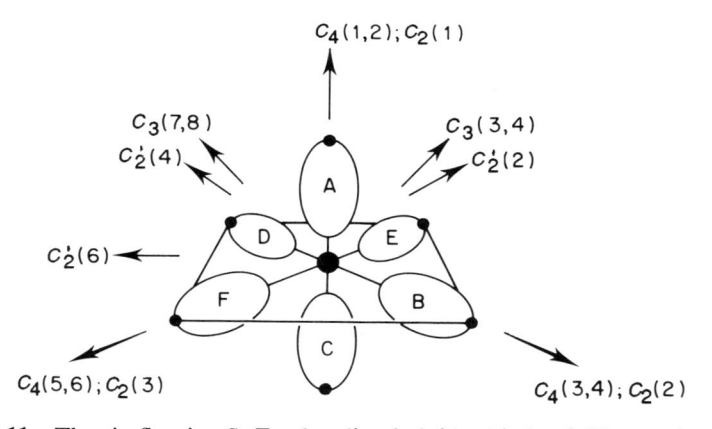

Figure 7.11   The six fluorine S–F σ-bonding hybrid orbitals of SF_6 together with the labels used for them in the text. Some of the axes used in obtaining Table 7.3 are indicated.

contribution to the character associated with the corresponding operation. It follows, therefore, that because the fluorine atoms are located on the fourfold axes of the octahedron it is only under the fourfold and corresponding twofold rotation operations (and, of course, the identity operation) that any of the fluorine σ orbitals can be left unchanged. Remembering that there are two fluorines on each C_4 axis, it follows that the reducible representation generated by the transformation of the fluorine σ orbitals is:

E	$8C_3$	$6C_4$	$3C_2$	$6C_2'$
6	0	2	2	0

This is a sum (sums such as this are sometimes called a *direct* sum) of the $A_1 + E + T_1$ irreducible representations.

> **Problem 7.14**   Use Table 7.1 to show that the above reducible representation has A_1, E and T_1 components.
> *Hint:* This is similar to, but more difficult than, Problems 5.14 and 6.5.

This result was obtained in the group O. There are several ways in which we could proceed to obtain the g and u nature of the A_1, E and T_1 combinations in O_h—the most obvious way is to repeat the above sequence again but using the full O_h group. However, we should still be without the explicit forms of the combination of σ orbitals which transform as each irreducible representation, and this information is something that will be needed later. If these explicit forms were available at the present point in the text, a detailed study of them would show their g or u nature. It is therefore simplest next to use the projector operator method to obtain these combinations (working in the point group O) and subsequently to ask which are g and which are u in nature. To some extent the result may be anticipated. The fluorine σ orbitals used to generate the reducible representation were neither inherently symmetric nor antisymmetric with respect to inversion in the centre of symmetry—because they were not located at this centre they were always permuted by the inversion operation. In such a situation this indifference is reflected by an equal number of combinations of g and u symmetries being generated. That is, in the present case there must be three linear combinations of fluorine orbitals which are g and three which are u. There are, then, two possibilities. Either we have $A_{1g} + E_g + T_{1u}$ in O_h or, alternatively, $A_{1u} + E_u + T_{1g}$. Physically, only the first choice makes any sense, because the irreducible representations generated by the transformation of the sulfur valence shell s and p orbitals are included in this set whereas they are not in the second. That is, if the first set is correct then there can be interactions between the fluorine σ orbitals and the sulfur orbitals—and so the existence of the molecule explained—whereas for the second set there would be no interactions and the molecule SF$_6$ would not exist!

As just mentioned, in order to obtain the fluorine σ orbital combinations transforming as the A_1, E and T_1 irreducible representations the projection operator method will be used—a method that has been met several times before. Because the present case provides a particularly good example of the general method it will be given in detail, bringing together the techniques developed in previous chapters.

First, each ligand σ orbital is given a label, A → F, as shown in Figure 7.11. The transformation of one of these orbitals under the operations of the group is then considered in detail. Table 7.4 lists the 24 operations of the group O, and beneath each is the ligand σ orbital into which A is transformed by the particular operation. Within each set of operations, $8C_3$ for example, the order in which the operations are considered is unimportant; what matters is that all are included.

Table 7.4

E	$C_4(1)$	$C_4(2)$	$C_4(3)$	$C_4(4)$	$C_4(5)$
A	A	A	F	E	B
$C_4(6)$	$C_2(1)$	$C_2(2)$	$C_2(3)$	$C_3(1)$	$C_3(2)$
D	A	C	C	D	E
$C_3(3)$	$C_3(4)$	$C_3(5)$	$C_3(6)$	$C_3(7)$	$C_3(8)$
E	B	B	F	D	F
$C_2'(1)$	$C_2'(2)$	$C_2'(3)$	$C_2'(4)$	$C_2'(5)$	$C_2'(6)$
D	B	E	F	C	C

Problem 7.15 Use Figures 7.3(a) and 7.11 to obtain Table 7.4.
Hint: Good diagrams are important—it may be necessary to sketch out parts of Figure 7.3(a) several times to retain clarity in distinguishing the

different effects of the various operations. Note that because the are only six fluorine σ orbitals but 24 operations, each orbital label appears four times in Table 7.4. To help with this problem as many axes as is consistent with graphic clarity have been indicated in Figure 7.11.

The orbitals in Table 7.4 are now multiplied by the characters appropriate to the listed operations. The products and then added together. The sum is either the required ligand group orbital or is simply related to it. For the A_1 group orbital, multiplying each of the orbitals by 1 (the value of each of the A_1 characters) and adding the products together gives

$$4A + 4B + 4C + 4D + 4E + 4F$$

Normalizing, the A_1 orbital is obtained

$$\psi(a_1) = \frac{1}{\sqrt{6}} \ (A + B + C + D + E + F)$$

Turning to the E orbitals, the sum obtained after multiplication is

$$4A + 4C - 2B - 2D - 2E - 2F$$

which after normalizing is

$$\psi e(1) = \frac{1}{\sqrt{12}} \ (2A + 2C - B - D - E - F)$$

Problem 7.16   Derive $\psi e(1)$.
Hint: This problem is quite similar to that solved in Table 6.2 and the associated discussion.

Now a problem which is closely related to one met in Chapter 6—how to obtain the second E function. Using either B or E as the generating orbital in Table 7.4 the (un-normalized) combinations which would have been obtained are:

from B: $4B + 4D - 2A - 2E - 2C - 2F$
from E: $4E + 4F - 2A - 2B - 2C - 2D$

Neither of these can be the second E function for they are different and the choice between them is arbitrary; they cannot both be correct and there cannot be three different E functions. The method described in Section 6.2 may be used to systematically obtain the second function. Study of the results obtained there suggests that the difference between the functions given above should be taken. This difference is

$$6B + 6D - 6E - 6F$$

which on normalizing gives the second function

$$\psi e(2) = \tfrac{1}{2}(B + D - E - F)$$

Problem 7.17 Show by squaring and adding the coefficients with which the fluorine σ orbitals appear in $\psi e(1)$ and $\psi e(2)$ that each orbital makes an equal contribution to the E set.

Hint: This problem resembles Problem 6.7. Note that the sum of squares of coefficients is equal to the ratio

$$\frac{\text{Number of } E \text{ functions}}{\text{Total number of } \sigma \text{ orbitals}} = \frac{2}{6}$$

In Section 6.2 the argument used leading to the generation of a second E function depended on the fact that the functions which we were seeking to generate had vector-like properties. Those that have just been obtained do not behave like axes (as the basis functions given at the right-hand side of Table 7.1 show; they transform like sums of products of axes). The method deduced in Section 6.2 clearly has a wider generality than could have been anticipated.

The T_1 functions are readily obtained. The transformations of A in Table 7.4 when multiplied by the T_1 characters and added gives $4A - 4C$ which, normalized, gives

$$\psi t_1(1) = \frac{1}{\sqrt{2}} (A - C)$$

Similarly, the transformation of B (or D) and E (or F) give

$$\psi t_1(2) = \frac{1}{\sqrt{2}} (B - D)$$

and

$$\psi t_1(3) = \frac{1}{\sqrt{2}} (E - F)$$

respectively. Because these three functions each involve different fluorine orbitals they are clearly independent of each other.

Problem 7.18 Derive the three T_1 functions listed above.

The complete list of ligand group orbitals† is given in Table 7.5. In order to determine their symmetries in O_h the behaviour of these functions under the operation of the inversion centre of symmetry has to be determined. Because this operation interchanges the fluorine σ orbitals as follows:

$$A \leftrightarrow C$$
$$B \leftrightarrow D$$
$$E \leftrightarrow F$$

---

† This is the name commonly used in transition metal chemistry.

Table 7.5

Symmetry		Symmetry-adapted function
In the group O	In the group O_h	(ligand group orbitals)
A_1	A_{1g}	$\dfrac{1}{\sqrt{6}}$ (A + B + C + D + E + F)
E	E_g	$\begin{cases} \dfrac{1}{\sqrt{12}} (2A + 2C - B - D - E - F) \\ \dfrac{1}{2} (B + D - E - F) \end{cases}$
T_1	T_{1u}	$\begin{cases} \dfrac{1}{\sqrt{2}} (A - C) \\ \dfrac{1}{\sqrt{2}} (B - D) \\ \dfrac{1}{\sqrt{2}} (E - F) \end{cases}$

the effect of this operation is obtained by making these substitutions (A for C, C for A etc.) in the functions given in Table 7.5. When this is done, the A_1 and E's are left unchanged but each T_1 function changes sign. It is concluded that they transform in O_h as A_{1g}, E_g and T_{1u}, respectively. These labels have been included in Table 7.5. This means that symmetries of the sulfur 3s and 3p orbitals(A_{1g} and T_{1u} respectively) are matched within the fluorine σ orbital symmetries. The bonding molecular orbital of A_{1g} symmetry is shown in Figure 7.12(a), a representative T_{1u} bonding molecular orbital in Figure 7.12(b) and one of the two E_g functions in Figure 7.12(c). Because the two E_g functions have different mathematical forms, it should perhaps be commented that it is the second that was generated which is shown in Figure 7.12(c).

Problem 7.19   Sketch a diagram of the interaction of the first-generated E_g function with the appropriate fluorine group orbital.
Hint: If in doubt, first sketch the first-generated E_g orbital.

There is no doubt that the bonding orbitals have bonding energy stabilities in the order

$$A_{1g} > T_{1u} > E_g.$$

This order of orbital energies is also the order in terms of the number of nodes. The A_{1g} bonding molecular orbital is nodeless, the T_{1u} has one nodal plane and

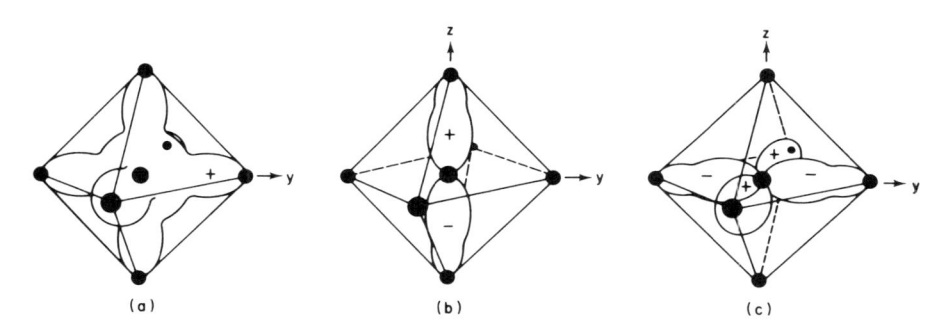

Figure 7.12   (a) The A_{1g} bonding molecular orbital.
(b) One of the T_{1u} bonding molecular orbitals (that involving the sulfur p$_z$ orbital).
(c) One of the fluorine σ orbital symmetry-adapted combinations of E_g symmetry.

the E_g two. These nodal patterns are implicit in the expressions given in Table 7.5 and are also evident in Figure 7.12.

This discussion has assumed that only σ bonding is involved in the interaction between the central sulfur atom and the surrounding fluorines. Although this is quite a good approximation for SF$_6$ it is useful, none the less, to extend the discussion to include those p$_\pi$ orbitals on the fluorine atoms which so far have been ignored. This is because they are of relatively high energy and they will be seen in the photoelectron spectrum. These p$_\pi$ orbitals transform as a degenerate pair of E symmetry under the local C_{4v} symmetry of each fluorine atom. It follows from the discussion of this symmetry in Chapter 5 that there is no unique specification of the direction of the local x and y axes but the choice and notation in Figure 7.13 prove to be convenient in practice.

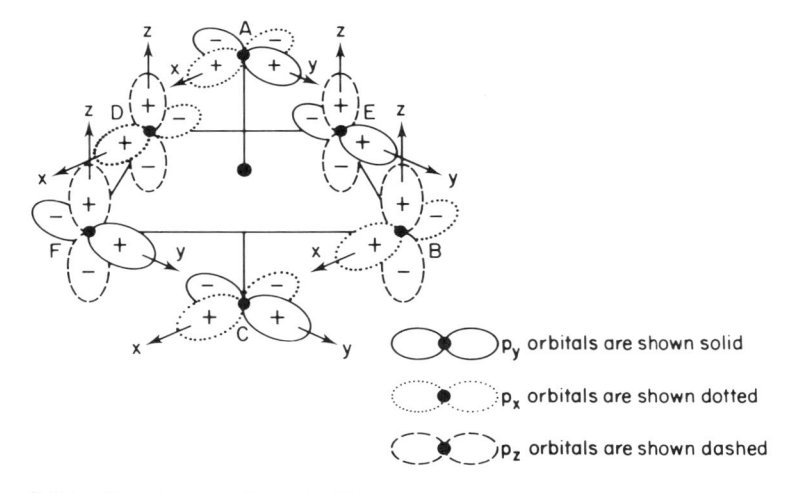

Figure 7.13   Fluorine p$_\pi$ orbitals in SF$_6$.

Problem 7.20   Figure 7.13 appears rather complicated. Show that it has an internal consistency in that all p_x orbitals are all oriented in the same x direction, the p_y in the same y and the p_z in the same z direction.

The next step is to determine the reducible representation generated by the transformation of these twelve p_π orbitals. As for the case of the σ orbitals it is only possible to obtain non-zero characters for the identity operation, for the C_4 rotation operations and the corresponding C_2 rotation operation (because the fluorine atoms lie on the fourfold axes of the octahedron). Of these, the character for the C_4 rotation operation is zero (because, for those fluorine atoms left unshifted by the operation, the p_x and p_y orbitals are interchanged) so that only the identity and $3C_2$ classes contain non-zero characters. The reducible representation obtained is

E	$8C_3$	$6C_4$	$3C_2$	$6C_2'$
12	0	0	-4	0

which has components $2T_1 + 2T_2$ (under O_h symmetry these become $T_{1g} + T_{1u} + T_{2g} + T_{2u}$).

Table 7.6   Symmetry-adapted combinations of fluorine p_π orbitals

---

T_{1g} orbitals
$$t_{1g}(1) = \tfrac{1}{2}[p_x(A) - p_x(C) + p_z(E) - p_z(F)]$$
$$t_{1g}(2) = \tfrac{1}{2}[p_y(A) - p_y(C) - p_z(B) + p_z(D)]$$
$$t_{1g}(3) = \tfrac{1}{2}[p_x(B) - p_x(D) + p_y(E) - p_y(F)]$$

T_{1u} orbitals
$$t_{1u}(1) = \tfrac{1}{2}[p_z(B) + p_z(D) + p_z(E) + p_z(F)]$$
$$t_{1u}(2) = \tfrac{1}{2}[p_y(A) + p_y(C) + p_y(E) + p_y(F)]$$
$$t_{1u}(3) = \tfrac{1}{2}[p_x(A) + p_x(B) + p_x(C) + p_x(D)]$$

T_{2g} orbitals
$$t_{2g}(1) = \tfrac{1}{2}[p_x(A) - p_x(C) - p_z(E) + p_z(F)]$$
$$t_{2g}(2) = \tfrac{1}{2}[p_y(A) - p_y(C) + p_z(B) - p_z(D)]$$
$$t_{2g}(3) = \tfrac{1}{2}[p_x(B) - p_x(D) - p_y(E) + p_y(F)]$$

T_{1u} orbitals
$$t_{2u}(1) = \tfrac{1}{2}[p_z(B) + p_z(D) - p_z(E) - p_z(F)]$$
$$t_{2u}(2) = \tfrac{1}{2}[p_y(A) + p_y(C) - p_y(E) - p_y(F)]$$
$$t_{2u}(3) = \tfrac{1}{2}[p_x(A) + p_x(B) - p_x(C) - p_x(D)]$$

$p_y(A)$ means the p_y orbital on atom A as indicated in Figure 7.13.

Problem 7.21 Show that the twelve fluorine p_π orbitals of Figure 7.13 generate the above reducible representation and that it has $2T_1$ and $2T_2$ components in the point group O.

Appropriate linear combinations are obtained by the usual projection operator method but a difficulty arises because, in O symmetry, two quite independent sets of functions transform as T_1 and two other sets as T_2. The problem of distinguishing between them is readily solved by working, instead, in O_h symmetry—where all sets are symmetry-distinguished—but this is a rather tedious task because this group has 48 symmetry operations, each of which has to be separately considered. In Appendix 4 an alternative, short-cut, method of obtaining these linear combinations is described. This method, depending on an ascent-in-symmetry, is a most useful one for high symmetry systems in which a large number of basis functions has to be handled.

   The appropriate linear combinations are given in Table 7.6 and one of each symmetry species shown in Figures 7.14–7.17. Our interest in these

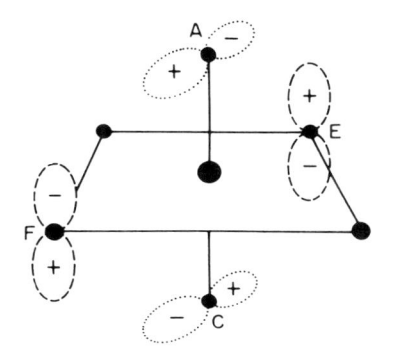

Figure 7.14   One of the T_{1g} fluorine p_π combinations.

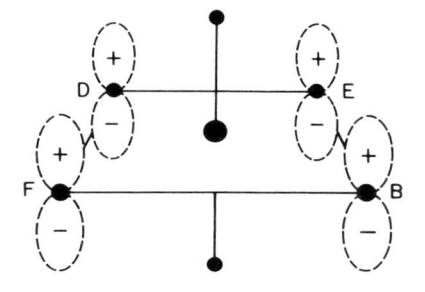

Figure 7.15   One of the T_{1u} fluorine p_π combinations.

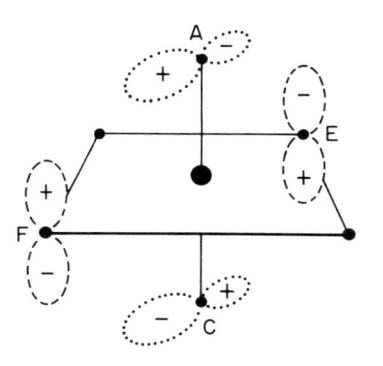

Figure 7.16   One of the T_{2g} fluorine p_π combinations.

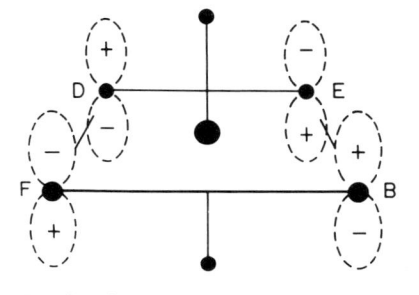

Figure 7.17   One of the T_{2u} fluorine p_π combinations.

combinations lies in the interactions between adjacent fluorine p_π orbitals for we shall use these interactions to predict relative energies for comparison with the results of photoelectron spectroscopy. Figures 7.14–7.17 show an interesting situation. The interactions between the fluorine p_π orbitals are of two types. For the T_{1u} (Figure 7.15) and T_{2u} (Figure 7.17) sets the component p_π orbitals are arranged parallel to each other; their interactions are therefore of π-type. For the T_{1g} (Figure 7.14) and T_{2g} (Figure 7.16) orbitals the axes of adjacent atomic p_π orbitals are at right angles to each other so that their interaction is a mixture of σ and π-types (as shown in Figure 7.18 the p_π orbitals may be treated as vectors and the neighbouring interactions resolved into σ and π components). Qualitatively, σ-interactions are usually greater than π and so it is reasonable to expect that the energy difference between the T_{2u} and T_{1g} orbitals would be greater than that between T_{1u} and T_{2g}, provided that the interactions are comparable in other respects. The other important factor is relative nodality. As is evident from Figures 7.14–7.17, the T_{1u} and T_{2g} orbitals are no-node combinations (apart from the nodes inherent in the p_π orbitals themselves)—the positive lobe of a p orbital is adjacent to the positive lobe of an adjacent p orbital—whereas the T_{1g} and T_{2u} orbitals each have two

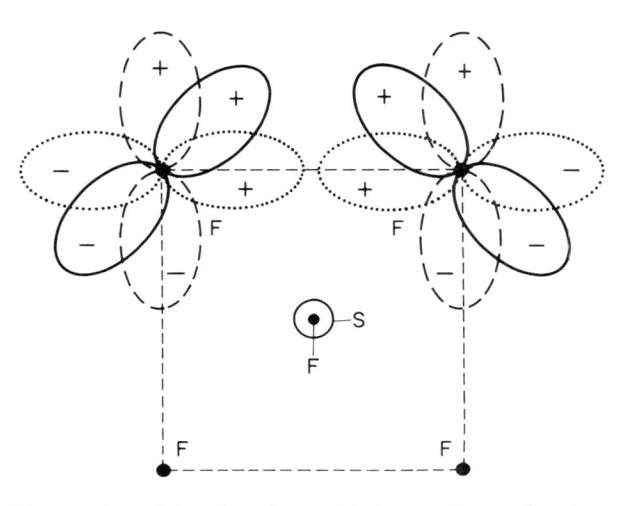

Figure 7.18 The overlap of 'coplanar' p$_\pi$ orbitals on adjacent fluorine atoms (shown solid) may be expressed as a sum of a π overlap (between orbitals shown dashed) and a σ overlap (between orbitals shown dotted). This diagram shows SF$_6$ viewed down a C_4 axis.

additional nodes—the positive lobe of one p orbital is adjacent to the negative lobe of its neighbour. We have:

T_{1g}: 2 nodes, σ and π interactions
T_{2u}: 2 nodes, π interaction only
T_{1u}: 0 nodes, π interaction only
T_{2g}: 0 nodes, σ and π interactions

Our discussion leads us to expect that the stability of these orbitals varies in the order

$$T_{2g} > T_{1u} > T_{2u} > T_{1g}$$

In arriving at this order any possible bonding of the fluorine p$_\pi$ orbitals with the sulfur atom has been ignored. This defect is easily remedied. The only symmetry in common with the sulfur valence shell orbitals is T_{1u}. The latter are involved in S–F σ bonding and so interactions between the p$_\pi$-derived T_1 set and the (more stable) S–F σ-bonding T_{1u} set have to be considered (these are the only ones that will affect the photoelectron spectrum). Any interaction will lead to a further stabilization of the S–F σ bonding set and a corresponding destabilization of the p$_\pi$-derived T_{1u} orbitals. This destabilization could change the p$_\pi$ orbital energy sequence; although the stability order

$$T_{2g} > T_{2u} > T_{1g}$$

seems clear enough, the T_{1u} set could slot in between the T_{2g} and T_{2u}—as before—or between the T_{2u} and T_{1g} (assuming that the destabilization is not too great). In practice, the stability observed sequence for SF_6 seems to be†

$$T_{2g} > T_{2u} > T_{1u} > T_{1g}$$

or, including the orbitals associated with σ-bonding $(A_{1g} > T_{1u} > E_g)$:

$$A_{1g} > T_{1u}(1) > T_{2g} > E_g > T_{2u} > T_{1u}(2) > T_{1g}$$

This order is in good agreement with that given by the qualitative model developed above, although this was not able to predict that the σ-interaction energy level sequence would overlap with the π levels.

Problem 7.22   Draw an orbital energy level diagram for SF_6 (cf. for instance, Figure 6.8).

Sulfur hexafluoride is the last, and most complicated, molecule of which the electronic structure will be considered in detail in this book. The molecules that were selected were chosen more because they enabled particular aspects of group theory to be introduced, rather than for their own intrinsic interest. However, the methods developed are of general applicability and can be used to gain insight into the electronic structure of quite complicated molecules. It is particularly useful here to assume the highest reasonable symmetry for a molecule or molecular fragment and to consider the effects of a reduction in symmetry to the real-life level as a minor perturbation. Such reductions in symmetry will be covered in Chapter 8. First, however, the discussion of octahedral molecules will be extended to transition metal complexes of this symmetry.

7.3   OCTAHEDRAL TRANSITION METAL COMPLEXES

It is probably true that a majority of transition metal complexes have octahedral symmetry, at least approximately. Entire books have been written on this subject but only the more important features will be described here. At the simplest level an octahedral transition metal complex may be regarded as built up from a transition metal ion, M^{n+}, surrounded by six atoms or ions arranged at the corners of a regular octahedron. The six surrounding atoms may indeed be single atoms or they may be an atom through which a molecule is attached to the transition metal ion. In the simplest picture the metal ion is bonded to the six surrounding *ligands* (a collective noun covering both bonded atoms and molecules) by pure electrostatic attraction. This simple model leads to *crystal field theory* and it is this which will now be discussed in outline. Although

---

† W. von Niessen, W. P. Kraemer and G. H. F. Diercksend, *Chemical Physics Letters*, **63** (1979), 65. The order given above is of calculated orbital energies; in the vertical ionization potentials (observed in the photoelectron spectrum) the T_{2u} and T_{1u} are identical.

simple, it provides the basis for all other, more detailed, models and so time is well spent studying it.

Transition metals are characterized by the fact that they exhibit variable valencies in their salts. The corresponding transition metal cations have different numbers of d electrons, the number of d electrons varying with the valence state of the cation. Loosely speaking, if a transition metal ion is oxidized then it loses a d electron; if it is reduced it gains one. Attention is therefore focused on the d electrons and on the d orbitals. In an octahedral ML_6 molecule a set of d orbitals on the central metal atom divides into two sets. One, consisting of the d_{xy}, d_{yz} and d_{zx} orbitals, has T_{2g} symmetry, as indicated in Tables 7.1 and 7.2. Figures 7.19–7.22 illustrate the transformations of members of this set and their individual contributions to the characters detailed.

Problem 7.23   Check the transformations of the d_{xy}, d_{yz} and d_{zx} orbitals shown in Figures 7.19–7.22 and thus show that these orbitals transform as T_2 in O. Because they are centrosymmetric orbitals, it follows that they transform as T_{2g} in O_h.

The second set of d orbitals,† $d_{x^2-y^2}$ and $d_{(1/\sqrt{3})(2z^2-x^2-y^2)}$ transform together as the E_g irreducible representation. Their transformations are illustrated in

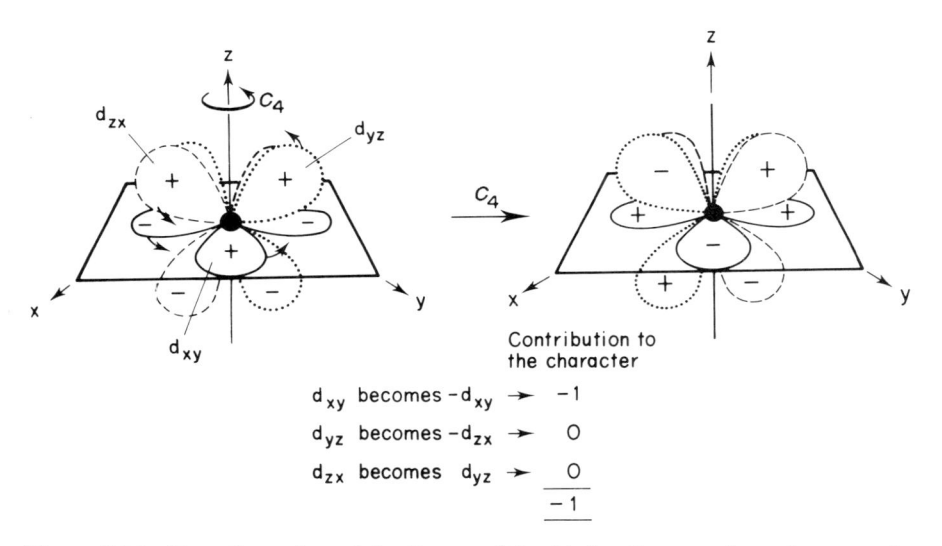

Contribution to
the character

d_{xy} becomes $-d_{xy}$ $\rightarrow$ -1

d_{yz} becomes $-d_{zx}$ $\rightarrow$ 0

d_{zx} becomes d_{yz} $\rightarrow$ $\underline{\quad 0 \quad}$

$\quad -1$

Figure 7.19   Transformation of the T_{2g} set of d orbitals of a central metal atom under a C_4 rotation operation of the octahedron.

---

† These orbitals are usually called $d_{x^2-y^2}$ and d_{z^2}. In the present context we have to recognize that the label z^2 is a shorthand symbol for $(1/\sqrt{3})(2z^2 - x^2 - y^2)$.

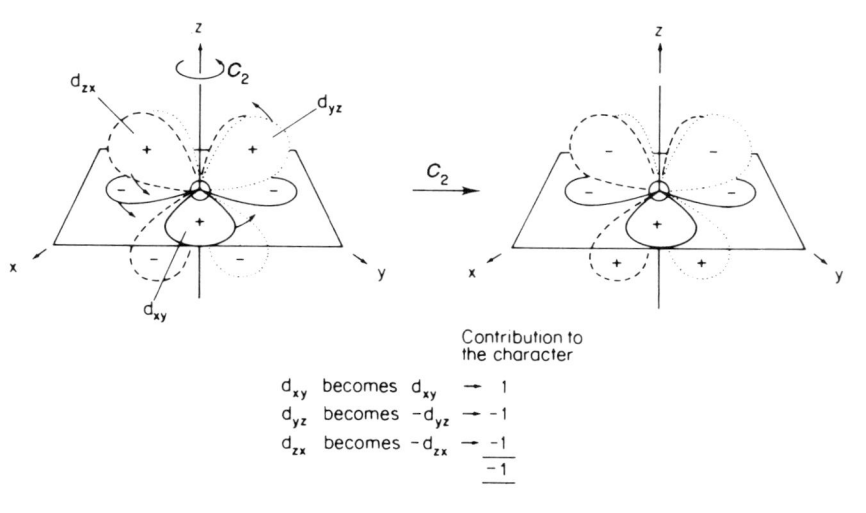

Contribution to
the character

d_{xy} becomes d_{xy} → 1
d_{yz} becomes $-d_{yz}$ → -1
d_{zx} becomes $-d_{zx}$ → -1
$\overline{-1}$

Figure 7.20 Transformation of the T_{2g} set of d orbitals of a central metal atom under a C_2 rotation operation of the octahedron.

Figures 7.23–7.26 where their individual contributions to the characters are also given. The only point of difficulty arises in connection with the C_3 rotation operations and resembles that discussed in detail in Section 6.1. There, too, a doubly degenerate irreducible representation gave a character of -1 under a C_3 rotation. In the present case it is helpful to write the E_g orbitals as $d_{(x^2-y^2)}$ and $d_{(1/\sqrt{3})[(z^2-x^2)-(y^2-z^2)]}$, because this helps to demonstrate that rotation of the pair

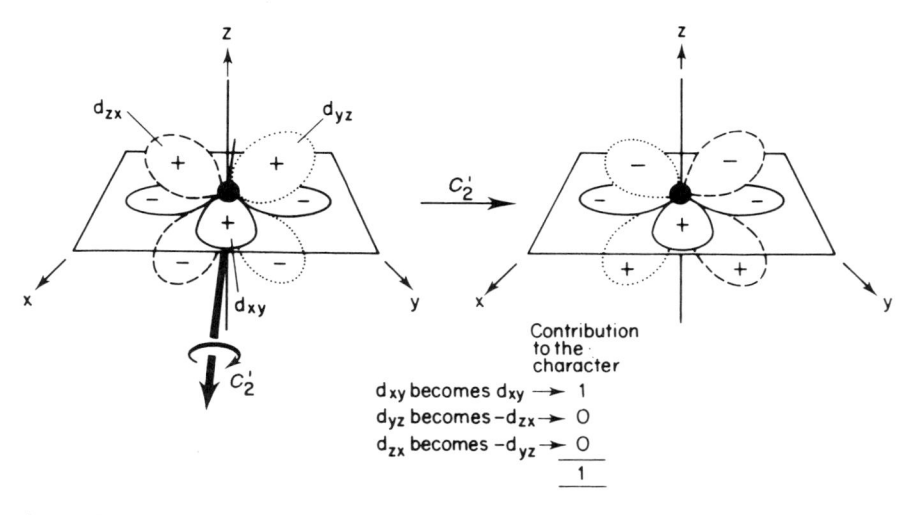

Contribution
to the
character

d_{xy} becomes d_{xy} → 1
d_{yz} becomes $-d_{zx}$ → 0
d_{zx} becomes $-d_{yz}$ → 0
$\overline{1}$

Figure 7.21 Transformation of the T_{2g} set of d orbitals of a central metal atom under a C_2' rotation operation of the octahedron.

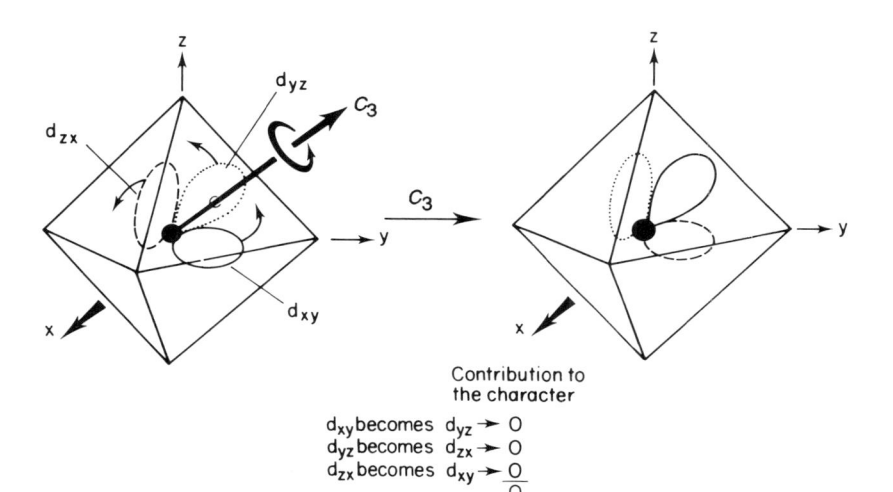

Contribution to
the character

d_{xy} becomes $d_{yz} \rightarrow 0$
d_{yz} becomes $d_{zx} \rightarrow 0$
d_{zx} becomes $d_{xy} \rightarrow \underline{0}$
$\phantom{d_{zx}\text{ becomes }d_{xy} \rightarrow }\overline{0}$

To simplify the diagram only one lobe of each of d_{xy}, d_{yz} and d_{zx} is shown

Figure 7.22 Transformation of the T_{2g} set of d orbitals of a central metal atom under a C_3 rotation operation of the octahedron.

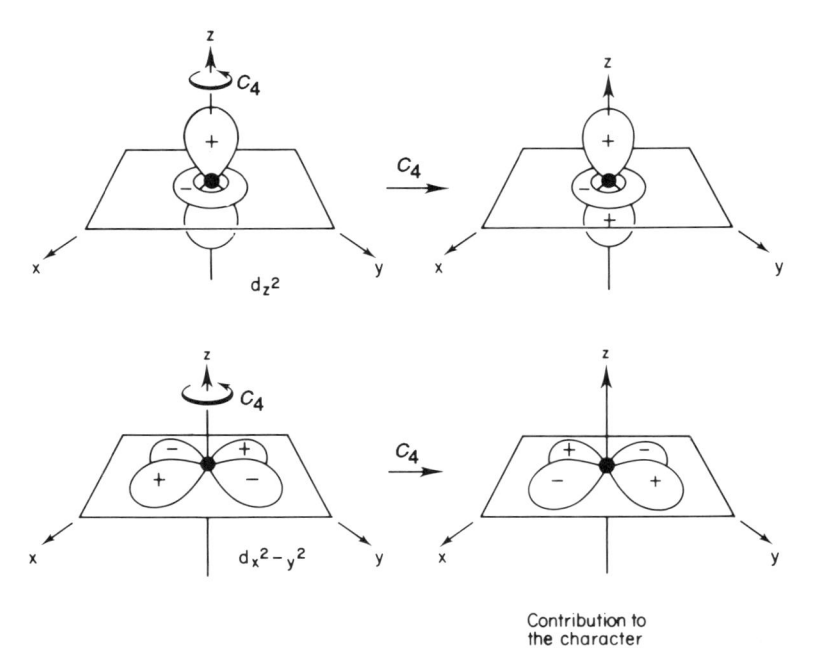

Contribution to
the character

d_{z^2} becomes $d_{z^2} \rightarrow 1$
$d_{x^2-y^2}$ becomes $-d_{x^2-y^2} \rightarrow \underline{-1}$
$\phantom{d_{x^2-y^2}\text{ becomes }-d_{x^2-y^2} \rightarrow }\overline{0}$

Figure 7.23 Transformation of the E_g set of d orbitals of a central metal atom under a C_4 rotation operation of the octahedron.

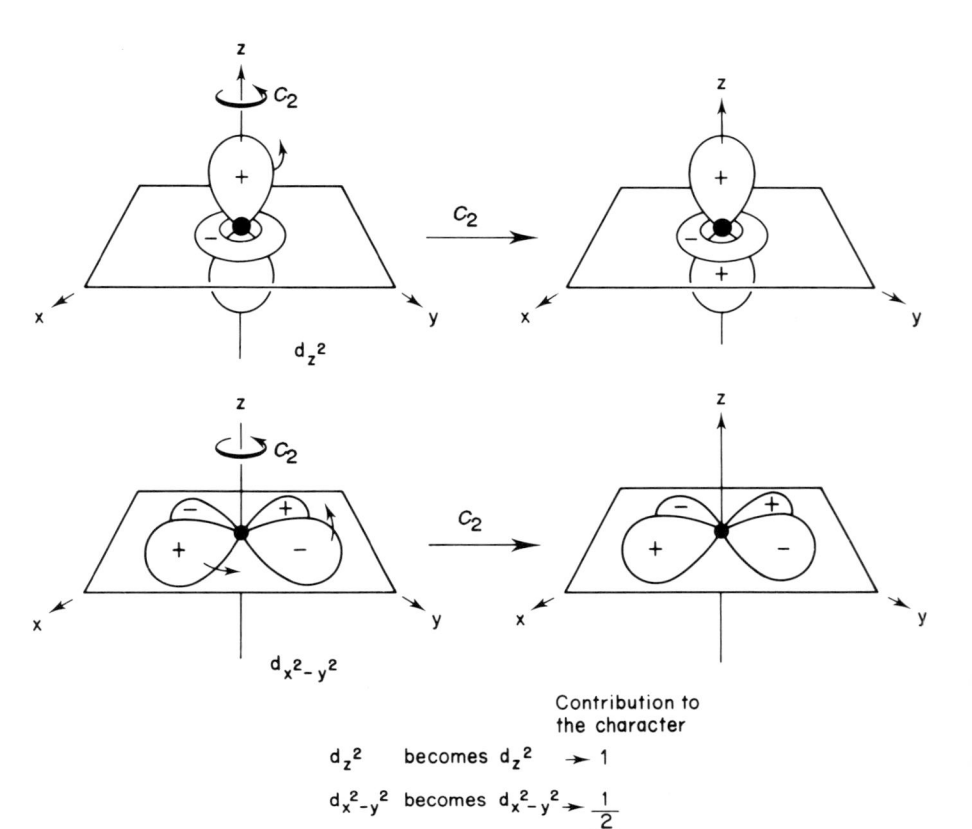

d_{z^2} becomes d_{z^2} → 1

$d_{x^2-y^2}$ becomes $d_{x^2-y^2}$ → $\dfrac{1}{2}$

Contribution to the character

Figure 7.24 Transformation of the E_g set of d orbitals of a central metal atom under a C_2 rotation operation of the octahedron.

by 120° to give, for instance, $d_{(y^2-z^2)}$ and $d_{(1/\sqrt{3})[(x^2-y^2)-(z^2-x^2)]}$ (i.e. $x \to y \to z \to x$) leads to functions which may be expressed in terms of the original. It is easy to show by expansion of the coefficients that, for instance,

$$d_{(y^2-z^2)} = -\frac{1}{2} d_{(x^2-y^2)} - \frac{\sqrt{3}}{2} d_{(1/\sqrt{3})[(z^2-x^2)-(y^2-z^2)]}$$

so that the coefficient with which $d_{x^2-y^2}$, the 'starting' orbital, appears in this expression $(-\frac{1}{2})$ is its contribution to the character under the C_3 rotation. The contribution of $d_{(1/\sqrt{3})(2z^2-x^2-y^2)}$ to $d_{(1/\sqrt{3})(2x^2-y^2-z^2)}$ is similarly shown to be $-\frac{1}{2}$ so that the aggregate character is -1.

Problem 7.24 Check the transformation of the $d_{x^2-y^2}$ and d_{z^2} orbitals given in Figures 7.23–7.26 and thus show that these orbitals transform as

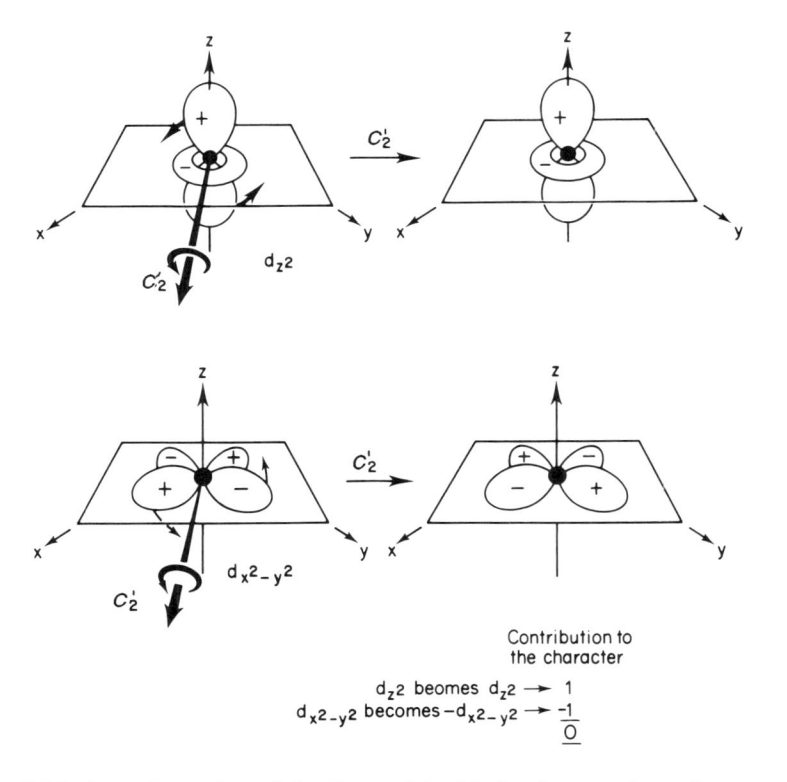

Contribution to
the character

d_{z^2} beomes $d_{z^2} \rightarrow 1$

$d_{x^2-y^2}$ becomes $-d_{x^2-y^2} \rightarrow \underline{\underline{\dfrac{-1}{0}}}$

Figure 7.25 Transformation of the E_g set of d orbitals of a central metal atom under a C_2' rotation operation of the octahedron.

E in O. Because they are centrosymmetric orbitals it follows that they transform as E_g in O_h.

Crystal field theory, being a purely electrostatic theory which does not admit of the existence of bonding and antibonding molecular orbitals, asserts that since the d electrons (like all other metal electrons) are non-bonding, they will occupy preferentially that arrangement in which electrostatic repulsion with the ligands (most simply represented as point negative charges) and with each other is a minimum. It is convenient to consider these two factors separately. Consider first the requirement of minimum electrostatic repulsion between the metal d electrons and the negatively charged ligands. Figure 7.27 shows a representative E_g orbital (the $d_{x^2-y^2}$) and a representative T_{2g} (the d_{xy}). Symmetry ensures that whatever we conclude about these also holds for the other member(s) of their respective sets. As Figure 7.27 suggests, it is in the E_g orbitals that an electron gets closest to the ligands and so experiences the greatest electrostatic repulsion. This conclusion, which is confirmed by detailed

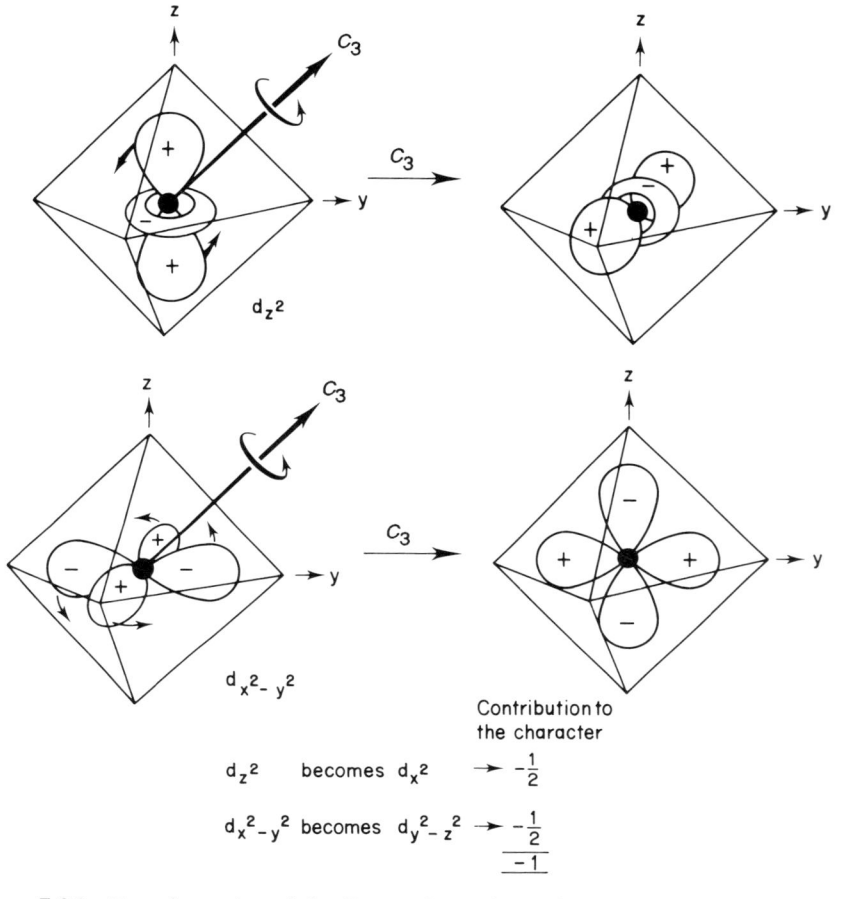

d_{z^2} becomes d_{x^2} $\rightarrow \; -\frac{1}{2}$

$d_{x^2-y^2}$ becomes $d_{y^2-z^2}$ $\rightarrow \; -\frac{1}{2}$

$\overline{\qquad\qquad\qquad\quad -1}$

Figure 7.26 Transformation of the E_g set of d orbitals of a central metal atom under a C_3 rotation operation of the octahedron. It will help to understand this diagram if it is recognized that the C_3 operation shown has the effect of converting $z \rightarrow x$, $x \rightarrow y$ and $y \rightarrow z$.

calculations, means that the T_{2g} set of d orbitals has a lower energy than the E_g set. The energy splitting between the two sets is usually denoted by either Δ or 10 Dq. If d electron–ligand repulsion were the only factor to be considered then the d electrons in octahedral transition metal complex ions would occupy the lower, T_{2g}, set of d orbitals until these were filled up. However, this preference is opposed by the effects of electron repulsion between the d electrons themselves. This electron repulsion is minimized if, as far as possible, the d electrons occupy different d orbitals with parallel spin. That is, occupation of the E_g orbitals will start as soon as the T_{2g} set is half full. We

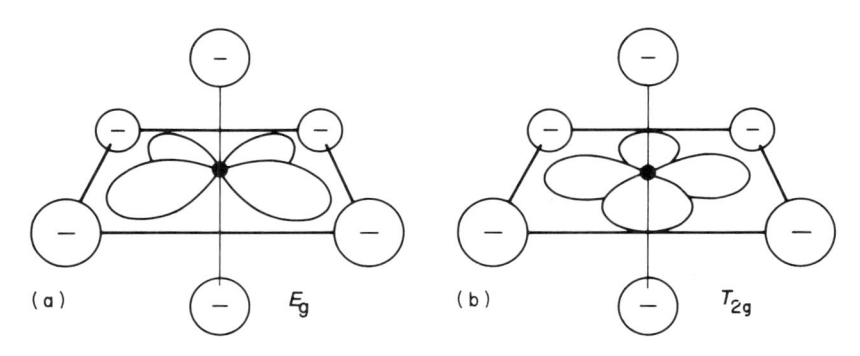

Figure 7.27 Representative (a) E_g and (b) T_{2g} orbitals of a central metal atom in an octahedral metal complex. In this figure ⊖ indicates negative electrical charge; the d orbitals are envisaged as also containing electron density so that electron–electron repulsion occurs.

have here a straight conflict between two opposing forces. When the d electron–ligand repulsion wins we have the so-called 'strong field' case' when the d–d electron repulsion dominates we have the so-called 'weak field' case. Alternative, but not quite equivalent (see below), names are to talk of 'low spin' and 'high spin' complexes, the names originating from the fact that some weak field complexes have more unpaired electrons—a higher resultant spin—than do the corresponding strong field complexes.

In summary, in crystal field theory, the relative magnitudes of Δ (10 Dq) and the d electron–repulsion energies—the so-called 'pairing energy'—determine the way that the set of d orbitals is occupied. This is illustrated in Figure 7.28, where the clear differences between high spin and low spin electron arrangements for ions with between four and seven d electrons is evident (the names 'high' and 'low' spin really only apply to these electron configurations). Small orbital occupation differences also exist for ions with two, three and eight d electrons but these differences are rather subtle and are not manifest in obvious orbital occupancies. Consequent upon these differences between high and low spin cases is a variety of associated spectral, magnetic, structural, kinetic and thermodynamic differences.

Inclusion of covalent bonding, along the lines discussed earlier in this chapter for sulfur hexafluoride, in the interaction between metal ion and ligands in a transition metal complex leads to *ligand field theory*. It differs from crystal field theory in that quantities which are well defined in crystal field theory become less well defined in ligand field theory (and are generally treated as parameters to be deduced from experiment). Qualitatively, Figure 7.28 remains appropriate except that, as will be explained, the E_g set of d orbitals is now identified as the antibonding counterpart of the E_g set involved in metal–ligand σ bonding.

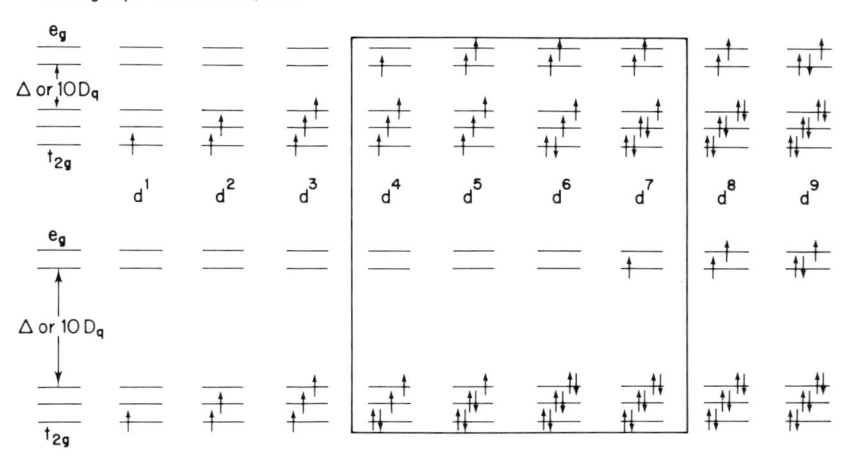

Figure 7.28 Differences in arrangement of electrons in the d orbitals of a metal atom in an octahedral complex occur for the d^4–d^7 configurations (those within the box).

In contrast to the discussion of SF_6 earlier in this chapter, the valence shell of the central atom in transition metal complexes consists of s, p and d atomic orbitals. This means that the nine available metal orbitals span the A_{1g}, T_{1u}, E_g and T_{2g} irreducible representations. It will be recalled that the σ orbitals of the surrounding six atoms—be they fluorine in SF_6 or ligands in a complex— span $A_{1g} + T_{1u} + E_g$. In a transition metal complex these ligand orbitals are full. This is evident if the ligand is a closed shell anion such as F^-, Cl^- etc. and is equally true if it is a molecule such as H_2O or NH_3, where the ligand σ orbital is identified as a lone pair of electrons on the electronegative atom. This means that the interaction with the metal orbitals can be regarded as stabilizing the ligand orbitals—lowering their energy. In this case the metal orbitals are to be regarded as being correspondingly destabilized by virtue of the same interactions and the e_g orbitals, which in crystal-field theory are pure d orbitals, to be regarded as antibonding combinations of ligand σ and metal d orbitals. Here the common practice of using lower case letters before the word 'orbital' is followed. To avoid possible disruption of the logistic sequence, this convention has not been applied earlier in this chapter but use of the labels e_g and t_{2g} is so common in transition metal chemistry that they have to be introduced at some point. Before concluding this section on transition metal ions it is of interest to note that in ligand field theory the d orbitals of T_{2g} symmetry may also interact with ligand orbitals. It will be recalled that the fluorine p_π orbitals in SF_6 transform as $T_{1u} + T_{1g} + T_{2u} + T_{2g}$. The p_π orbitals of any ligand, L, in an octahedral ML_6 complex will have the same symmetries.

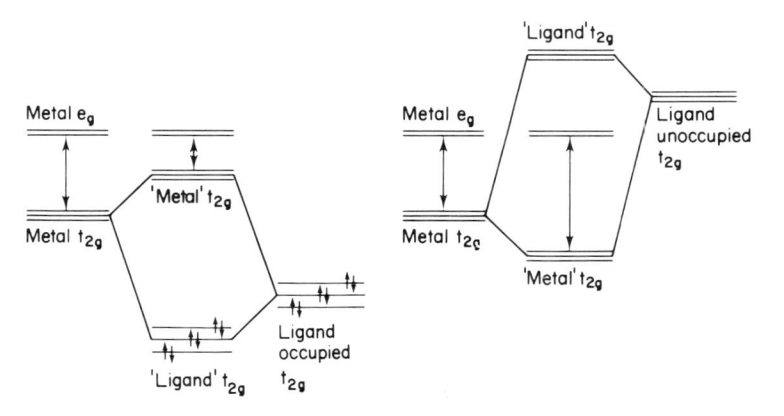

Figure 7.29   The effect of ligand π orbitals on the e_g-t_{2g} splitting (indicated by the double-headed arrows) depends on the relative energies of the metal and ligand t_{2g} orbitals.

Evidently, in transition metal complexes the metal d orbitals of T_{2g} symmetry may interact with the T_{2g} set of ligand π orbitals. If the relevant ligand π orbitals are empty—and therefore of high energy—then the effect of ligand metal T_{2g} interactions will be to depress (stabilize) the lower T_{2g} orbitals. These are those corresponding to the T_{2g} d orbitals shown in Figure 7.28—and to raise the energy of the (empty) T_{2g} ligand π orbitals. The effect on the molecular orbitals corresponding to the d orbitals of Figure 7.28 will be to increase the splitting Δ. If the ligand π orbitals are filled—and therefore of relatively low energy—the effect will be to decrease the e_g-t_{2g} splitting. These two cases are illustrated in Figure 7.29. The π bonding that has just been described seems to be of importance because it is found that it is ligands with available but empty π orbitals that give large values of Δ, and thus strong field complexes, while those with filled π orbitals give small values of Δ and so weak field complexes. Examples of the former are the CN^- and CO ligands (the empty π orbitals being C–N or C–O π antibonding) and of the latter, Br^- and Cl^-. For these halide anions the filled π orbitals are the atomic p_π orbitals.

7.4   THE BONDING IN TETRAHEDRAL MOLECULES

Early in this chapter it was mentioned that it is often possible to discuss octahedral and tetrahedral transition metal complexes together because their geometries are both derived from a cube. Tetrahedral complexes, of general formula ML_4, are of widespread occurrence but are not as common as octahedral. Together, species with geometries which approximate to either octahedral or tetrahedral account for at least 80 per cent of all coordination compounds. In organic chemistry, of course, it is the tetrahedral geometry which is important. Clearly, it is appropriate that at least an outline discussion of tetrahedral molecules should be included in this text.

The Cartesian coordinate axes which were used for an octahedron (Figure 7.3(b)) are those of the corresponding cube. Similarly, it is advantageous to use cube-derived axes for a tetrahedron, notwithstanding the fact that this choice is not in accord with taking the z axis as the axis of highest rotational symmetry (which would mean C_3 for the tetrahedron). Because a tetrahedron is derived from a cube it is not surprising that the symmetry operations which turn a tetrahedron into itself are also symmetry operations of the cube (but the converse is not true). The corresponding symmetry operations are:

Cube (and octahedron) $E, 8C_3, 6C_4, 3C_2, 6C_2', i, 8S_6, 6S_4, 3\sigma_h, 6\sigma_d$
Tetrahedron $E, 8C_3,   3C_2,   6S_4   6\sigma_d$

The group of operations of the tetrahedron is given the shorthand label T_d (pronounced 'tee-dee').

Problem 7.25 Draw diagrams to show all of the symmetry operations of a tetrahedron.
Hint: Figure 1.4 is helpful. The mid-point of each cube face in this figure corresponds to a corner of the octahedron shown in Figures 7.3 and 7.4.

It will be noticed that, although there exists a group of the pure rotations of the tetrahedron ($E, 8C_3, 3C_2$, a group called T), the group T_d is not a direct product group of T with any other group (if it were there would be three, not two, additional classes of T_d compared with T and they would have 1, 8 and 3 operations in them).

Problem 7.26 Detail the arguments behind the assertion just made.

The character table of the T_d group is given in Table 7.7.

Table 7.7

T_d	E	$8C_3$	$3C_2$	$6S_4$	$6\sigma_d$
A_1	1	1	1	1	1
A_2	1	1	1	-1	-1
E	2	-1	2	0	0
T_1	3	0	-1	1	-1
T_2	3	0	-1	-1	1

The bonding in tetrahedral molecules will not be discussed in detail but the essentials are given in Table 7.8, which summarizes the ways in which the various orbitals transform.

Table 7.8

		Symmetry
Orbitals of an atom at the centre of the tetrahedron	s	A_1
	(p_x, p_y, p_z)	T_2
	$(d_{x^2-y^2}, d_{(1/\sqrt{3})(2z^2-x^2-y^2)})$	E
	(d_{xy}, d_{yz}, d_{zx})	T_2
Orbitals of the four atoms at the apices of the tetrahedron	σ	$A_1 + T_2$
	π	$E + T_1 + T_2$

Problem 7.27 Check that the transformations given above are correct. The generation of the correct reducible representation for the transformation of the apical atom π orbitals is not a trivial task and the reader who gets the correct answer is to be congratulated. Success depends on choosing the orientation of the π orbitals mindful of the symmetry operations under which they are to transform (this problem is best tackled by the techniques described in Appendix 4).

Problem 7.28 Use the projection operator method to derive explicit forms for the σ orbitals of four atoms arranged at the apices of a tetrahedron.

Problem 7.29 Use the data in Table 7.8 to describe the bonding in methane, CH_4.

It will be noted that double and triple degeneracies exist in a tetrahedral environment and, rather important, that the p and three of the d orbitals of a central atom both transform as T_2. This means that the t_2 d orbitals in a tetrahedron will be mixed with some of p, and vice versa. In this lies, ultimately, the explanation of the fact that tetrahedral transition metal complexes tend to be more highly coloured than do octahedral. Because d and p orbitals mix, this mixing makes some electronic transitions more allowed in a tetrahedron than they are in an octahedron (pure d–d transitions are forbidden, but d–p are allowed).

Just as for an octahedron, in a tetrahedral environment the d orbitals of a transition metal split into two sets; $d_{x^2-y^2}$ and $d_{(1/\sqrt{3})(2z^2-x^2-y^2)}$ are of E symmetry and, as has been commented, d_{xy}, d_{yz}, and d_{zx} of T_2. If a diagram analogous to

Figure 7.2 is drawn for a tetrahedron then it is concluded that in this geometry splitting the T_2 set is of higher energy than the E—the inverse of the splitting found for an octahedron.

> **Problem 7.30**   Draw diagrams analogous to Figure 7.2 for a tetrahedral transition metal complex and thus show that it is reasonable that the d_{xy}, d_{yz} and d_{zx} orbitals are of higher energy than $d_{x^2-y^2}$ and $d_{(1/\sqrt{3})(2z^2-x^2-y)}$. *Hint*: The lobes of the $d_{x^2-y^2}$ orbital point towards the mid-points of the faces and of d_{xy} towards the mid-points of the edges of the cube corresponding to a tetrahedron.

The splitting of the d orbitals in a tetrahedron is only about one half of that for the corresponding ligands arranged octahedrally (more accurately, $\frac{4}{9}$). This reduction in separation means that strong field tetrahedral complexes are virtually unknown; almost all are weak field.

In both a tetrahedral and octahedral environment the d orbitals of a transition metal split into a set of two (E in T_d, E_g in O_h) and a set of three (T_2 in T_d, T_{2g} in O_h). It is the fact that the d orbitals of E symmetry in T_d are the same as those of E_g in O_h (and similarly for the T_2 and T_{2g} orbitals) which enables a common discussion of the two symmetries in specialized texts. In this common discussion the orbital sets are referred to as E and T_2 (one can think of the discussion of octahedral molecules taking place in the group O for there these are the correct symmetry labels). The two geometries are then distinguished by the fact that the E–T_2 splittings are of opposite signs. One warning, however; a warning signalled in the discussion above. Although a set of three p orbitals of a central atom have T_{1u} symmetry in O_h, they have T_2 symmetry, *not* T_1, in T_d. The moral is clear—be careful not to extrapolate without checking.

> **Problem 7.31**   Show that the p orbitals of an atom at the centre of a tetrahedron have T_2 symmetry.

7.5   THE DETERMINATION OF THE POINT GROUP OF A MOLECULE

The stage has now been reached at which all of the different types of point group operations have finally been met. It is therefore a convenient time at which to briefly review the chemically important point groups and the allocation of a molecule to the correct one. That is, we shall tackle the question 'How do I decide what the symmetry of a particular molecule is?' Unless a correct answer can be guaranteed, it would be only too easy to end up trying to use the incorrect character table.

When a molecule has a single rotational axis, C_n, this is usually quite evident (it seems to become easier as n increases). It may be that this C_n axis is the only symmetry element, in which case the point group is C_n. Frequently, however, there will be other symmetry elements. If the only ones are mirror planes which contain the C_n axis (there must be n such σ_v planes because if there was only a single one, the existence of the C_n axis would create the other $n-1$), then the point group is C_{nv}. Were there just a single mirror plane perpendicular to the C_n axis (a σ_h plane), it would be a C_{nh} point group. The simultaneous existence of n σ_v mirror planes and a σ_h plane means that there must be further symmetry elements, in particular, n C_2 axes perpendicular to the original C_n axis, equally spaced around it (and symmetrically related to the n σ_v mirror planes). Such pointgroups are D_{nh} (D for Dihedral). Can these additional n C_2 axes exist without the n σ_v and σ_h? The answer is that they can; the point group produced is D_n. What of the simultaneous existence of the C_n, the n σ_v's and the n C_2's? These combinations also exist and lead to the point groups D_{nd}. The n σ_v axes now symmetrically interleave the n C_2 axes and so are, more correctly, referred to as σ_d mirror planes. In similar fashion the 'σ_v' mirror planes in the D_{nh} point groups should be called σ_d. Unfortunately, most authors do not use this (correct) notation. In such groups with n even the vertical mirror plane reflection operations invariably fall into two classes. In most texts, one of these is usually—arbitrarily—denoted $(n/2)\sigma_v$ and the other $(n/2)\sigma_d$ (in Appendix 3 the notation $(n/2)\sigma_d$ and $(n/2)\sigma_d'$ has been used). The combination of just C_n, σ_h and n C_2 does not exist—the existence of these elements requires the co-existence of n σ_v and we are back to D_{nh}.

In the present chapter S_n operations were met for the first time. A set of such operations, together with the identity, can comprise a group, provided that n is even. Such groups are called S_n, although the case of $n=2$, S_2, is usually called C_i because the operation S_2 is precisely equivalent to inversion in a centre of symmetry (Figure 7.30).

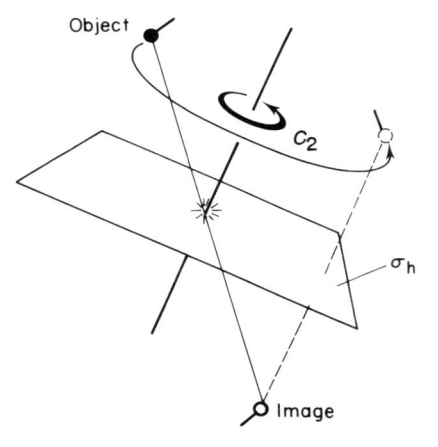

Figure 7.30 The operation i is equivalent to 'rotate about an arbitrary C_2 axis containing the i and reflect in a σ_h containing the i'. That is, $i \equiv S_2$.

If it is clear that a molecule has non-coincident C_n and $C_{n'}$ axes, where n and n' are both greater than two (n can be equal to n') then the point group of the molecule is one of those for which the x, y and z axes are interchanged by some of the operations of the group. The Cartesian axes then transform together as a triply degenerate irreducible representation. If the molecule contains a C_5 axis (it would actually have several) then the point group would be an icosahedral one—I, or I_h (pronounced 'eye aich'). An icosahedron—of I_h symmetry—is shown in Figure 7.31. If the molecule contains a C_4 axis (there would be three of them in all) then the point group would be cubic (or, equivalently, octahedral)—O or O_h. Finally, if its pure rotation axis of highest symmetry is a C_3 axis (there would be a total of four of these) then its symmetry would be that of a tetrahedral group—T, T_d or T_h. The distinction between each of the two icosahedral, the two octahedral or three tetrahedral groups is quite simple. Groups with no suffix are groups with only pure rotation operations—they have no mirror planes and no centre of symmetry, for example. They are rare. Groups with suffixes contain improper rotation operations. The distinction between T_d and T_h is that the latter contains a centre of symmetry, whereas the former does not. The octahedral and tetrahedral groups, together, are often referred to as 'cubic' groups.

Linear molecules all have an axis—the molecular axis—about which rotation by any angle, no matter how large or small, is a symmetry operation. The 'fundamental' rotation operation, from which all others may be built up, is therefore a rotation by an infinitesimally small angle. It takes an infinite number of these rotations to return the molecule to its original position (rather than an equivalent, rotated, one). This axis is therefore a C_∞ axis. If they have a centre of symmetry they are of the $D_{\infty h}$ point group; if they have not, they are $C_{\infty v}$. Because they each have an order of infinity, the reduction of reducible

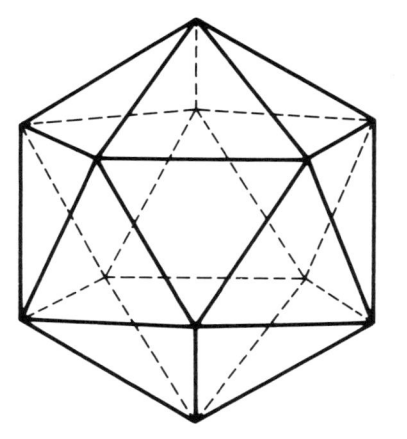

Figure 7.31 An icosahedron. A fivefold rotational axis passes through each pair of opposite apices and a threefold through the mid-points of each pair of opposite faces.

representations in these groups has to be handled differently to the method developed in this text. The problem is one that is seldom encountered and is discussed in Appendix 5.

One point group remains. It is that which, in addition to the identity element, contains only the operation of reflection in a single mirror plane. It is denoted C_s ('cee ess').

The way that most experienced workers identify the point group of a molecule is by a spontaneous knee-jerk type of reflex (most common) or to list as many symmetry operations as they can immediately see (much less common). Such a list is usually mental but the beginner may prefer to use pencil and paper. Even if incomplete, this list may at once identify the point group; if not, it will certainly reduce the number of possibilities to two or three. A glance at the list of operations across the head of the character tables of the possible groups (Appendix 3) will reveal the operations in which the possible groups differ. These operations (or, rather, the corresponding elements) are then explicitly looked for and thus the correct group selected. This procedure of scanning likely character tables is strongly recommended to the beginner as the best way forward; it requires intelligent comparisons to be made of different point groups and this can be a very enlightening process. An alternative, the one recommended in most texts, is to mount a more systematic search for symmetry elements. Several schemes for such a search exist (although experienced workers never use them!) and one is given in Figure 7.32. One starts at the top and traces a path by answering the questions on the way, which ends with the correct point group (provided that no mistakes have been made).

Problem 7.32   Determine the point groups of the following molecules. The answers are given on page 179.

C_2H_6 (eclipsed)
C_2H_6 (staggered)
C_6H_6
PF_5 (trigonal bipyramid)
Hg_2^{2+}
$[PtCl_4]^{2-}$ (a planar molecule)
$TeCl_4$ (two pairs of equivalent chlorines)
ClF_3 (two equivalent fluorines)
The environment of the Na^+ ion in crystalline NaCl
CO_2
CO
SO_4^{2-}
NO_3^- (planar)
NO_2^-
C_3H_6 (cyclopropane)

Fe(C$_5$H$_5$)$_2$ (ferrocene, eclipsed)
Fe(C$_5$H$_5$)$_2$ (ferrocene, staggered)

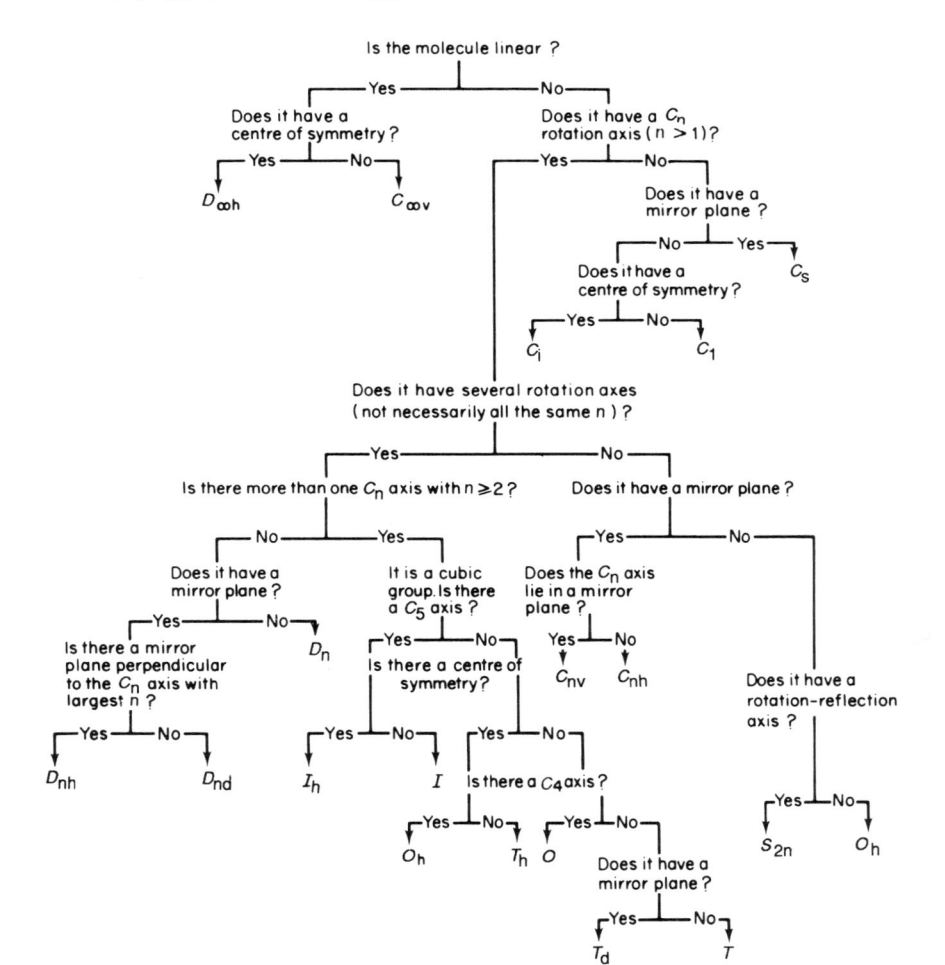

Figure 7.32 A 'yes'–'no' response table which is one of the many variants available which may be used to assign a molecule to the correct point group.

7.6 SUMMARY

In this chapter several cubic points have been met, along with the symmetry-enforced double and triple degeneracies associated with them (pp. 141, 172). This high level of symmetry can compensate for quite high molecular complexity and discussions of octahedral and tetrahedral transition metal complexes are now invariably symmetry-based, the d orbitals of the central metal atom being split into a degenerate pair (E) and a degenerate trio (T_2) (pp. 164, 173). Finally, the identification of the point of a molecule was discussed and a diagnostic scheme given in Figure 7.32 (p. 178).

Answers to Problem 7.32

D_{3h}, D_{3d}, D_{6h}, D_{3h}, $D_{\infty h}$, D_{4h}, C_{2v}, C_{2v}, O_h, $D_{\infty h}$, $C_{\infty v}$, T_d, D_{3h}, C_{2v}, D_{3h}, D_{5h}, D_{5d}

8
Groups and Subgroups

8.1   INVARIANT AND NON-INVARIANT SUBGROUPS

So far in this book we have been concerned with particular molecules and, more important, particular symmetries. It is the purpose of the present chapter to adopt a less selective viewpoint and to examine, first, the relationships which exist between similar symmetries and, second, some of the consequences of low symmetry.

A molecule in which a central atom M is surrounded by six identical atoms or groups L, ML_6, is usually of octahedral symmetry, O_h, as was seen in Chapter 7. Suppose one of the L is now replaced by a chemically similar, but different, atom or group X (for instance, both L and X could be halogen atoms). The molecule is then ML_5X and has, at most, C_{4v} symmetry (assuming that the change from L to X does not lead to a gross structural change in the molecule—no C_5 axis is introduced, for example). Strictly, then, a discussion of the ML_5X molecule should follow the general pattern developed in Chapter 5, where the electronic structure of BrF_5 was considered, due allowance being made for the presence of X. However, the difference between L and X could be negligible (if they are isotopic variants of the same element, for instance). In such a case the difference in conclusions between a discussion based on O_h symmetry and one based on C_{4v} should also be negligible. There must be a continuity between the C_{4v} and O_h cases because even if the difference between L and X is large, it could be broken down into a series of small, hypothetical, steps. Similar arguments will apply whenever there is a similar relationship between two groups.

Just what is this relationship between groups? In the above example, it is clear that some of the symmetry elements (and, therefore, operations) of the O_h point group are not present in C_{4v}. However, the existence of operations common to the two groups means that there will be some relationship between their group multiplication tables and, very important, between their character tables. The group which has the smaller number of operations is referred to as a *subgroup* of the other; there are many fascinating relationships which exist between a group and its subgroups, some of which will be met in this chapter.

It is evident from group multiplication tables that the symmetry operation of a point group are not, in general, independent of one another—if one symmetry operation is removed then usually as a consequence others will be removed also. So, for example, if in the case of the C_{2v} point group one mirror plane reflection operation is to be deleted then this can only be done if a second operation is also deleted. This is shown in Table 8.1 where it is seen that deletion of the σ_v column and row still leaves a σ_v entry (as a product of C_2 and σ_v'). We can only totally remove σ_v entries from the table if we remove either σ_v and σ_v' or σ_v and C_2. Deletion of the former pair leaves the point

Table 8.1

C_{2v}	E	C_2	$\cancel{\sigma_v}$	σ_v'
E	E	C_2	$\cancel{\sigma_v}$	σ_v'
C_2	C_2	E	$\cancel{\sigma_v'}$	σ_v
$\cancel{\sigma_v}$	$\cancel{\sigma_v}$	$\cancel{\sigma_v'}$	$\cancel{E}$	$\cancel{C_2}$
σ_v'	σ_v'	σ_v	$\cancel{C_2}$	E

group C_2 (operations E, C_2) as a subgroup of C_{2v} and deletion of the latter pair gives C_s (operations E, σ). Because there is only one possible way of producing the subgroup C_2 from C_{2v}, C_2 is said to be an *invariant subgroup* of C_{2v}. More rigorously, an invariant subgroup contains only complete classes of the parent group. It is thus understandable that the point group C_s is also an invariant subgroup of C_{2v}, despite the fact that it could be derived from either σ_v or σ_v'. Key is the fact that these two mirror plane reflection operations are not in the same class in C_{2v}. The existence of an invariant subgroup can be of key importance. In Chapter 13, for instance, the problem of the enormous size of space groups will be encountered. If the faces of a crystal are ignored then it is, effectively, infinite because there is an infinity of translation operations in the group. Fortunately, it will not prove necessary to work with a group of this size. This is because the (infinite) group of all translations is an invariant subgroup of the space group. This sounds formidable when first encountered, but all that it really means is that the group of all translations can only be obtained from the space group in one way. As will be seen in Chapter 13, it makes good physical sense to ignore the group of all translations and to consider only operations of the point group variety, those that have been the subject of this book so far. Instead of being an infinite problem, it is reduced to the sort that have been met many times. It may make good physical sense but it also has to make mathematical sense. It does, and the reason is that the group of all translations is an invariant subgroup of the space group. All this will be explained in detail in Chapter 13.

Not all subgroups are invariant. The C_{3v} point group, which was the subject of Chapter 6, provides an example. The multiplication table for this group is given in Table 8.2; the operations are those indicated in Figure 6.1. A clockwise

Table 8.2 The first operation is listed along the top and the second down the left-hand side

C_{3v}	E	C_3^+	C_3^-	$\sigma_v(1)$	$\sigma_v(2)$	$\sigma_v(3)$
E	E	C_3^+	C_3^-	$\sigma_v(1)$	$\sigma_v(2)$	$\sigma_v(3)$
C_3^+	C_3^+	C_3^-	E	$\sigma_v(2)$	$\sigma_v(3)$	$\sigma_v(1)$
C_3^-	C_3^-	E	C_3^+	$\sigma_v(3)$	$\sigma_v(1)$	$\sigma_v(2)$
$\sigma_v(1)$	$\sigma_v(1)$	$\sigma_v(3)$	$\sigma_v(2)$	E	C_3^+	C_3^-
$\sigma_v(2)$	$\sigma_v(2)$	$\sigma_v(1)$	$\sigma_v(3)$	C_3^+	E	C_3^-
$\sigma_v(3)$	$\sigma_v(3)$	$\sigma_v(2)$	$\sigma_v(1)$	C_3^-	C_3^+	E

rotation by 120° is denoted by C_3^+, an anticlockwise rotation by C_3^-; the mirror plane reflections are $\sigma_v(1)$, $\sigma_v(2)$ and $\sigma_v(3)$. The multiplication table in Table 8.2 differs from all the other multiplication tables that have been explicitly given in this book. It is not symmetric about the leading diagonal (top left to bottom right). Put another way, for some combinations of operations the result depends on the order in which the operations are applied. Thus,

$$C_3^+ \sigma_v(1) = \sigma_v(2)$$
but
$$\sigma_v(1)C_3^+ = \sigma_v(3)$$

Care therefore has to be taken to specify that the operations at the head of the columns in the multiplication table are on the *right* in expressions such as those above. Equivalently, they are the *first* operation. This may seem strange but if so, it is only because we are accustomed to reading from left to right so that in the first example above we *read* C_3^+ before $\sigma_v(1)$. However, if the operations operate on some function, ψ say, then we have

$$C_3^+ \sigma_v(1)\psi$$

and, clearly, here $\sigma_v(1)$ must operate *before* $C_3{}^+$.

Problem 8.1 Using Figure 6.1 check that the C_{3v} group multiplication table given in Table 8.2 is correct.

In the multiplication table, Table 8.2, the complete deletion of a single σ_v operation requires that the other two σ_v's are also deleted, to give the group C_3 as an invariant subgroup, a subgroup that can only be obtained in one way. However, deletion of the two C_3 operations causes the multiplication table to break up into three disconnected multiplication tables. This is because we can only remove the C_3^+ and C_3^- entries from Table 8.2 by deleting the C_3^+ and C_3^- columns and rows and then deleting one of the pairs [$\sigma_v(1)$ and $\sigma_v(2)$] or [$\sigma_v(2)$ and $\sigma_v(3)$] or [$\sigma_v(3)$ and $\sigma_v(1)$]. For each of these three choices we arrive at C_s as a subgroup. That is, there are three different but equivalent ways that C_s can be obtained as a subgroup, so it is *not* an invariant subgroup of C_{3v}—i.e. one that can be obtained in only one way—(although, as has been seen, it is an invariant subgroup of C_{2v}).

Problem 8.2 Check the above assertions by deleting from Table 8.2.
(a) all C_3 operations. Is it possible to just delete C_3^+ but leave C_3^-?
(b) one σ_v operation.
(c) one σ_v operation and the C_3 operations.

The distinction between invariant and non-invariant subgroups may seem rather academic. In fact, it has quite a variety of consequences, as two examples will show. The first example concerns molecular dynamics. Suppose that a molecule of C_{3v} symmetry is momentarily distorted—by a molecular vibration, for instance—to give a molecule of C_s symmetry. Thus, in the ammonia molecule, one N–H bond might be momentarily longer (or shorter) than the other two. Because the symmetry of the molecule has been reduced to that of a non-invariant subgroup there exists other different but equivalent distortions (in the case of our ammonia molecule there are two such equivalent distortions, corresponding to distortion of one of the two other N–H bonds to give a different but equivalent arrangement of C_s symmetry). That is, because there are three different C_s subgroups of C_{3v} there will be three equivalent distortions; the molecule would be of the same energy in each of the three equivalent configurations. In this situation, the distortion can 'rotate' from one bond to the next with no nett cost in energy. That is, the presence of non-invariant subgroups means that a molecule may indulge in some unexpected gymnastics. A full discussion of this is well beyond the scope of the present book, although there is more detail in Chapter 9. It is clear that special care has to be taken in a detailed analysis of the vibrational and rotational properties of molecules with symmetries which have non-invariant subgroups.

The second example is concerned with the character tables of invariant subgroups. When the operations of a point group can be written as a product of the operations of two of its invariant subgroups, then its character table can also be derived from those of these subgroups. Consider the C_{2v} point group. We have seen that it has two invariant subgroups, C_2 and C_s. It follows that all of the operations of C_{2v} can be derived from those of these two subgroups. Take each of the operations of one invariant subgroup and combine it, in turn, with all of the operations of the other invariant subgroup. Therefore, in our case, carry out the steps shown in Table 8.3.

Table 8.3 shows that the operations of C_{2v} are products of the operations of C_2 and C_s. Using the language of Section 4.3, the group C_{2v} is said to be the direct product of the groups C_2 and C_s, a relationship usually written as

$$C_{2v} = C_2 \times C_s$$

(more strictly, the symbol $\otimes$ should be used in place of the multiplication sign).

Table 8.3 The combination of operations
of the invariant subgroups of C_{2v}

C_2	combines with	C_s	to give	C_{2v}
E	combines with	E	to give	E
E	combines with	σ	to give	σ_v
C_2	combines with	E	to give	C_2
C_2	combines with	σ	to give	σ_v'

In Section 4.3 it was also seen that a similar property holds for the corresponding character tables. Thus, in the present case the character table for C_2 is taken and the whole of it multiplied by the characters of the C_s table, to give a table four times the size of that of C_2. This is shown in Table 8.4 where, for simplicity, the C_2 character table has been written out four times on the left. Each one is then multiplied by the corresponding C_s character to give the C_{2v} table.

The C_{2v} character table given in Table 8.4 is the same as that met in Chapter 2 (Table 2.4), with the A_2 and B_1 irreducible representations interchanged in position.

Problem 8.3   Check through the individual steps in Tables 8.3 and 8.4.

Examples of this relationship between character tables have already been met. In Chapter 4 the fact that D_2 and C_i are both invariant subgroups of D_{2h} was exploited (Tables 4.3 and 4.4). In Chapter 7 the fact that O_h has invariant subgroups O and C_i was used in Table 7.2 and the preceding discussion.

Problem 8.4   Show that the operations of the group C_{3v} are the product of operations of the groups C_3 (E, C_3^+, C_3^-) and C_s (E, σ). However, because C_s is not an invariant subgroup of C_{3v}, the character table of C_{3v} is *not* the direct product of the character tables of C_3 and C_s. This is immediately seen when the character tables of C_{3v} and C_3 are compared (Appendix 3).

Table 8.4

C_2	E	C_2	E	C_2
A	1	1	1	1
B	1	-1	1	-1
A	1	1	1	1
B	1	-1	1	-1

$\times$

C_s	E	σ
A'	1	1
A''	1	-1

$=$

C_{2v}	E	C_2	σ_v	σ_v'
A_1	1	1	1	1
B_1	1	-1	1	-1
A_2	1	1	-1	-1
B_2	1	-1	-1	1

At the beginning of this chapter it was recognized that C_{4v} is a subgroup of O_h. It is not an invariant subgroup because there are eight C_4 operations in O_h but only two in C_{4v}. It is also evident that the character table of O_h is not a direct product of that of C_{4v} with any other group because that of O_h contains triply degenerate irreducible representations whereas C_{4v} does not. This is another illustration of the rule that the character table of a group is never the direct product of the character table of a non-invariant subgroup with that of another group.

8.2 CORRELATION TABLES

Having discussed how the character table of a group may be related to that of its subgroups we now consider the opposite problem: how is the character table of a subgroup related to that of the parent group? Again, the general form of the relationship is best seen by considering an example. The example which we choose corresponds to the physical situation described earlier in this chapter, that in which a molecule of C_{3v} symmetry is distorted to give a structure with C_s symmetry (i.e. a distortion leading to the loss of the threefold axis). The character tables of the C_s and C_{3v} point groups are given in Table 8.5. Note that in the C_s character table a single prime as a superscript indicates something which is symmetric with respect to a mirror plane reflection and a double prime indicates antisymmetry. This use (and meaning) of primes reappears in other point groups—see Appendix 3. In the C_{3v} character table in Table 8.5 the loss of the C_3 axis has been indicated by deleting the column associated with the corresponding operations. Since loss of this axis also leads to the loss of two σ_v mirror planes (those generated by C_3 operations acting on the 'first' σ_v) the figure 3 has also been deleted from the $3\sigma_v$ entry. It is clear from Table 8.5 that the remaining characters of the A_1 irreducible representation of the C_{3v} point group are those of the A' irreducible representation of the C_s point group. One says that the 'A_1 irreducible representation of C_{3v} correlates with the A' irreducible representation of C_s'. This means that any function or object which transforms as A_1 in C_{3v} must transform as A' in C_s when the molecular symmetry changes. Similarly, Table 8.5 shows that the A_2 irreducible representation of C_{3v} correlates with A'' of C_s.

The E irreducible representation of C_{3v} is both interesting and important for it does not correlate uniquely with a single irreducible representation of C_s.

Table 8.5

C_{3v}	E	$2C_3$	$3\sigma_v$	C_s	E	σ
A_1	1	1	1	A'	1	1
A_2	1	1	-1	A''	1	-1
E	2	-1	1			

Rather, it gives rise to a reducible representation, one which is readily seen to have $A' + A''$ components. In summary, then, we have the correlations shown in Table 8.6.

Table 8.6

C_{3v}	C_s
$A_1 \rightarrow A'$	
$A_2 \rightarrow A''$	
$E \rightarrow A' + A''$	

This example illustrates the general theorem that each irreducible representation of a group gives rise to a representation, which may be either reducible or irreducible, of each of its subgroups. Tables showing these correlations—so-called *correlation tables* are available, but it is very easy to work them out using the example given above as a model. Working-out sometimes has an advantage over the use of tables. The D_{2h} group was described in Chapter 4 (its character table is given in Table 4.1). This group has C_{2v} as a subgroup and correlation of the irreducible representations of the two groups seems very easy.

Problem 8.5   Use either Tables 4.1 and 2.4 or Appendix 3 to correlate the irreducible representations of the D_{2h} and C_{2v} groups.
Hint: If you find this problem more difficult than expected, read the next part of this section.

As the reader may have discovered when tackling Problem 8.5, while the problem is not a difficult one, there is a catch in correlating from D_{2h} to C_{2v}. The D_{2h} group has three different C_2 axes. The precise correlation between the two groups depends on which of the three twofold axes is retained in going from D_{2h} to C_{2v}. This does not indicate any fundamental problem, rather that it may be necessary to relabel coordinate axes (and associated basis functions) in moving between the two groups. The twofold axis retained in C_{2v} may not be that labelled z in D_{2h}, although it would be called z in C_{2v}. In compilations of correlation tables it is usual to indicate all three possible $D_{2h} \rightarrow C_{2v}$ correlations but one still has to decide which correlation is appropriate before using the tables. In such cases even experienced workers may find that they are less likely to make a mistake by working out the correlation for themselves rather than by using the tables!

There is another way of showing correlations, and that is by use of a diagram. That for the $C_{3v} - C_s$ correlation is given in Figure 8.1.

Such diagrams emphasize another aspect of the consequences of a decrease in symmetry. Figure 8.1 shows, for example, that a function transforming as

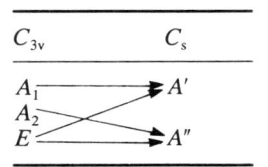

Figure 8.1 The correlation between the groups C_{3v} and C_s.

A_1 in C_{3v} and one of the two functions transforming as E have a common symmetry in C_s, that described by the A' irreducible representation. This means that in C_s symmetry these two functions can interact with each other, an interaction which is symmetry-forbidden in C_{3v} symmetry.

Another aspect of a reduction in symmetry, equally evident from either Table 8.6 or Figure 8.1, is that a decrease in symmetry may lead to a decrease in degeneracy. In the example above, the degeneracy of functions transforming as E in C_{3v} is lost in C_s. A particularly important case is that of octahedral transition metal coordination compounds, discussed in Chapter 7. Although much of the basic theory of such compounds is conveniently developed assuming full octahedral symmetry (O_h), real-life examples usually show some minor distortion. The most important cases are those in which such a distortion is either along a fourfold or a threefold axis (either distortion therefore destroying all other fourfold and threefold rotation axes), as shown in Figure 8.2. The appropriate correlation table is given in Table 8.7.

> **Problem 8.6** Use the character tables of the O_h, D_{4h} and D_{3d} point groups in Appendix 3 to check the correlations given in Table 8.7.

It is seen from Table 8.7 that in D_{4h} symmetry all degeneracies present in O_h symmetry are at least partially removed. One important consequence of this is

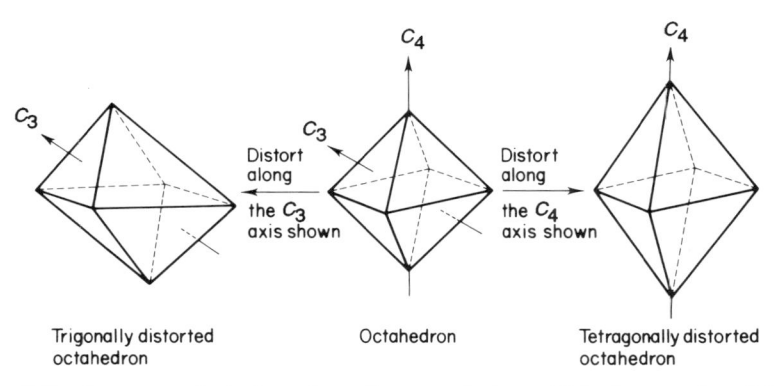

Figure 8.2 A symmetrical distortion of an octahedron (O_h) along a threefold axis gives a figure with D_{3d} symmetry while a symmetrical distortion along a fourfold axis gives a D_{4h} figure.

Table 8.7

	trigonal		tetragonal	
D_{3d}	$\xleftrightarrow{\text{distortion}}$ (along C_3)	O_h	$\xleftrightarrow{\text{distortion}}$ (along C_4)	D_{4h}
A_{1g}		A_{1g}		A_{1g}
A_{2g}		A_{2g}		B_{1g}
E_g		E_g		$A_{1g} + B_{1g}$
$A_{2g} + E_g$		T_{1g}		$A_{2g} + E_g$
$A_{1g} + E_g$		T_{2g}		$B_{2g} + E_g$
A_{1u}		A_{1u}		A_{1u}
A_{2u}		A_{2u}		B_{1u}
E_u		E_u		$A_{1u} + B_{1u}$
$A_{2u} + E_u$		T_{1u}		$A_{2u} + E_u$
$A_{1u} + E_u$		T_{2u}		$B_{2u} + E_u$

that when a single spectral band is predicted in the electronic absorption spectrum of an octahedral transition metal complex this band would be expected to show a splitting if the real symmetry is D_{4h} and a degeneracy is involved in the transition (for instance, the excited state might be triply degenerate). Such a splitting could take the form of the observation of a separate peak, a shoulder or an asymmetry on the band. The D_{3d} case shows, however, that it is not always true that a reduction in symmetry causes all degeneracies to be relieved to (i.e. a splitting to occur); thus the E_g and E_u irreducible representations of O_h persist in D_{3d}. There is, however, a trap for the unwary. In the point group O_h convention was followed in choosing a C_4 axis as the z axis; one would do the same in D_{4h}. In D_{3d}, the axis of highest symmetry is a C_3 axis and *this* is the z axis. It follows that, although E_g of O_h becomes E_g of D_{3d} it is NOT true that the basis functions for E_g in O_h, $x^2 - y^2$ and $(1/\sqrt{3})(2z^2 - x^2 - y^2)$, are basis functions for E_g in D_{3d}. A detailed analysis, using the methodology of Appendix 2, is needed to describe the correlations between basis functions in O_h and D_{3d}.

In practice, the correlations which exist between groups are quite important for two reasons. First, as indicated above, they enable the properties of low symmetry molecules to be related to those of high symmetry species. Another aspect of this occurs when a molecule is high symmetry but is trapped in a low symmetry environment—an octahedral molecule on a low symmetry lattice site in a crystal, for example. Second, some of the problems of degenerate representations—and some were met in the last chapter—can often be neatly side-stepped by pretending that a molecule has a lower symmetry than is in fact the case—so that the degeneracy is split (or 'relieved')—and, after working in the low symmetry group, using a correlation relationship to apply the result to the high symmetry case.†

Another interesting aspect of the relationship of a group to its subgroups is that the number of symmetry operations in a group (the order of the group) is a simple multiple of the number of symmetry operations of any of its subgroups. The multiplication factor—which is always an integer—is called the *index* of the subgroup (relative to the particular parent group). Thus, the C_3 group (of order 3) is subgroup of index 2 of the point-group C_{3v} (of order 6). However, the same C_3 group (of order 3) is a subgroup of index 40 of the point group I_h (of order 120).

Problem 8.7 Use Appendix 3 to determine the index of each of the following subgroups of O_h.

$$D_{4h}, C_{4v}, D_{3d}, C_{3v}, D_{2h}, C_{2v}$$

An important application of the concept of index concerns rotational subgroups. A point group may only contain operations which are proper rotation operations (such as C_2, C_3 and so on) or it may contain some operations which are pure rotations and others which are improper rotations (such as σ_v, i, S_4). By deleting all of the improper rotations one can always obtain a subgroup of a group which itself contains both proper and improper rotations. What remains is the *pure rotational subgroup* of the parent group. This subgroup is always of index 2. The importance of rotational subgroups is their relationship to the (infinite) group consisting of all the pure rotation operations associated with a sphere. A rotation of any angle about any radial axis is a symmetry operation of a sphere. In particular, infinitesimally small rotations are symmetry operations. This property is closely associated with the importance of angular momentum in the theory of atomic structure. When there is an atom at the centre of mass of a molecule then all the pure rotational symmetry elements associated with the molecular point group pass through it. The corresponding rotation operations are all that remain of the infinity of rotation operations which would have turned this atom into itself if the rest of the molecule were not present. What remains of the consequences of the angular momentum in the free atom are therefore manifest in the molecule in its pure rotational sub-group. In particular, this group provides a method of determining how the degeneracies which may be associated with the free atom (and these degeneracies may be quite large) are split up when the atom is placed in the molecule. For the metal atom at the centre of a transition metal complex, in particular, this is quite invaluable information. As was mentioned in Chapter 7, these compounds often contain unpaired electrons. Besides

---

† The particular attraction of a reduction in symmetry is that interactions can be caused to become 1:1—ambiguities about just which functions interact no longer exist.

behaving like tiny bar magnets themselves (this is the physical meaning of their spin), the orbital motion of these (negatively charged) electrons can have a magnetic effect, just like an electrical current in a solenoid. This additional, orbital, magnetism is closely connected with any angular momentum possessed by the unpaired electrons (the angular momentum of the electrons is the electron density in circulation and so is akin to the current flowing round a solenoid). To understand this orbital magnetism it is necessary to know something about the residual angular momentum and this information is contained within the rotational subgroup.

> **Problem 8.8** Use the symmetry operations listed at the top of the character tables in Appendix 3 to show that deletion of improper rotation operations in the following point groups in each case leads to a pure rotational subgroup of index 2.
>
> $$I_h, T_h, D_{5h}, C_{2h}, D_{3d}$$
>
> *Note:* In several of these examples it is possible to obtain subgroups by deletion of all improper and some proper rotations. Such subgroups are not of index 2. The statements made in the text refer to the largest pure rotational subgroup of a given group.

8.3 SUMMARY

There are relationships between a group and its subgroups. The operations of a group can immediately be obtained from the operations of its subgroups (p. 184) as can its character table (p. 183) provided that the subgroups are invariant (p. 181). Correlations exist between the irreducible representations of a group with its subgroups and are useful in discussions associated with molecules which approximate to high symmetry species (p. 185). Groups containing improper rotation operations always have a pure rotational subgroup of index 2 (p. 189).

9
Molecular Vibrations

So far in this book we have been largely concerned with the techniques of group theory and their application to the problem of the electronic structure of molecules. There are a few additional techniques which we have yet to meet but we are already in a position to discuss one important application of group theory to quite a different area, the analysis of molecular vibrations. Although a chemist may be interested in vibrational spectra in a qualitative way—some molecular fragments reveal their presence by characteristic 'fingerprint' peak patterns—the concern of the present chapter is with a more detailed analysis of the relationship between spectra and structure. Vibrational spectra provide a way of determining the geometrical arrangement of groups attached to an atom both quickly and with reasonable accuracy. Because such geometrical arrangements have often been a particular problem in inorganic chemistry, it is in inorganic chemistry that the methods of this chapter find the greatest application. This chapter, then, is concerned with the prediction of the number of infrared and Raman peaks (i.e. spectroscopic features resulting from the excitation of molecular vibrations) that is expected for a particular molecular geometry. The predictions generally vary with geometry and so provide a method of distinguishing between alternatives. However, the spectral activity of vibrations is a subject which must be deferred until Chapter 10 because this topic is one that requires a further development of basic ideas. The present chapter will be concerned solely with determining the symmetries of the normal modes of vibration of a molecule. Having determined these symmetries, it will be found in Chapter 10 that the selection rules follow immediately.

9.1 NORMAL MODES

A normal mode is a 'natural' vibration of a molecule. Just as a tuning fork has a 'natural' frequency and motion (or mode), so too has a molecule, a difference being that all molecules (except diatomic molecules) have more than one natural frequency. More precisely, a normal mode is one which has the property that if each atom in a molecule is displaced from its equilibrium position by a displacement which corresponds to its maximum amplitude in the normal mode, then, when the atoms are simultaneously 'let go' the atoms will all undergo a motion at the same frequency. Further, once having been 'let go' they will all *simultaneously* pass through the equilibrium configuration and, later, simultaneously again reach their positions of maximum amplitude.† It follows that the motions of symmetry-related atoms in the molecule will be simply related to each other, so that it is possible to place a symmetry label on each normal mode. Our concern, then, is with the determination of these symmetry labels.

The description just given of the vibrations of molecules differs from that which seems intuitively more realistic. It seems reasonable to expect the motion of individual atoms in a molecule to be much more complicated than this. A motion in which atoms move in apparently random directions, amplitudes and phases seems more likely. However, provided the amplitudes are not too great, such a complicated motion can be regarded as a sum of normal motions occurring with different phases, just as the sound from a musical instrument can be regarded as a similar sum of different harmonic frequencies. Clearly, normal modes of vibration are of key importance in the description of molecular vibrations. These normal modes are quantized—just like the harmonics of a vibrating stretched string—and it is possible to add a further quantum of vibrational energy to any mode. For some of the modes this quantum can be added by infrared radiation and for some it can be added by a Raman mechanism. These excitation processes are the basis of infrared and Raman spectroscopies. In adding an additional quantum it is a good approximation to ignore the other vibrations already occurring within a molecule. That is, we can pretend that the molecule is not vibrating at all! There is more than one justification for this step. One particular justification for this pretence is seldom met but it is not very difficult and—because it is both good fun and rather fundamental—it is included as an optional section at the end of the present chapter.

There are two distinct methods which may be adopted in tackling the problem of determining the symmetries of the normal modes of vibration of a molecule. First, the molecule may be regarded as made up of convenient fragments and the vibrations of each fragment considered in turn. The normal vibrations cannot usually be obtained in this way because these vibrations will probably involve the movement of atoms outside the fragment under consideration. An additional step is needed to obtain the normal vibrations. Consider an example. For chloromethane, CH_3Cl, one would consider separately the C–H bond stretches, the C–Cl bond stretch, the H–C–H angle change and the H–C–Cl angle change vibrations. Such a breakdown is often a useful starting point for the study of molecular vibrations because each motion is often associated with a characteristic spectral region (which, correctly, suggests that, despite what was said above, the vibrations of fragments are often not too different from normal vibrations). This approach to molecular

---

† This statement needs modification when the vibrational motion concerned has the symmetry species of an irreducible representation which has complex characters. The motion is then that of a travelling wave. Such irreducible representations will be dealt with in Chapter 11.

vibrations, which regards them as a superposition of the vibrations of fragments, requires care. The method depends on an ability to spot all of the different vibrators (or 'internal coordinates' as they are usually called), and it is all too easy to miss some. The twists and turns of one part of a complicated molecule against another or the puckering motion of a ring are all too easy to miss, for example. It is here that the second method is useful since it provides a check on the first. In it one determines all of the symmetries of the normal modes together. It has no place for chemical experience or intuition about the motion of the atoms involved in each normal vibration. But, it has an advantage—nothing gets missed out! In this chapter both methods will be considered. That which divides the vibration of the molecule into several smaller problems, that of the vibration of fragments, is the simpler and will be considered first.

9.2   SYMMETRY COORDINATES

It is easiest to work with a particular example and so we shall consider the C–H stretching vibrations of chloromethane, CH_3Cl, using the C_{3v} character table given in Table 9.1. Although the frequencies of the A–B stretching vibrations in an AB_3 unit will be dependent on the masses of the atoms involved, the symmetries of the vibrations will not. Our discussion is therefore equally appropriate to other molecules of C_{3v} symmetry, such as ammonia. The problem is easier to visualize if this generality is exploited. This is readily done by thinking about displacements (e.g. a bond stretch or a bond angle change) rather than the motion of a particular set of atoms. Displacements may be symbolically represented by arrows as shown in Figure 9.1. In chloromethane there are three C–H bond stretches to consider and, so, there are three arrows in Figure 9.1.

It is easy to show that the three arrows of Figure 9.1 form the basis for a reducible representation, the components of which are the $A_1 + E$ irreducible representations of the C_{3v} point group. What does this mean? As has been mentioned, normal vibrations may be labelled by that irreducible representation of the point group which describes the phase relationship between the motions of symmetry-related internal coordinates (bond stretches in the present

Table 9.1   The character table of the C_{3v} point group

C_{3v}	E	$2C_3$	$3\sigma_v$	
A_1	1	1	1	z
A_2	1	1	-1	R_z
E	2	-1	0	(x, y), (R_y, R_x)

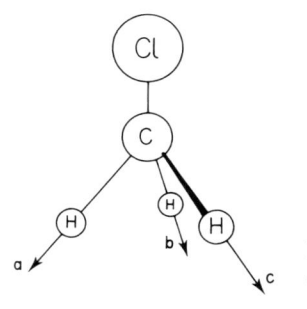

Figure 9.1 Three independent C–H stretch internal coordinates in CH_3Cl.

context). A diagram such as Figure 9.1 merely indicates the existence of some independent vibrators (of which there are three in the present case). The diagram could be redrawn with the direction of some of the arrows reversed if we chose and this would not change the fundamental problem. The projection operator technique automatically corrects for arbitrary choices of initial phase. Once it has been used to generate a symmetry-adapted combination of internal coordinates (a so-called 'symmetry coordinate') the relative phases are all fixed. There is nothing new in this—analogous situations have been met in previous chapters. However, a particular confusion arises here because, as will be seen in Figure 9.2(a), a diagram such as Figure 9.1 may be indistinguishable from one used to depict a symmetry coordinate. The moral is—read text and captions carefully! The application of the projection operator technique to the present problem will not be detailed because this application is indistinguishable from that described in Chapter 6. In Figure 6.5 the 1s orbitals of the hydrogen atoms in ammonia were labelled a, b, and c, and the A_1 and E combinations were derived. They were found to be:

$$\psi(A_1) = \frac{1}{\sqrt{3}} (a + b + c)$$

$$\psi_1(E) = \frac{1}{\sqrt{6}} (2a - b - c)$$

$$\psi_2(E) = \frac{1}{\sqrt{2}} (b - c)$$

where it has been assumed that a, b and c all have the same phase. The present, vibrational, problem is mathematically just the same provided that a, b and c are taken to represent the bond extensions shown in Figure 9.1.

Problem 9.1 Repeat the derivation described in Chapter 6 but without choosing the same phase for a, b and c. Show that, although the mathematical form of the functions generated differ from those given above, the actual motions deduced are the same.

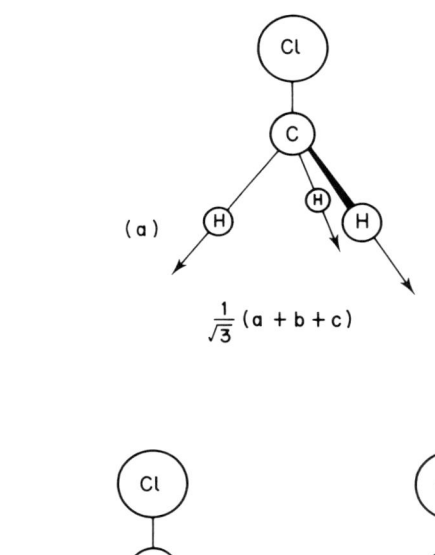

$$\frac{1}{\sqrt{3}}\,(a+b+c)$$

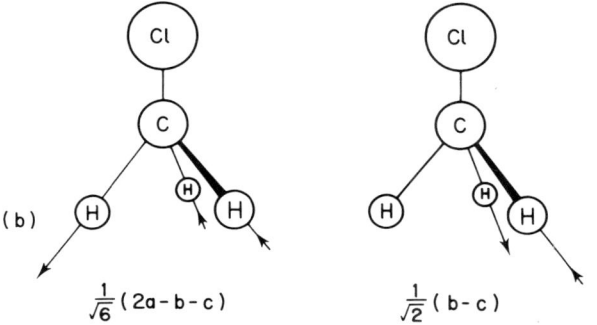

$$\frac{1}{\sqrt{6}}\,(2a-b-c)\qquad\qquad\frac{1}{\sqrt{2}}\,(b-c)$$

Figure 9.2 (a) The A_1 combination of the C–H stretching internal coordinates in CH_3Cl.
(b) The E combinations.

The analogy just drawn between the group theoretical analysis of molecular bonding and molecular vibrations is quite general. When the members of two different basis sets—for instance, the atomic orbitals in a bonding problem and the internal coordinates in a vibrational problem—transform isomorphously (that is, there is a 1:1 correspondence between members of the two sets) then their transformations lead to symmetry-adapted combinations of identical symmetries and mathematical form. Corresponding to the symmetry-adapted group orbitals of bonding theory are the symmetry coordinates of vibrational theory. Similarly, corresponding to the molecular orbitals of bonding theory (which are combinations of symmetry-adapted group orbitals of the same symmetry species) are the normal coordinates in vibrational theory. These normal coordinates are linear combinations of symmetry coordinates of the same symmetry species. However, because neither is symmetry-determined, the coefficients with which symmetry coordinates appear in a normal coordinate have no relationship to the coefficients with which symmetry-adapted orbitals appear in molecular orbitals.

Problem 9.2   Use the results of the sections indicated to write down the symmetry species and symmetry coordinates of:
(a) The C–H stretching vibrations of ethene (Section 4.5).
(b) The Br–F (equatorial) stretching vibrations of BrF_5 (Section 5.5).
(c) The S–F stretching vibrations of SF_6 (Section 7.2).
Note: Angle change vibrations can be handled similarly. However, there is a hidden problem associated with them which is discussed towards the end of the present section.

In Chapter 6, the molecular orbitals of the ammonia molecule were expressed as combinations of orbitals of the same symmetry species. Thus, the A_1 N–H bonding molecular orbital of ammonia was taken as an in-phase sum of the A_1 symmetry-adapted combination of hydrogen 1s orbitals with a nitrogen 2s–2p combination of A_1 symmetry. As just mentioned, in just the same way a normal coordinate (of a particular symmetry species) is taken as a sum of contributions from symmetry coordinates of this symmetry species. Thus, if there is only one symmetry coordinate of a particular symmetry type then this symmetry coordinate *is* the normal coordinate.

The A_1 and E symmetry coordinates derived from the C–H bond stretching coordinates shown in Figure 9.1 are drawn in Figure 9.2. Evidently, a crucial difference between A_1 and E symmetry coordinates is that in the latter a C–H bond may contract as its neighbour stretches, whereas for the former all of the C–H vibrators stretch and contract together. Provided, then, that a C–H bond is sensitive to whether its neighbour is contracting or stretching the spectral bands associated with the A_1 and E modes will have different energies and will therefore appear at different frequencies. The greater the sensitivity (or, as it is usually put, the greater the coupling between) the C–H vibrators the greater the separation between the spectral peaks associated with the different vibrations. With strong coupling, two peaks would be expected in the C–H stretching region of the spectrum subject to two provisos. These are that the normal modes are not too different from symmetry coordinates (otherwise, conclusions may not be transferable) and that both modes are spectrally active. In Chapter 10 it will be shown that the A_1 and E modes are both infrared and Raman active. Further, it turns out that the normal modes are closely approximated by the symmetry coordinates above. It is the direct connection between the analysis and prediction for a particular spectral region which makes the molecular fragment approach so useful. However, it is not without its problems. Suppose we are considering the H–C–H bond angle change vibrations of chloromethane. This basis set is pictured in Figure 9.3, where each bond angle change is represented by a double-headed arrow. It is simple to show that this basis gives rise to a representation with $A_1 + E$ components. What is the form of the A_1 vibration? In this vibration, all H–C–H bond angles must increase and decrease in phase. It is easy to see that the only way that they can do this is as shown in Figure 9.4. But, as this figure shows, this vibration seems to be more evidently associated with changes in the H–C–Cl bond

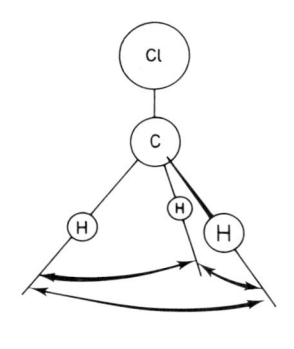

Figure 9.3 Three independent HCH bond angle change internal coordinates in CH₃Cl.

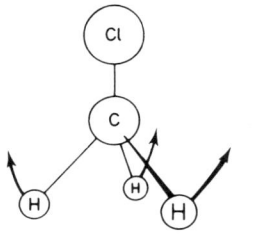

Figure 9.4 The A_1 combination of HCH (or HCCl) bond angle change internal coordinates in CH₃Cl.

angles! Not surprisingly, the H–C–Cl bond-angle change set of internal coordinates also gives rise to an A_1 vibration, which is also that shown in Figure 9.4! This figure makes clear that when we fragment a molecule we may find ourselves inadvertently duplicating vibrations. This problem is not confined to angle change vibrations. In a cyclic system, for instance, vibrations in which some bond lengths increase while others decrease usually have a simultaneous change in at least one bond angle. In this situation there is not a 'right' and a 'wrong' vibration—a vibration has been duplicated and such duplication is inseparable from the method.

The above discussion of the vibrations of CH₃Cl is summarized and completed in Table 9.2. In this table convention has been followed by denoting bond-stretching internal coordinates by the symbol ν and bond angle change (deformation) coordinates by the symbol δ. The total number of symmetry coordinates given in this table is ten (the E's are doubly degenerate) and this is to be compared to the number of normal vibrations predicted by the $3N - 6$ rule. In CH₃Cl, N, the number of atoms, is five and so the $3N - 6$ rule requires

Table 9.2

Internal coordinate	Symmetry coordinate species
ν(C–H)	$A_1 + E$
ν(C–Cl)	A_1
δ(H–C–H)	$A_1 + E$
δ(H–C–Cl)	$A_1 + E$

that this molecule has just nine normal vibrations. As has been seen, the disparity arises from the fact that an A_1 symmetry coordinate has been duplicated in the two deformation sets. We conclude that the normal vibrations of CH_3Cl are of $3A_1 + 3E$ symmetries.

Although this conclusion is correct, it is not entirely justified. That this is so would become evident if we were to consider a large and complicated molecule. In such cases it is easy to overlook some internal coordinates or to be uncertain whether or not they have already been included—as when one part of a molecule rocks or twists relative to another part, for instance. Equally, duplication of symmetry species such as that which occurred in CH_3Cl could easily go undetected. Clearly, a more systematic and reliable method is needed and this is given in the next section.

From what has just been said, the value of the fragment model might appear dubious. This is not so; in practice, the model is not used to predict the symmetries of the normal modes of a molecule. Rather, it is used to predict the number of bands expected in the spectral regions associated with individual internal coordinates such as those listed on the left hand side of Table 9.2. Even so, care has to be taken to spot duplication and here qualitative diagrams of symmetry coordinates such as those of Figures 9.2–9.4 are of great help. The generation of such diagrams is usually not difficult, the projection operator method being used.

Problem 9.3 Associated with the equilateral triangular arrangement of three carbon atoms in cyclopropane, C_3H_6 are three $\nu(C–C)$ stretching vibrations and three $\delta(C–C–C)$ angle change vibrations. Derive the symmetry coordinates associated with these internal coordinates and suggest how they may be brought into conformity with the requirements of the $3N – 6$ rule.
Note: Although the symmetry of cyclopropane is D_{3h} it is simplest work in the point group C_{3v}—the results are equivalent.

Problem 9.4 Draw diagrams akin to those of Figures 9.2–9.4 for the three $\nu(C–C)$ stretching vibrations of cyclopropane.

9.3 THE WHOLE-MOLECULE METHOD

The alternative technique, that of considering the entire molecule and generating all of the vibrations, is particularly useful as a check on the results obtained from a fragment analysis such as that described above. Not only does it correctly give the symmetry species of the molecular vibrations but vibrations duplicated (or others inadvertently omitted!) in a fragment analysis can usually be detected. In the entire-molecule method one starts by consider- ing the total motional freedom within the molecule. That is, each atom is allowed to move in any direction. Of course, with this amount of freedom we are, implicitly, allowing the molecule to translate and rotate as well as vibrate but it is a simple matter to select the vibrations from the totality of allowed molecular motions.

The method is illustrated for the chloromethane case in Figure 9.5. Each atom is allowed to move in any direction by having as a basis set a set of translations. Each atom is allowed to move independently in three mutually perpendicular directions and therefore contributes three translational displacements to the basis set. The most obvious—and sometimes used—arrangement of displacement vectors (arrows) is that shown in Figure 9.5(a). In it, corresponding displacements of all atoms are parallel. However, the three displacements on each atom can be in any direction in space without in any way changing the final answer. A considerable simplification results if the arrows are chosen to point in symmetry directions (along rotational axes or in, or perpendicular to, mirror planes). The directions of the arrows shown in Figure 9.5(b) have been chosen with this in mind. Remembering that only arrows on atoms which are unshifted by a symmetry operation can contribute to the character, and bearing in mind the discussion of x and y axes in Section 6.1 (a discussion which applies equally to displacements along x and y) it is a simple matter to show that this set of arrows gives rise to the reducible representation

$$
\begin{array}{ccc}
E & 2C_3 & 3\sigma_v \\
15 & 0 & 3
\end{array}
$$

and that this representation has $4A_1 + A_2 + 5E$ components.

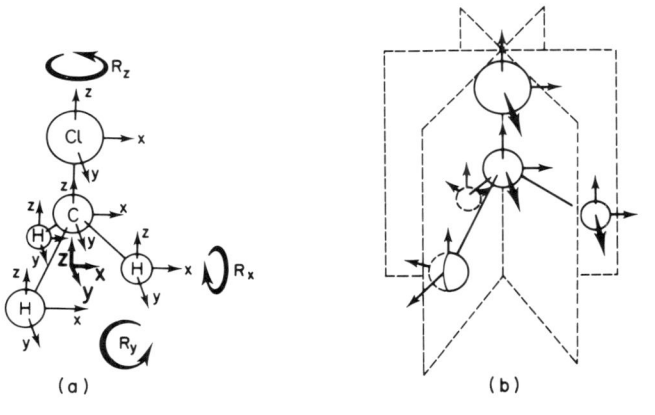

(a)   (b)

Figure 9.5   (a) The three independent translational displacements of each atom in CH_3Cl. In this diagram the three rotational motions of the entire molecule are also indicated (these rotations must be some combination of the atomic translational displacements).
(b) A more educated choice of atomic displacements than that shown in (a). It proves best to place the three independent translations (arrows) of each atom as far as possible along rotational axes or in, or perpendicular to, mirror planes.

Problem 9.5 Check the reducible representation given above using the choice of displacement vectors shown in Figure 9.5(b). The advantage of using this set of vectors over a set such as that shown in Figure 9.5(a) can be seen if an attempt is made to generate the reducible representation using this latter set of vectors. Note that if the vectors of Figure 9.5(a) are modified by rotating the x and y displacements on each hydrogen atom about the local z axis so that the local σ_v mirror plane bisects the angle between local x and y displacements then the reducible representation may again be readily generated.

The irreducible representations spanned by bodily translations and rotations of a molecule are usually to be found in the relevant character table, the symbols T_x, T_y and T_z (or just x, y, z) indicating the transformation of the translations and R_x, R_y and R_z the corresponding rotations. This has been done in Table 9.1 (the character table of the C_{3v} point group); in the character tables in Appendix 3 both x, y, z and T_x, T_y, T_z are indicated.

If one wishes to check on the symmetry species spanned by T_z, T_x and T_y this can be done by the usual method of investigating their transformational properties; this is conveniently done by representing them by the solid arrows drawn at the centre gravity of the three hydrogens of Figure 9.5(a). Rotations may be similarly treated, representing them by the curved arrows in Figure 9.5(a). Usually this is found to be more difficult but, fortunately, there is an alternative—and simpler—way. There is a general rule that

---

A rotation about an axis R_α ($\alpha = x$, y or z) has the same character as the corresponding translation T_α for all proper rotation operations. To obtain the character for R_α under an improper rotation operation, however, one simply has to change the sign of the character of T_α—that is, multiply it by -1.

Thus, it is easy to see from Figure 9.5(a) that T_z has A_1 symmetry in C_{3v}. That is, it has the characters

	E	$2C_3$	$3\sigma_v$
T_z	1	1	1

Application of the above rule shows that R_z has the characters

	E	$2C_3$	$3\sigma_v$
R_z	1	1	-1

That is, it is of A_2 symmetry. For the C_{3v} group we thus find (Table 9.1) that the three translations transform as $A_1 + E$ and the three rotations as $A_2 + E$.

Problem 9.6   Check that the three rotations R_x, R_y and R_z transform as $A_2 + E$ in C_{3v} by considering the behaviour of the three curved arrows in Figure 9.5(a) under the group operations.

Problem 9.7   By inspecting the character tables in Appendix 3 check that the above rule is invariably true. Having done this, return to Figure 9.5(a) and attempt to understand, pictorially, the origin of the rule.

As is well known, and as has already been used, the number of normal vibrations of a non-linear molecule is $3N - 6$, where N is the number of atoms in the molecule. The $3N$ is the total motional freedom of the atoms in the molecule and the 6 arises because the translations and rotations of the molecule as a rigid body are included in the $3N$ degrees of freedom. It has just been seen that these translations and rotations transform as $A_1 + A_2 + 2E$ for CH_3Cl. The symmetries of the normal vibrations can be obtained by subtracting these from those generated by the $3N$ degrees of freedom, $4A_1 + A_2 + 5E$. It follows that the vibrations transform as $3A_1 + 3E$, a result which agrees with the conclusions reached by the fragment analysis of the previous section.

9.4   THE VIBRATIONS OF ALREADY-VIBRATING MOLECULES (this section may be omitted at a first reading)

There is one final question to which we address ourselves. It arises when it is recognized, as at the beginning of this chapter, that at room temperature any molecule with low frequency vibrations is likely to have some of these vibrations thermally excited. If at least one such vibration is not totally symmetric then it must reduce the molecular symmetry (only totally symmetric vibrations maintain—or, in some situations, increase, molecular symmetry).

Problem 9.8   For each of the molecules of Problem 9.2 a totally symmetric vibration will have been obtained as part of the answer to that problem. Use the symmetry coordinates obtained to sketch out the form of the vibration and thus show that the vibration does not lead to a change in molecular symmetry. Similarly show that all other vibrations obtained in answer to Problem 9.2 lead to a reduction in molecular symmetry.
Hint: Exaggerate the vibrational distortion and determine the symmetry of the distorted molecule, using Figure 7.32 if necessary.

If there is a high probability that a molecule has a symmetry which is lower than that which has been assumed, how valid is an analysis based solely on the high symmetry case? The answer is a rather unexpected one. The symmetry group which we have been using is not the one that we think that we have been using! Consider the case of the chloromethane molecule but now subject the molecule to an arbitrary distortion such as that shown in Figure 9.6. To this

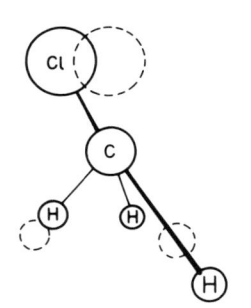

Figure 9.6 The distorted CH_3Cl molecule to be used in Figure 9.7. Dotted circles show the original 'atomic' positions. It is assumed that the distortion gives the molecule neither linear nor angular momentum.

distorted molecule we now apply the operations of the C_{3v} point group. The result of this is shown in the top half of Figure 9.7. Finally, apply a permutation operation to the labels on the hydrogen atoms, the permutation chosen being that which brings the hydrogens back to their original positions. Such a permutation is described by a symbol such as (123), a symbol which means 'replace the label 1 by label 2; replace the label 2 by the label 3 and replace the label 3 by the label 1'. Application of these permutation operations leads to the bottom half of Figure 9.7. The combined effect of point group and permutation operations has been that of giving us back the original molecule, but with the distortion differently related to the labels 1, 2 and 3. If the hydrogen atoms 1, 2 and 3 had been quite different atoms (e.g. if 1 were H, 2 were F and 3 were Br) then the six distorted molecules in the lower group of six would have been quite different. The fact that all three *are* hydrogen atoms, however, means that all these six distorted molecules have exactly the same energy. That is, they tell us something about the potential energy surfaces of the CH_3Cl molecule. For any one distorted arrangement there are always five others with precisely the same energy. But molecular vibrations explore potential energy surfaces, so this sixfold repetition is information which is clearly relevant to a vibrational analysis. It is a straightforward, if somewhat tedious, task to show that the six operations

$$E(1)(2)(3)\dagger$$
$$C_3(132)$$
$$C_3(123)$$
$$\sigma_v(1)(23)$$
$$\sigma_v(2)(31)$$
$$\sigma_v(3)(12)$$

---

† A symbol such as (1) means 'leave 1 alone'.

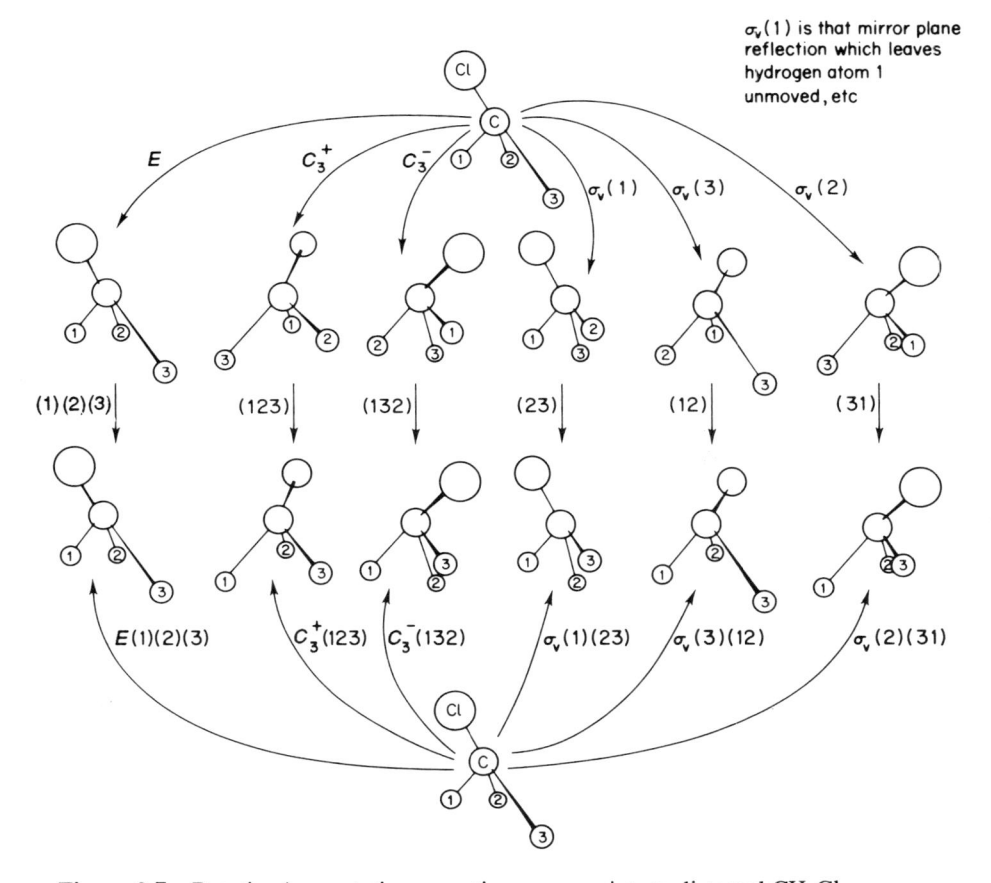

$\sigma_v(1)$ is that mirror plane reflection which leaves hydrogen atom 1 unmoved, etc

Figure 9.7 Rotation/permutation operations appropriate to distorted CH_3Cl.

multiply to form a group; further, these new operations are in a 1:1 correspondence with the operations of the C_{3v} point group and also multiply in the same way as their C_{3v} counterparts; the two groups are isomorphous.

Problem 9.9 Show by deriving the group multiplication table that the set of operations given above form a group and that this table is isomorphic to the C_{3v} multiplication table (Table 8.2).

Hint: Although diagrams such as Figure 9.7 or a model may be of help in tackling this problem it is perhaps easiest to treat it as algebraic, using Table 8.2 for the operations and treating the permutations separately.

Note: As indicated in the footnote, the permutation (1)(23) means '2 and 3 interchange while 1 remains unchanged'. It follows that, for example (1)(23) followed by (123) is equal to (3)(12), whereas (123) followed by

(1)(23) gives (2)(31). Such permutations are most easily worked out as follows, using (123) followed by (1)(23) as an example.

Start (Identity)	1	2	3
Application of (123) gives	2	3	1
Followed by (1)(23) gives	3	2	1

which, on comparison with the identity, is seen to be (2)(31).

It cannot be too strongly stressed that the two groups that we have been discussing are isomorphous and give rise to the same character table (or, more strictly, they have isomorphic character tables). It is this close connection between the 'correct' group—that containing combined point group and permutation operations— and the more immediately accessible C_{3v} point group which enabled us to use the latter in our discussion of the vibrations of the chloromethane molecule.

9.5   SUMMARY

In this chapter the problem of the determination of the symmetry species of the normal modes of vibration of a molecule has been studied. Such a classification is an essential prelude molecular to the prediction of vibrational spectra. A fragment analysis (p. 192) is of particular utility to the chemist but a complete treatment (p. 198) is useful as a check on errors that may have been introduced in a fragment analysis. For this, knowledge of the transformational properties of the bulk translations and rotations of the molecule is essential. Although point groups are invariably used in vibrational analyses, a more detailed study shows that they are actually used because they are isomorphic to the correct vibrational symmetry group (p. 201).

10

Direct Products

10.1 THE SYMMETRY OF PRODUCT FUNCTIONS

All of the previous chapters have been concerned with a discussion of the symmetries of individual objects, such as orbitals. Very commonly, however, the chemist is interested in products of such quantities. So, in a many-electron atom or molecule the electronic wavefunctions will be a product wavefunction which, at the simplest level, takes the form

$$\psi = \phi_1 \phi_2 \phi_3, \dots, \phi_n$$

where $\phi_1, \dots, \phi_n$ are individual one-electron orbitals and ψ is the single wavefunction which describes all n electrons.

Even in a one-electron problems products of one-electron wavefunctions —products of orbitals—have to be discussed as soon as we become interested in overlap between orbitals. The overlap integral between two orbitals ϕ_1 and ϕ_2 is given by

$$S_{12} = \int \phi_1 \phi_2 \, \delta v$$

where the integral is over all space. We know how to place symmetry labels on $\phi_1, \phi_2, \dots$, etc., but how do we place a symmetry label on a product function such as $\phi_1 \phi_2$? The present chapter is concerned with the answer to this question and with some important consequences which stem from it.

It is easiest to progress by considering a specific example and we will take one from the C_{2v} point group, the character table which is given in Table 10.1.

Consider a product function $\phi_1 \phi_2$ and suppose that ϕ_1 has symmetry A_2; we write this $\phi_1(A_2)$. Similarly, take ϕ_2 to be of B_2 symmetry and write it $\phi_2(B_2)$.

Table 10.1

C_{2v}	E	C_2	σ_v	σ_v'
A_1	1	1	1	1
A_2	1	1	-1	-1
B_1	1	-1	1	-1
B_2	1	-1	-1	1

We have to determine the symmetry of their product ψ; that is, we have to fill in the empty bracket in:

$$\psi(\ ) = \phi_1(A_2) . \phi_2(B_2)$$

where a dot has been placed between the two functions on the right to facilitate separate consideration of these two functions. In principle, the method is simple—it is the one used many times before in this book. We subject ψ to all of the operations of the group and obtain a set of characters by relating the transformed function to the original. The application of the C_2 operation, for example to ψ means that we are really applying it to $\phi_1(A_2)$ and to $\phi_2(B_2)$, simultaneously. Now, from Table 10.1, under this operation $\phi_1(A_2) \rightarrow \phi_1 A(_2)$ because the A_2 irreducible representation has a character of 1 for this operation. Similarly, $\phi_2(B_2) \rightarrow -\phi_2(B_2)$ and a character of -1. Putting these two results together, we have that under the C_2 rotation operation

$$\phi_1(A_2) . \phi_2(B_2) \rightarrow -\phi_1(A_2) . \phi_2(B_2)$$

That is, $\psi(\ ) \rightarrow -\psi(\ )$, so the character generated by the transformation of $\psi(\ )$ under this operation is -1. It is clear that this -1 really occurs because the products of the characters of the A_2 and B_2 irreducible representations under the C_2 operation is -1. Similarly, because the A_2 and B_2 characters under the σ_v operation are -1 and -1, respectively, their product, 1, is the character of $\psi(\ )$ under this operation. Summarizing this, and extending it to include the other operations, we have:

	E	C_2	σ_v	σ_v'
Characters generated by the transformation of $\phi_1(A_2)$, i.e. the A_2 irreducible representation				
	1	1	-1	-1
Characters generated by the transformation of $\phi_2(B_2)$, i.e. the B_2 irreducible representation				
	1	-1	-1	1
Characters of the transformation of $\psi(\ )$, i.e. the products of the two rows of characters above				
	1	-1	1	-1

The representation generated is the B_1 irreducible representation. That is, $\psi(\ )$ can now be identified as $\psi(B_1)$.

Problem 10.1 Using the procedure described above, fill in the empty brackets in the product functions:

$$\psi(\quad) = \phi_1(B_1) \cdot \phi_2(B_2)$$
$$\psi(\quad) = \phi_1(B_1) \cdot \phi_2(A_2)$$
$$\psi(\quad) = \phi_1(A_2) \cdot \phi_2(A_1)$$

Your answers can be checked by reference to Table 10.2.

In this example the general method of determining the symmetries of product functions was that used; multiply together the characters of the irreducible representations which describe the transformation of the individual functions. The act of multiplying two irreducible representations in this way is said to give rise to the *direct product* of the two individual representations; if we multiply three irreducible representations we form a triple direct product, and so on. The name 'direct product' is not new—it was first met in Section 4.3 where the operations of the group D_{2h} were described as the direct product of the operation of the groups D_2 and C_i. These two usages of 'direct product' are related; the connection may be seen in the discussion of Section 2.3 where the close relationship between the way that operations of a group multiply and the way that the corresponding irreducible representations multiply became evident. It is, then, not surprising that the same name, direct product, should be applicable to each type of multiplication. The connection between the two multiplications is described more fully in Appendix 2.

As will be seen in the remainder of this chapter, direct products are very important in the application of symmetry to chemistry. For these applications, all that is needed is a list—a table—of two-function direct products. Triple and higher direct products can readily be deduced from such a table. The (two-function) direct product table for the C_{2v} point group is given in Table 10.2. In this table an obvious, and conventional, symbolism has been used. The entry at a particular point in the table is the symmetry of the direct product of the species that label the column and row in which the entry falls.

Table 10.2 Direct products of the irreducible representations of the C_{2v} group

C_{2v}	A_1	A_2	B_1	B_2
A_1	A_1	A_2	B_1	B_2
A_2	A_2	A_1	B_2	B_1
B_1	B_1	B_2	A_1	A_2
B_2	B_2	B_1	A_2	A_1

Problem 10.2 Check that Table 10.2 is correct—this will provide useful additional practice in the formation of direct products.

Problem 10.3 Use Table 10.2 to obtain symmetry labels for the following product functions

$$\psi(\quad) = \phi_1(A_1)\phi_2(B_1)\phi_3(B_2)$$
$$\psi(\quad) = \phi_1(A_2)\phi_2(B_1)\phi(B_2)$$
$$\psi(\quad) = \phi_1(A_2)\phi_2(B_2)\phi(B_1)$$
$$\psi(\quad) = \phi_1(A_2)\phi_2(B_1)\phi(A_1)\phi(B_1)$$
$$\psi(\quad) = \phi_1(A_2)\phi_2(B_1)\phi(A_2)\phi(B_1)$$
$$\psi(\quad) = \phi_1(A_2)\phi_2(A_2)\phi(B_1)\phi(B_1)$$

What do your results indicate about the importance of the order in which functions are listed on the right-hand side of these expressions?

Note that Table 10.2 is symmetric about the leading diagonal (top left to bottom right). Thus, the result obtained for the example considered earlier in this chapter is

$$A_2 \otimes B_2 = B_1$$

where the symbol $\otimes$, which is that conventionally used to denote the direct product between two irreducible representations, has been used in preference to the $\times$ which might have been expected.

It is equally true that:

$$B_2 \otimes A_2 = B_1$$

This equivalence follows because sets of numbers are being multiplied together and the result obtained is independent of the order in which they are multiplied—the origin of the diagonal symmetry of Table 10.2 is at once evident.

This method of obtaining direct products is entirely general. However, another result in Table 10.2—that the product of two irreducible representations is always another irreducible representation—is not general. Direct products involving two or more degenerate irreducible representations invariably give rise to a reducible representation as a product. Consider the C_{4v} point group. The C_{4v} character table is given in Table 10.3 (it was first met as Table 5.6).

Table 10.3

C_{4v}	E	$2C_4$	C_2	$2\sigma_v$	$2\sigma_v'$
A_1	1	1	1	1	1
A_2	1	1	1	-1	-1
B_1	1	-1	1	1	-1
B_2	1	-1	1	-1	1
E	2	0	-2	0	0

It is evident that the direct product $E \otimes E$ must be reducible because the number that will appear in the identity column (4) is larger than any character in the table. The direct product $E \otimes E$ is:

	E	$2C_4$	C_2	$2\sigma_v$	$2\sigma_v'$
$E \otimes E$	4	0	4	0	0

which is readily seen to be a representation with components

$$A_1 + A_2 + B_1 + B_2$$

Problem 10.4   Show that the direct product table for the C_{4v} group is:

C_{4v}	A_1	A_2	B_1	B_2	E
A_1	A_1	A_2	B_1	B_2	E
A_2	A_2	A_1	B_2	B_1	E
B_1	B_1	B_2	A_1	A_2	E
B_2	B_2	B_1	A_2	A_1	E
E	E	E	E	E	$(A_1 + A_2 + B_1 + B_2)$

10.2   CONFIGURATIONS AND TERMS

It is instructive to consider the meaning of the $E \otimes E$ direct product in the C_{4v} point group in more detail. Suppose that there are two electrons, one of which is to be placed in the degenerate pair of orbitals of E symmetry denoted individually e_1 and e_2. The second electron is to be placed in a different degenerate pair of orbitals of E symmetry which we individually denote by E_1 and E_2. The possible two-electron functions are

$$e_1 E_1$$
$$e_1 E_2$$
$$e_2 E_1$$
$$e_2 E_2$$

That is, they are four in number (in agreement with the number 4 which appears in the identity column when the $E \otimes E$ direct product is formed). Group theory tells us that it is possible to take linear combinations of these four functions such that one combination has A_1 symmetry, one has A_2 symmetry, one has B_1 symmetry and one has B_2 symmetry. These symmetry-adapted functions may be obtained by the projection operator method described in Chapter 4 and, more particularly—because it deals with a non-Abelian group—Chapter 5. We first simply choose one function—$e_1 E_1$ for instance—and work out how it transforms under the operations of the group. For this, we need to know how the individual functions e_1 and E_1 transform. This information is detailed in Table 5.7 (where it is necessary to replace p_x by e_1 or E_1 and p_y by e_2 or E_2). In this way Table 10.4 is obtained.

Table 10.4

				Operation			
E	C_4^+	C_4^-	C_2	$\sigma_v(1)$	$\sigma_v(2)$	$\sigma_v'(1)$	$\sigma_v'(2)$
e_1	$-e_2$	e_2	$-e_1$	$-e_1$	e_1	e_2	$-e_2$
E_1	$-E_2$	E_2	$-E_1$	$-E_1$	E_1	E_2	$-E_2$
$e_1 E_1$	$e_2 E_2$	$e_2 E_2$	$e_1 E_1$	$e_1 E_1$	$e_1 E_1$	$e_2 E_2$	$e_2 E_2$

Multiplication by the A_1 characters and adding, in the usual projection operator method, leads to the conclusion that

$$\psi(A_1) = \frac{1}{\sqrt{2}} (e_1 E_1 + e_2 E_2)$$

Problem 10.5 Use Table 10.4 to show that

$$\psi(B_1) = \frac{1}{\sqrt{2}} (e_1 E_1 - e_2 E_2)$$

Problem 10.6 Derive a table similar to Table 10.4 but appropriate to the function $e_1 E_2$. Use it to show that

$$\psi(A_2) = \frac{1}{\sqrt{2}} (e_1 E_2 - e_2 E_1)$$

and

$$\psi(B_2) = \frac{1}{\sqrt{2}} (e_1 E_2 + e_2 E_1)$$

In the belief that a specific example would help the reader, the above discussion was concerned with electronic wavefunctions. The method, however, is not limited to such wavefunctions. Thus, the pairs (e_1, e_2) and (E_1, E_2) could equally have been vibrational wavefunctions, in which case the product wavefunctions would have been the ones relevant to a discussion of combination bands in a vibrational spectrum (vibrational excitations in which two different vibrations are excited by a single quantum of energy). Indeed, the discussion is of general applicability, with one exception. Note that the members of the two pairs of E functions are different—(e_1, e_2) and (E_1, E_2), not (e_1, e_2) and (e_1, e_2). This is not to say that the case in which the members of the pairs are identical is not important. In vibrational spectroscopy, for instance, overtone bands arise from double excitations—where (e_1, e_2) is combined with (e_1, e_2). Such cases are not immediately covered by the discussion because when products are formed within the members of a doubly degenerate set only *three* product functions can be distinguished. Two quanta of vibrational energy can be excited in e_1 or e_2 or one quantum can be excited in each so that the distinguishable excited states are of the form:

$$e_1 e_1$$
$$e_2 e_2$$
$$e_1 e_2$$

The concept of a direct product can be developed further to deal with this problem but this development is not particularly simple. The interested reader will find the electronic case developed in Ballhausen[1] and the vibrational in Wilson, Decius and Cross.[2] Not surprisingly, the two derivations are closely related.

There is another simple application of direct products that it is important to consider. As has been seen, if, in C_{2v} symmetry, we have a molecule with the electron configuration $a_2^1 b_2^1$ [or, as it was previously expressed, the product wavefunction is $\phi_1(A_2)\phi_2(B_2)$], then it is said that this configuration gives rise to a *term* of B_1 symmetry [or, in the form used earlier, there is a product wavefunction $\psi(B_1)$], the direct product $A_2 \otimes B_2$ being B_1. Note that throughout the present discussion electron spin will be ignored, although the electron would normally be indicated by a pre-superscript; thus a triplet spin term would be $^3B_1$, a singlet $^1B_1$, and so on. It is possible to extend group theoretical concepts to include electron spin but this is an advanced topic. This neglect of spin is formally expressed by saying that we are only concerned with orbital terms. Thus, in C_{2v} the orbital configuration $a_2^1 b_2^1$ gives rise to the orbital term B_1. Similarly, for the C_{4v} example considered above, the electron configuration $e^1 E^1$ gives rise to the terms A_1, A_2, B_1 and B_2. More correctly, and following the notation used in earlier diagrams in this book, one says that the electron configuration $1e^1 2e^1$ gives rise to the terms A_1, A_2, B_1 and B_2.

When a singly degenerate orbital is occupied by two electrons the product wavefunction describing this situation is totally symmetric—because the direct product is totally symmetric (see, for example, Table 10.2 and the table given in Problem 10.4). It is not so easy to see that the same result follows when a set of degenerate orbitals is completely occupied by electrons because simply forming direct products leads, apparently, to a large number of terms. However, the Pauli exclusion principle eliminates all but one of these terms. There is only one way of filling all orbitals of a degenerate set and that is by putting two electrons into each orbital. There is, then, only a single wavefunction and so there must be a singly degenerate term. This simple argument does not tell us whether or not this term is totally symmetric. It seems intuitively likely that it will be, and this is confirmed by following the transformations of the product wavefunction under the operations of the point group in the way that was done for e_1E_1 above. This leads to a general and very valuable conclusion:

Closed shells of electrons are invariably totally symmetric.

Here, by 'closed shell' is meant configurations like a_1^2, b_{1u}^2, e^4, t_{2g}^6 and so on. This conclusion means that for a many-electron molecule the possible terms arising from a configuration can be obtained simply by considering those orbitals which are partially filled. Those that are totally filled are ignored— unless these are the only ones present, in which case the orbital term is totally symmetric.

Problem 10.7 Show that in the C_{4v} group the electron configuration e^4 gives rise to a term of A_1 symmetry.
Hint: It may be helpful to write this configuration, using the notation adopted earlier in this chapter, as $e_1^1 e_1^1 e_2^1 e_2^1$. The table constructed as part of Problem 10.6 can then be modified to be used in the present problem.

It might be thought that, having determined that a totally symmetric term results from a closed shell—that is, that the many-electron wavefunction is totally symmetric—that this would be the end of the matter. This is not the case. Consider the situation shown in Figure 10.1, in which for a C_{2v} molecule in addition to a filled A_2 orbital there is an empty B_2 orbital at higher energy. Both the configurations a_2^2 and the configuration in which both electrons are promoted into the b_2 orbital b_2^2 have orbital symmetry A_1. In general, it is found by detailed calculation that although the ground term wavefunction is well represented as one derived solely from a configuration such as a_2^2, this wavefunction is improved if there is mixed in with it a contribution from the excited term configuration b_2^2, which also gives rise to a term of orbital symmetry A_1. Such *configuration interaction* is an important step in most detailed calculations of molecular properties, although more than one excited

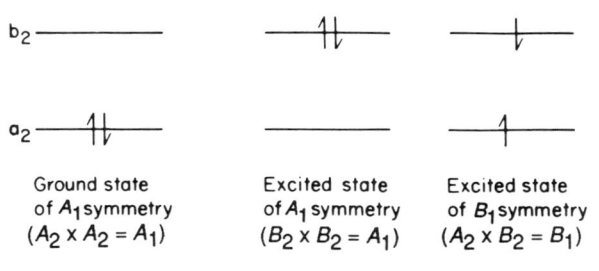

Figure 10.1 Ground and two excited configurations for a C_{2v} molecule.

term is usually involved in mixing with the term arising from the ground term configuration. So, if, in the present example, the ground term configuration were one in which a doubly occupied orbital of A_1 symmetry had above it a doubly occupied orbital of A_2 symmetry followed by empty orbitals of B_2 and B_1 symmetries (Figure 10.2), then configuration interaction would be expected between the A_1 terms $a_1^2 a_2^2$, $a_1^2 b_2^2$, $a_1^2 b_1^2$, $a_2^2 b_2^2$, $a_2^2 b_1^2$, $b_2^2 b_1^2$ and $a_1^1 a_2^1 b_2^1 b_1^1$ and also configuration interaction between the excited B_1 terms arising from the configurations $a_1^2 a_2^1 b_2^1$, $a_1^1 a_2^2 b_1^1$, $a_1^1 b_2^2 b_1^1$ and $a_2^1 b_2^1 b_1^2$.

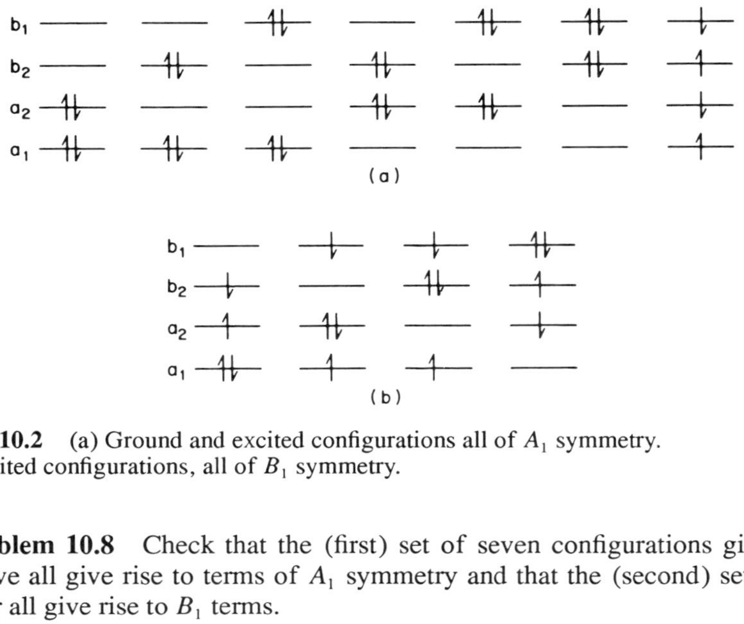

Figure 10.2 (a) Ground and excited configurations all of A_1 symmetry. (b) Excited configurations, all of B_1 symmetry.

Problem 10.8 Check that the (first) set of seven configurations given above all give rise to terms of A_1 symmetry and that the (second) set of four all give rise to B_1 terms.

As has just been mentioned, the inclusion of configuration interaction is usually an important step in accurate calculations on the electronic structure of molecules—for instance, in obtaining those results which have been used at several points in this book. Two points should be made. First, just as for orbital interactions, so too for configuration interactions; only terms of the same symmetry species interact (they also have to be of the same spin multiplicity). Second, it is evident that as the number of orbitals included in a molecular problem—and so the number of configurations that arise increases—so the number of terms of a given symmetry species which may interact under configuration interaction increases. In that the improvement that results in the description of the ground term (and, usually, the lowest excited terms) is often considerable, an upper limit on the improvement is usually set by the capacity of the computer available, its ability to handle the enormous number of integrals that has to be calculated.

10.3 DIRECT PRODUCTS AND QUANTUM MECHANICAL INTEGRALS

This is a very important section of this book. It contains the heart of the simplification provided by group theory to chemical problems—a reduction in the number of integrals that arise in quantum chemistry. This statement may not impress the student who hopes to avoid such integrals altogether—although it indicates how this avoidance may be achieved! In fact, group theory indicates integrals which are zero, without any need to evaluate them. Such zero integrals reappear as spectroscopic selection rules, for instance. A forbidden transition is one for which the predicted spectral intensity is zero. Let us look at this in more detail.

It is all too easy for quantum mechanics to appear formidable because of the large number of rather unpleasant looking integrals which it seems to involve. In practice, these integrals are found to be rather less objectionable because if they cannot be evaluated algebraically they can be evaluated numerically, either by hand or by a digital computer. Even so, a great deal of work can be saved by the intelligent use of group theory. Let us first consider what is meant by an integration over all space (which is the integration which is usually involved in quantum mechanics). Integration may be pictorially regarded as the adding together of an infinite number of infinitesimally small fragments. As a consequence of this it is sometimes possible to see the result of an integration without actually carrying out the calculation. Consider the p_z orbital shown in Figure 10.3. What is the value of the integral over all space of the p_z orbital? That is, what is the value of $\int p_z \, \delta v$, where δv is an infinitesimally small volume element? Treat this integral as $\sum p_z \, \delta v$, where the summation is over an infinity of minute volume elements. In order to perform this summation—this integration—one has to collect into one box, as it were, all of the infinitesimally small fragments which comprise this wavefunction. We must pay due

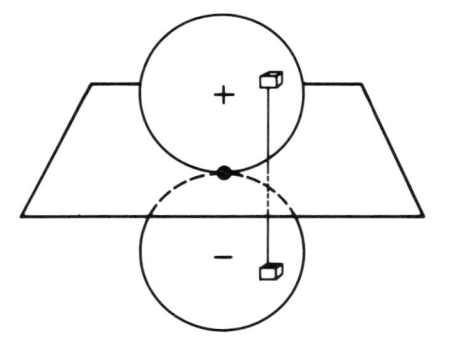

Figure 10.3 There is an exact cancellation of the contributions of the two boxes (at equivalent positions in the lobes of the p_z orbital shown) to an integral over all space of the p_z orbital.

regard to the signs of the fragments—some fragments are from that part of space in which the wavefunction has a positive amplitude and others from that part in which it has a negative amplitude. From the shape of the orbital it is evident that for every volume element that makes a positive contribution to the integral, there is a corresponding volume element which makes a negative contribution. A pair of such mutually cancelling volumes is shown in Figure 10.3, the positive contribution from the top volume being cancelled by the negative contribution from the bottom. By adding together pairs of points in this way until the whole of space is included it is seen that the value of the integral $\int p_z \, \delta v$ is zero—even though it has not been explicitly evaluated. It is the fact that arguments such as this can be cast, very simply, in the language of symmetry that makes group theory so valuable. Thus, an alternative way of stating the above argument is to recognize that the 'top' and 'bottom' of the p_z orbital are (a) symmetry-related by reflection in the mirror plane shown in Figure 10.3 and (b) of opposite phase. These two facts, taken together, establish that the integral must be zero. Can this procedure be generalized? Is there a general rule to replace the two specific points made above, which are relevant only to the p_z orbital (and any other functions that behave similarly).

To establish the general rule it is helpful to qualitatively consider the corresponding integral involving an s orbital:

$$\int s \, \delta v$$

It is clear, from Figure 10.4, that reflection in the mirror plane now interrelates two volume elements which make *identical* contributions to the integral. The character of the s orbital under the mirror plane reflection operation is 1 and so the contribution to the integral coming from volume elements related by this operation do not cancel. This is in contrast to the p_z orbital which has a character of -1 under reflection in the mirror plane. Clearly, the integral over all space of a function which transforms as an irreducible representation which has all its characters equal to $+1$ will not be equal to zero by symmetry. That is, integrals over all space of functions transforming under the totally symmetric irreducible representation of a point group may be non-zero. Does the contrary rule hold: can one say, in general, that the integral over all space of a function transforming as a non-totally symmetric irreducible representation must be

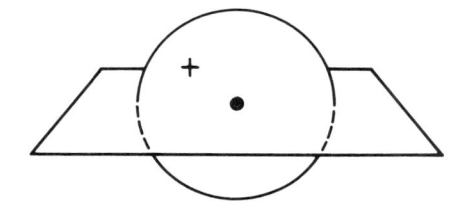

Figure 10.4   Integration over all space of an s orbital-type quantity is non-zero.

zero? The answer is 'yes', as the following argument shows. In Chapter 5 a number of character table theorems were introduced. Of these, consider Theorem 3:

> Take any two different irreducible representations and multiply together the two characters associated with each class. Then, in each case, multiply the product by the number of operations in the class. Finally, add the answers together. The result is always zero.

Consider the case in which one of the irreducible representations chosen is the totally symmetric. Multiplication by its characters is always multiplication by the number 1. The second irreducible representation cannot be a totally symmetric one because we are working with two different irreducible representations. The theorem tells us that for this second irreducible representation the sum of (the products of the character in a class multiplied by the number of operations in that class) is equal to zero. This means that an integral having the symmetry of this second irreducible representation will also have the value of zero—positive and negative contributions to the integral will cancel. We know the value of the integral without ever evaluating it. This is particularly easy to see for Abelian groups in which only the numbers 1 and -1 appear in the character table (C_{2v} provides an example). For all but the totally symmetric irreducible representation there are equal numbers of 1's and -1's; the number of positive contributions to an integral exactly match the number of negative contributions of the same magnitude. This example is simple because the magnitude of the contributions from all symmetry-related points in space is identical. The above rule also applies when the characters involved are other than 1 and -1.† As an example consider the T_{2u} irreducible representation of the O_h point group discussed in Chapter 7, and given in Table 7.2. The point group O_h is of order 48; that is, the operations of the point group relate a general point in space to 47 equivalent points. For a non-totally symmetric singly degenerate irreducible representation such as A_{1u} 24 of these points have a phase which is the opposite of the phase of the other 24. For a degenerate irreducible representation, such as the triply degenerate T_{2u}, the picture is more complicated because there are several basis functions simultaneously under study (three for T_{2u}). In such a case it is the integral over all space of the several functions, together, which is zero. However, the symmetry-equivalence of the functions ensures that the zero value applies to each individual.‡ For the T_{2u}

---

† The development in the text assumes that no complex characters are involved. A small change, along the lines of Chapter 11, is needed to cover this case.
‡ As explained in Chapter 9, if there is a particular interest in one of the three functions a useful trick is to work in a subgroup in which that function transforms as a singly degenerate irreducible representation.

irreducible representation of the O_h point group we have:

O_h:	E	$8C_3$	$6C_4$	$3C_2$	$6C_2'$	i	$3S_6$	$8S_4$	$3\sigma_h$	$6\sigma_d$
T_{2u}:	3	0	-1	-1	1	-3	0	1	1	-1

Product of T_{2u} characters with the number of operations in the class:

3	0	-6	-3	6	-3	0	6	3	-6

It is easy to see that the sum of the numbers in the final row is zero. This means that an integral over all space of a set of T_{2u} functions is zero—the negative contributions exactly cancel the positive—and that the same is true of each individual T_{2u} function.

Problem 10.9   Select any of the character tables discussed in a previous chapter. Select a non-totally symmetric irreducible representation and show that the sum of all (character multiplied by the number of operations in the corresponding class) products is zero.

From this argument it follows that the earlier statement about non-zero integrals may be replaced by a stronger one:

> Only integrals of functions transforming under the totally symmetric irreducible representation of a point group may give rise to non-zero integrals over all space.

It is this theorem which leads to the simplifications introduced by group theory; one knows immediately which integrals must be zero without ever having to actually evaluate them. Amongst other things, this is the basis of all spectroscopic selection rules. The reader may have noted that the word 'may' was used in the generalization above, rather than the stronger 'will'. The reason for this will become clear from a comparison of Figures 10.3 and 10.5. The latter shows the p_z orbital of Figure 10.3 but now in C_{2v} symmetry. None of the operations corresponding to the symmetry elements of Figure 10.5 interrelate the bottom and top of the p_z orbital—the mirror plane of Figure 10.3 is not a symmetry element of C_{2v}. As a consequence, and as was seen in Chapter 2, the p_z orbital is totally symmetric in C_{2v}. This does not alter the fact that the integral

$$\int p_z \, \delta v$$

remains equal to zero. We see, then, that integrals over all space of functions of A_1 symmetry can be equal to zero, and so we must say 'may' rather than 'will'. In the literature one sometimes encounters use of the phrase 'hidden symmetry' to indicate that all of the nodal properties of a function are not explicitly revealed. In such cases, integrals are zero even though, like p_z in C_{2v}, the function is totally symmetric.

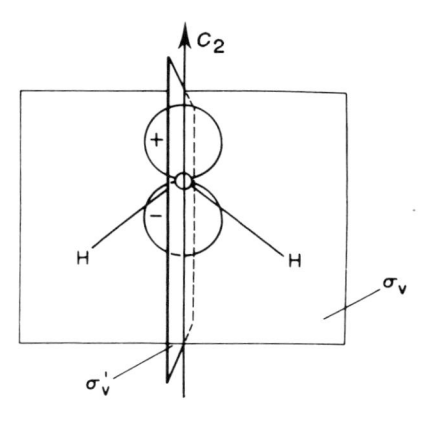

Figure 10.5 An integration over all space of p_z is zero, even when it transforms as the totally symmetric irreducible representation (here, in C_{2v}).

The discussion on non-zero integrals which has just been developed in the context of simple functions, like p_z, can immediately be extended to product functions using the arguments developed earlier in this chapter. Just as for simple functions, only product functions which are totally symmetric may give non-zero integrals over all space. Already in this chapter we have seen how to obtain the symmetry species of such product functions from the symmetries of their component functions—one simply forms the appropriate direct product. If this direct product either is or contains the totally symmetric irreducible representation then the integral is non-zero, or, if the direct product is reducible and contains the totally symmetric symmetry species, the integer contains a non-zero component. In the latter case, the non-zero component can be investigated further by applying the projection operator method to obtain the explicit form of the totally symmetric product function.

As an example of this application the assertion made in Section 3.5 when discussing the bonding in the water molecule that 'interactions between orbitals of different symmetry species are always zero' will now be justified. 'Interactions' in this context means, specifically, that both the overlap integral

$$\int \psi_a \psi_b \, \delta v$$

and the energy integral

$$\int \psi_a \mathcal{H} \psi_b \, \delta v$$

where $\mathcal{H}$ is the so-called Hamiltonian operator for the system (fortunately, a detailed expression is not needed for $\mathcal{H}$ in the present context)—are only non-zero when ψ_a and ψ_b are of the same symmetry species. Consider the overlap integral first. Direct products of the symmetry species of ψ_a and ψ_b have to be formed and those which give rise to the totally symmetric irreducible representation (A_1 in C_{2v}) selected. It is clear from Table 10.2 that A_1 only appears along the leading diagonal. That is, A_1 only results, in the C_{2v} group, when ψ_a and ψ_b are of the same symmetry species. It follows that only overlap between orbitals of the same symmetry species may give rise to the non-zero overlap integral in the C_{2v} group.

What of the energy integrals

$$\int \psi_a \mathcal{H} \psi_b \, \delta v \; ?$$

The Hamiltonian operator $\mathcal{H}$ expresses all of the energies—be they attractive, repulsive, kinetic or potential—present in the molecule. At a particular point in the molecule there will be a particular blend of the forces corresponding to each of these energy components. However, at all symmetry related points these blends must be equivalent. It follows that $\mathcal{H}$ must have the symmetry of the molecule. That is, $\mathcal{H}$ is totally symmetric, A_1, and inclusion of it in a direct product will not change the final direct product. It follows that the answer to the question of whether the above energy integral is symmetry-required to be zero is determined by the direct product of the symmetry species of ψ_a and ψ_b. But this problem has already been dealt with when we discussed the overlap integral between these orbitals. The energy integral will also be zero unless ψ_a and ψ_b have the same symmetries. The validity of the statement made in Chapter 3 that 'interactions between orbitals of different symmetry species are always zero' at once follows.

It is clear from the above discussion that the occurrence of the totally symmetric irreducible representation in tables of direct products is of particular interest. There is a relevant general conclusion which is implicit in that discussion. In Table 10.2 and also in the table given in Problem 10.4 it is seen that the totally symmetric irreducible representation occurs along the leading diagonals (top left to bottom right) and nowhere else. Further, it occurs in *every* entry along the leading diagonals. It is found that these two statements are true for all point groups:

> The totally symmetric irreducible representation always occurs when the direct product is formed between a particular irreducible representation and itself. It never appears when a direct product is formed between two different irreducible representations.

If nothing else, this generalization will save work when determining whether or not an integral is zero by symmetry. If, for instance, we have an integral over three functions of different symmetries (different irreducible representations) then we need only form the direct product of *two* of these irreducible representations and then compare the irreducible representation(s) contained in this direct product with that of the third function. If there is a matching, the integral may be non-zero. If there is no matching, then it certainly is zero. In this argument there was a switch from the word 'functions' to 'irreducible representations'. This is because the general argument holds no matter what the form of the functions involved. Indeed, we have just met a case where an ordinary function was not involved; the case of the energy integrals, where the concern was with the operator $\mathcal{H}$. However, we were able to put a symmetry label on this operator. This step is important because the general form of almost all the integrals of quantum mechanics is:

$$\int (\text{wavefunction})_2 (\text{operator})(\text{wavefunction})_1 \, \delta v$$

where the wavefunctions may be one-electron wavefunctions, many-electron wavefunctions, vibrational wavefunctions or many others. A wide range of operators occurs in quantum mechanics and the above discussion applies to all of them, although in the next section the discussion will be restricted to those associated with the spectroscopic properties of molecules.

10.4   SPECTROSCOPIC SELECTION RULES

It is a fundamental postulate of quantum mechanics that corresponding to every observable associated with a system there is a corresponding operator, a postulate justified by the fact that it has always been found to work! So, to apply quantum mechanics to the electronic spectrum of a molecule the appropriate ground and excited term electronic wavefunctions and the appropriate operator have to be inserted into the general expression given at the end of the previous section. If the transition represented by the integral is allowed, then the integral will be non-zero. The association of symmetry labels with the two wavefunctions should present no problems, given the discussion of the present chapter. If the symmetry species of the appropriate operator can be obtained then it will be a matter of simple group theory to determine the allowedness of the transition. There are limits, however. If we wished to predict the actual intensity of a transition we would have to evaluate the integral properly.

Fortunately, the task of working out the symmetry species of the operators appropriate to a particular form of spectroscopy is a very simple task compared with that of determining the detailed form of the operators themselves. In most forms of spectroscopy a beam of electromagnetic radiation is allowed to interact with the system under study (the term 'electromagnetic radiation' rather than the word 'light' is used because the wavelength of the radiation may be far from the visible region of the spectrum). Integrals such as those above then describe the consequences of this interaction between radiation and matter.

In the simplest (Maxwell) picture, electromagnetic, radiation is regarded as being composed of two mutually perpendicular oscillating fields, one an electric field and the other magnetic; both fields are perpendicular to the direction of propagation of the radiation. The most evident way, then, in which such radiation can interact with matter is by virtue of either or both of the electric and magnetic fields associated with it, so that the most common spectroscopic observations are of transitions which are either 'electric dipole' allowed or 'magnetic dipole' allowed (for the two to occur simultaneously the molecule has to be optically active; indeed it is the simultaneous allowedness of electric and magnetic dipole mechanisms which produces optical activity). The electric field associated with the light wave induces an oscillating electric dipole $(+ -)$ in atoms and molecules. When this oscillation matches a natural frequency of the atom or molecule, resonance occurs and energy is transferred from the light wave to the atom or molecule. Similarly, the magnetic field induces an oscillating magnetic dipole (N–S) which can also cause excitations. Magnetic dipole allowed transitions are of particular importance in nuclear magnetic resonance and electron paramagnetic resonance spectroscopies. To determine the symmetries of the corresponding operators, then, it is only necessary to determine the symmetry species associated with an electric dipole or a magnetic dipole.

In order to obtain an electric dipole it is necessary to separate charges of opposite sign along an axis. In our three-dimensional world there are only three independent directions in which one may bring about such a charge separation and so there are just three electric dipole operators, one corresponding to the x, one to the y and one to the z axes. Further, because the Cartesian axes are dipolar —they have $+$ and $-$ regions—the transformations of the electric dipole operators mimic—are isomorphous to—those of the Cartesian axes of a molecule. Equally, they are isomorphous to the translations of the molecule along x, y or z axes. This isomorphism of x, y and z with T_x, T_y and T_z was noted in Section 9.3 when they were discussed in connection with vibrational analyses. A second use for them has now been found—they give the transformational properties of the three electric dipole operators—and a second reason why one or other set (and sometimes both) are included at the right-hand side of character tables.

Just as an electric dipole corresponds to a movement of electric charge along an axis, so a magnetic dipole corresponds to a rotation of charge about an axis (as in a solenoid carrying a current). There are three magnetic dipole operators, one for each of the three Cartesian axes. The symmetry species of the magnetic dipole operators will be the same as those of the rotations about these axes. These rotations (usually denoted R_x, R_y and R_z in a character table) were met and used in the discussion of molecular vibrations in Section 9.3. The entries R_x, R_y and R_z at the right-hand side of a character table tell us how the three magnetic dipole operators transform.

We are now in a position to make some general statements about whether or not an integral related to the intensity of a transition is required to be zero. That is, to state general selection rules for electric dipole and magnetic dipole allowed processes. This rule is derived from the integral given towards the end of the previous section by replacing wavefunctions and operator with the appropriate symmetry species.

A transition is electric dipole allowed if the triple direct product of the symmetry species of the initial and final wavefunctions with that of the symmetry of a translation contains the totally symmetric irreducible representation. Similarly, a transition is magnetic dipole allowed if the triple direct product of a rotation with the symmetry species of the initial and final wavefunctions contains the totally symmetric irreducible representation.

These rules are rather lengthy and triple direct products can be tedious to work out, so it is convenient to recast them into a simpler form by making use of the fact that the totally symmetric irreducible representation only arises in the direct product of an irreducible representation with itself. In compiling the triple direct product, first form the direct product of the symmetry species of the initial and final wavefunctions. The transition will only be allowed if this direct product contains within it the same symmetry species as that of the operator. This is a particularly useful way to state the selection rule because, as has been seen, there are commonly several alternative operators—dipole moment operators corresponding to T_x, T_y and T_z, for instance—and in this form a choice between the alternatives does not have to be made until the last step. One simply looks for a matching between the irreducible representation(s) arising from the direct product of wavefunction symmetries with those of the appropriate operators. If a matching exists, the transition is allowed. In summary, then, the general spectroscopic selection rule—of which all others are particular cases—is:

A transition is allowed only if the direct product symmetry species of the initial and final wavefunctions contains the symmetry species of the operator appropriate to the transition process.

All that is needed in order to make use of this rule is a list of those spectroscopic processes which normally arise as a result of electric dipole transitions and those which normally arise from magnetic dipole transitions. Such a list is given in Table 10.5 which also lists the simple functions usually contained in character tables which are of the same symmetry species as the operators relevant to Raman spectroscopy. The quadratic form of operator for the Raman process arises because in it one wavelength of light is incident on a molecule but a different wavelength is emitted. The symmetries of the operators relevant to Raman spectroscopy are therefore the same as those of products like $T_x T_y$ but these are never given in character tables and an equivalent form—such as xy—is listed instead.

Table 10.5

Form of spectroscopy and spectral region	Form of operator	Symmetry properties of the operator are the same as those of:
Electronic (visible and ultraviolet) Vibrational (infrared)	} Electric dipole	T_x, T_y, T_z (or more simply x, y, z)
Rotational (microwave) NMR (radiofrequency) EPR (microwave)	} Magnetic dipole	R_x, R_y, R_z
Raman (visible)	Polarizability (this resembles 'electric quadrupole' but is a little wider)	$x^2, y^2, z^2, xy, yz, zx$ (or combinations of these)

Problem 10.10 Confirm that the following transitions are electric dipole allowed:

Point group	Ground term symmetry	Excited term symmetry
C_{2v}	A_1	B_2
C_{2v}	B_1	B_1
C_{4v}	A_2	E
C_{4v}	E	E

Confirm that the following transitions are electric dipole forbidden:

Point group	Ground term symmetry	Excited term symmetry
C_{2v}	B_2	B_1
C_{4v}	A_1	B_1
C_{4v}	B_1	B_2
C_{4v}	A_2	A_1

Hint: These problems anticipate the discussion of the next section of the text.

This chapter is concluded by two illustrations of the application of the general selection rule to vibrational spectroscopy. The first is the example discussed in Chapter 9, that of the vibrational spectrum of CH_3Cl. The general principle will be sufficiently well illustrated if we confine our discussion to the C–H stretching vibrations, which, as has been seen in Section 9.4, have $A_1 + E$ symmetry in the molecular point group, C_{3v}. Molecular vibrational spectroscopy almost invariably admits of a further simplification to the selection rule problem. This is because the ground vibrational state can be assumed to be one in which no vibrations are excited. It is thus a totally symmetric state. The excited state has the symmetry of the vibration excited. The general selection rule therefore reduces to a very simple one:

> A vibration will be spectroscopically active if the vibration has the same symmetry species as the relevant operator.

Problem 10.11 Consider the water molecule; the normal modes have the symmetries A_1 and B_1. Suppose that the latter vibration is already excited in a water molecule, so that the symmetry of the vibrational wavefunction is B_1. Show that the above rule still applies.
Note: This is a particular example. Although it is not difficult to see that it is relevant to all cases in which a singly degenerate mode is previously excited, it could not immediately be applied to molecules in which a degenerate vibrational mode is excited.

The character table for C_{3v} is repeated in Table 10.6. From this table it is seen that a vibration of A_1 symmetry is infrared allowed (because T_z—or equivalently z—has A_1 symmetry). It is also a Raman allowed vibration because z^2 (and the sum $x^2 + y^2$) transforms as A_1. Similarly, because (T_x, T_y) or, equivalently, (x, y), transform as E, an E vibration is infrared allowed. It is also Raman allowed because products of coordinate axes also transform as E. We conclude that the CH_3Cl molecule is expected to have two infrared peaks and two coincident Raman peaks in the carbon–hydrogen stretching region of the spectrum. Apart from additional complications caused by the low moment of inertia about the C_3 axis, this is precisely what is seen. This spectral prediction is specific to C_{3v} geometry. If, for instance, CH_3Cl were a planar molecule with C_{2v} symmetry then three infrared and Raman-coincident peaks would be expected in the C–H stretching region of the spectrum.

Problem 10.12 If CH_3Cl were a planar molecule it could have C_{2v} symmetry provided that one hydrogen, the carbon and chlorine were co-linear. Use the C_{2v} character table to derive the spectral predictions given above.

Table 10.6

C_{3v}	E	$2C_3$	$3\sigma_v$	
A_1	1	1	1	T_z, z, z^2, $x^2 + y^2$
A_2	1	1	-1	
E	2	-1	0	(T_x, T_y) (x, y) (zx, yz) $(x^2 - y^2, xy)$

C_{2v}	E	C_2	σ_v	σ_v'	
A_1	1	1	1	1	z, T_z, x^2, y^2, z^2
A_2	1	1	-1	-1	xy
B_1	1	-1	1	1	y, T_y, yz
B_2	1	-1	-1	1	x, T_x, zx

There is an important subtlety hidden in the above discussion of selection rules; it is convenient to discuss it in the context of the vibrations of CH_3Cl but the ideas which will be introduced have a general validity. When talking of electric dipole selection rules the attention focused on the transformation properties of quantities like T_x, T_y, T_z. Effectively, an isolated molecule has been considered and its relationship with its environment ignored: T_x, T_y and T_z referred to *molecular* axes. In solution, the molecule will be tumbling and so these molecular axes will bear no fixed relationship to the axes within which we must, perforce, work—the laboratory-fixed axes. That is, even if plane polarized light is used (so that the electric and magnetic fields are in a fixed orientation) the molecular tumbling means that allowed transitions can involve the operators associated with T_x, T_y or T_z. However, if the molecular and laboratory axes can be brought into a fixed relationship with each other, then additional spectroscopic information can be obtained because transitions associated with T_x, T_y and T_z can be separately studied. First, the axes of all molecules must be brought into alignment and then held like this. Although there are several techniques for aligning axes—in some cases the application of a strong electric field produces an appreciable alignment; alternatively, molecules can be incorporated into a thin sheet of some transparent plastic which is then stretched—the simplest and most powerful is to crystallize the material. It sometimes happens that molecular axes persist within a crystal. For instance, it is possible that the threefold axis of a molecule such as CH_3Cl would lead to, and persist in, a crystal with a threefold axis. Such an axis can be identified by either microscopic or X-ray examination of the crystal. Suppose that such a crystal has been obtained and characterized. It is now used for an infrared experiment in which the infrared light is polarized (infrared polarizers are readily available). Call the direction of polarization of the light—that in which the electric vector lies—the direction p. This means that in laboratory axes, we are working with an operator which behaves like T_p. The crystal is now inserted into the beam of polarized infrared radiation so that, within the crystal—and so for molecular axes—p is coincident with the threefold axis (for the crystal previously discussed). This means that in terms of molecular axes the operator T_p is to be identified with the operator T_z, and *only* vibrations which are T_z allowed will be seen in the spectrum. That is, in a spectrum only A_1 bands would be seen (to return to the CH_3Cl example, ignoring all the obvious experimental difficulties). There would be no absorption at the frequencies of E bands. Conversely, rotating the crystal so that p is perpendicular to the crystal threefold axis, T_p, would, in the crystal, be T_x (or T_y—they transform as a pair so no distinction between them can be made). In this case only the E bands would be seen in the spectrum.

This discussion has been somewhat idealized—molecular axes do not usually persist in a crystal. The discussion can be modified to cover this case unless vibrational coupling occurs between the individual molecules in the crystal, when a rather different method of analysis has to be used. Further discussion of these aspects is given in Chapters 12 and 13. Even when molecular axes persist, real crystals are not perfect, alignment is not perfect, polarization is not perfect and so bands which, according to the above arguments, should not appear, in fact, usually do. However, they do so with very much reduced intensity and so studies such as those described above would, almost certainly, enable the determination of the actual symmetry species of a molecular vibration.

10.5 SUMMARY

This is an important chapter, one in which the main reason for the importance of group theory in chemical problems has been developed. Group theory is important because it simply and reliably gives answers of zero (p. 215). Knowledge of which integrals are zero simplifies discussions of molecular bonding and spectra (p. 217). The development falls into two parts. First, direct products. Here it was important to discover that only the direct product of an irreducible representation with itself gives the totally symmetric representation (p. 219). Second, molecular integrals, where it was found that only integrals over all space of totally symmetric functions are non-zero (p. 217). These two parts come together in the recognition that all integrals of any importance in quantum mechanics are integrals involving product functions (p. 220).

REFERENCES

1. C. J. Ballhausen, *Introduction to Ligand Field Theory* (McGraw-Hill, New York, 1962), p. 48.
2. E. B. Wilson, J. C. Decius and P. C. Cross, *Molecular Vibrations* (McGraw-Hill, New York, 1955), p. 152.

11

π-Electron Systems

11.1 SQUARE CYCLOBUTADIENE AND THE C_4 POINT GROUP

One of the areas of chemistry in which relatively simple quantum mechanical ideas have had a very important impact has been in the field of unsaturated organic molecules. When a molecule contains alternate single and double carbon–carbon bonds then it is found that those electrons involved in π-bonding can be considered on their own—that the σ electrons can be ignored. It seems that these π electrons largely determine the chemistry of such molecules, a recognition which has given an understanding of the chemical stability and reactions of these molecules and also of their spectroscopic properties. The distinction between σ and π orbitals was made in Section 4.4. It is important to recognize that when a molecule contains a series of atoms linked by alternate single and double bonds then on *each* atom in the series there is an orbital involved in the π bonding. It is usually the case that this orbital is a p orbital. The ready availability of detailed and accurate numerical calculations on simple organic molecules has shown that the idea of σ–π separability rests on less secure foundations than was once held to be the case. The orbital symmetry distinctions persist but configuration interaction of the type outlined in the last chapter serves to mix different electron configurations. None the less, there is no doubt that the predictions made by the simple theory are rather good, even if a detailed and general justification for this is not available. It is when the results of the simple model are symmetry-determined that the most evident justification occurs and it is such applications which will be the concern of this chapter.

The symmetry aspects of Hückel theory, the best known π-electron model, are most readily seen from an example. A simple molecule, but one which serves to illustrate all of the main points of the theory, will be considered. The molecule is a very unstable and fugitive one, cyclobutadiene, C_4H_4, which will be taken to be a planar molecule with its four carbon atoms arranged at the corners of a square. The carbon atoms are known to have this arrangement when the molecule is stabilized by complexing with a transition metal atom, as in the molecule $C_4H_4Fe(CO)_3$. Figure 11.1 shows square cyclobutadiene together with the four $2p_\pi$ orbitals that will be of interest (we suppose that the carbon 2s and the other carbon 2p orbitals are involved in the bonding of the σ framework). The molecular symmetry is D_{4h} and so this is the obvious group in which to work. However, we shall not. Although it is not particularly obvious

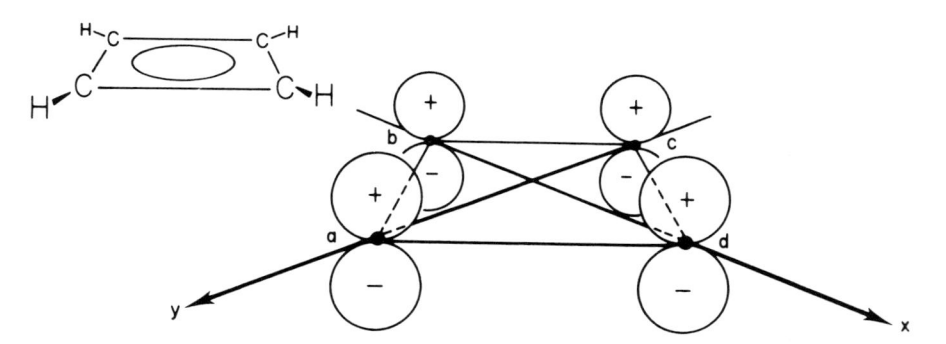

Figure 11.1   The four carbon $2p_\pi$ orbitals in cyclobutadiene, C_4H_4.

from the way that the D_{4h} character table is usually written (Appendix 3), the D_{4h} group is the direct product of $C_{4v} \times C_s$. That is, add a σ_h mirror plane to C_{4v} and other symmetry elements are at once generated so that the group becomes D_{4h}. Now, the problem that we are considering immediately defines the effect of this σ_h mirror plane. We are only interested in the p_π orbitals shown in Figure 11.1 and these, and anything derived from them, are antisymmetric with respect to reflection in the σ_h mirror plane. So, it might well be simpler to work in the C_{4v} point group and, at the end, move to D_{4h} by recognizing this σ_h antisymmetry. It is probable that most workers would be content to stop here and work in C_{4v}, but we shall press on!

> **Problem 11.1**   (a) Using Appendix 3 and Figure 11.1 show that square planar cyclobutadiene has D_{4h} symmetry.
> (b) Using Appendix 4, show that the D_{4h} group is the direct product of C_{4v} and C_s.

The C_{4v} group possesses two sorts of σ_v mirror planes; $2\sigma_v$ and $2\sigma_v'$. Either the σ_v or σ_v' mirror planes (it does not matter which we choose) cut vertically through the carbon p_π orbitals of cyclobutadiene. They therefore relate one side of each lobe of this orbital to the other side (Figure 11.2). But these sides must be of the same phase. So, the operation of reflection in these mirror planes gives no new information. The operations are superfluous and perhaps can be discarded. But, if they are discarded we must also discard the other mirror planes—point groups exist with both $2\sigma_v$ and $2\sigma_v'$ but there are no, and can be no, point groups with just one set. The most sensible thing to do would be to play safe and keep them all—after all, not much additional work is involved. We shall be more daring, however, and eliminate them because this will give us the opportunity to work in what seems a rather strange group—the C_4 point group. It is unusual to discuss a group of pure rotations such as the C_4 group in a text at the level of the present one and as a consequence these groups tend to be regarded as rather strange and difficult. However, the same problems that they present also occur in the theory of

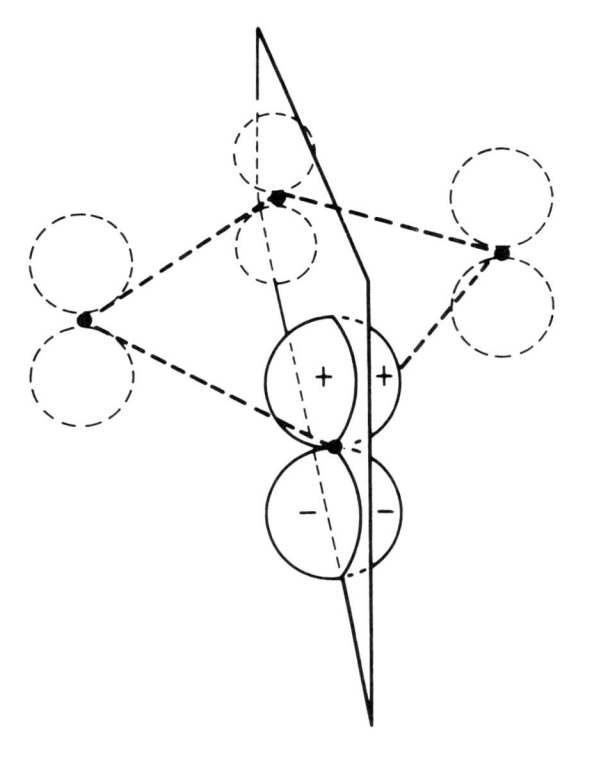

Figure 11.2 A σ_v (or σ_v') plane in C_{4v} cuts each carbon $2p_\pi$ orbital in half; the relationship between the two halves is determined by the orbital, not by the mirror plane.

space groups and so familiarity with the C_4 group is relevant to the theory of these groups. Space groups are the subject of the next two chapters and so we deliberately chose to work in the C_4 group in the present. It must be admitted, however, that the discussion which results is a little more difficult than would have been the case had we worked in an 'easier' group. As the reader may check for him or herself (Problem 11.9), we shall ultimately obtain the same answers as would have been obtained in C_{4v} (or D_{4h})!

The C_4 character table is given in Table 11.1. Note that it is an Abelian group—there is only one operation in each class. In particular, note that C_4 and C_4^3 (the C_4 operation carried out three times in the same sense)—are in different classes. In Chapter 5 the importance of the definition of 'class' was mentioned and this definition is given in Appendix 1. The proof that C_4 and C_4^3 are in different classes in the C_4 group is explicitly given in this appendix.

There are two apparently odd things about the C_4 character table. The appearance of i $(= \sqrt{-1})$ and the failure of the number 2 to appear against the E irreducible representation under the identity column. Note, that *if* the

Table 11.1

C_4	E	C_4	C_2	C_4^3
A	1	1	1	1
B	1	-1	1	-1
$E\ \big\{$	1	i	-1	$-i$
	1	$-i$	-1	i

$(i = \sqrt{-1})$†

number 2 did appear Theorem 2 of Chapter 5 would not be obeyed. The sum of squares of characters in the identity column would not be equal to the order of the group (which is 4, the number of operations in the group). The main reason for working in the C_4 point group is to give an opportunity to look at the E irreducible representation in some detail. In order to do this, it has to be remembered that the complex conjugate of $(a + ib)$, a and b being ordinary numbers, is $(a - ib)$. These complex conjugates have a special relationship to each other because when they are multiplied together a real number results:

$$(a + ib)(a - ib) = a(a - ib) + ib(a - ib) = a^2 - iab + iab - (i)^2b^2 = a^2 + b^2$$

because $-(i)^2 = -(-1) = 1$. In contrast, neither $(a + ib)^2$ nor $(a - ib)^2$ are free from i. Note that where, in the E irreducible representation, one component contains i, the other contains $-i$. These are complex conjugates (as is easily shown if one sets $a = 0$, $b = 1$ in the expressions earlier in this paragraph). This hints at what is, in fact, correct. In the previous chapter, in particular, an irreducible representation was sometimes multiplied by itself (when forming direct products). The way that reducible representations were decomposed into their irreducible components earlier in this book is very similar. The same procedure is followed when working with the C_4 point group for all irreducible representations except the E. For applications involving the E irreducible representation, it is complex conjugates that have to be multiplied. That is, one multiplies the first component of this doubly degenerate representation by the second and vice versa. In this way real, not complex, answers are obtained.

The E irreducible representation of the C_4 point group is said to be a *separable degenerate* representation. Some purists object to this name—holding that it is self-contradictory—but it is the name commonly used. The word 'degenerate' is used because functions transforming as this representation have the same energy—an example will be met shortly. 'Separable' because it

---

† It may be helpful to note that the complex conjugate of i, denoted i^*, is $-i$ ($ii^* = 1$). In some books i^* is found where in this text $-i$ has been used.

is possible to design an experiment on a molecule of C_4 symmetry which shows that all functions transforming as the E irreducible representations are not necessarily quite equivalent. In order to illustrate this we shall digress to give a brief discussion of optical activity.

11.2   OPTICAL ACTIVITY

Classically, a molecule is optically active when in an electronic transition there is a helical movement of charge density. Just as a left-hand screw thread is not superimposable on a right-hand thread, so there is an optical rotation difference between molecules in which otherwise identical charge displacements follow right-hand and left-hand helical paths. A characteristic of a helix is that it corresponds to a simultaneous translation and rotation and so, as was outlined in the discussion of Section 10.4, optically active molecules are those in which a transition is simultaneously both electric dipole (charge translation) and magnetic dipole (charge rotation) allowed. That is,

---

Molecules† may be optically active when they have a symmetry such that T_α and R_α ($\alpha = x$, y or z) transform as the same irreducible representation.

Comparison of this rule with the data given on the right-hand side of the character tables in Appendix 3 confirms the applicability of the commonly stated criteria for optical activity; optically active molecules possess neither a centre of symmetry nor a mirror plane. They do not have any improper rotation operations. As an alternative general statement, one can say that optically active molecules do not have any S_n axis, where n can assume any value ($n = 1$ corresponds to a mirror plane and $n = 2$ to a centre of symmetry).

Problem 11.2   The separation of the cobalt complex ion $[Co(en)_3]^{3+}$ into optical isomers is a common undergraduate experiment. The complex is, essentially, octahedral and en is the bidentate ligand ethylenediamine, $NH_2.CH_2.CH_2.NH_2$, which is bonded to the cobalt through the nitrogen atoms on adjacent (cis) coordination sites. Determine the symmetry of this molecule and thus show that it has no S_n axis.
Hint: The discussion of Section 7.5 should be helpful.

In the particular case of the C_4 point group, T_z and R_z both transform as A and the complex combination $T_x + iT_y$ transforms in the same way as $R_x + iR_y$.

---

† Note the word 'molecules' in this statement. It does not apply to crystals which, under some circumstances, can contain mirror planes of symmetry and yet be optically active.

Similarly, $T_x - iT_y$ transforms isomorphically with $R_x - iR_y$. The complex form of these latter combinations is a little off-putting, although it should be less so by the end of this chapter. Ignoring this, it is clear that T_α and R_α transform isomorphically in the C_4 group so that a molecule of C_4 symmetry is potentially optically active. A beam of polarized light incident on such a molecule down the fourfold axis might suffer a rotation. Clearly, this is not compatible with the isotropy which one normally associates with degeneracy in the x, y plane. The explanation lies, not surprisingly, in the appearance of complex coefficients in the character table.

Problem 11.3   Despite the discussion of optical activity in the context of cyclobutadiene in the text, it is believed that cyclobutadiene is not optically active. Why?

11.3 WORKING WITH COMPLEX CHARACTERS†

All of the character tables met in earlier chapters in this book contained simple integers as characters. Most people approach complex characters with some apprehension, expecting some strange twists. This apprehension is justified! One example of the different pattern is seen in the statement made towards the end of Section 10.3 that 'the totally symmetric irreducible representation always occurs when the direct product is formed between a particular irreducible representation and itself'. This statement is true for the C_4 point group but needs some elaboration. Consider the direct product of the first of the E irreducible representations of Table 11.1 with itself:

	E	C_4	C_2	C_4^3
$E(1)$	1	i	-1	$-i$
$E(1) \otimes E(1)$	1	-1	1	-1

this direct product is the B irreducible representation, not the A. To obtain the A, the direct product has to be formed of $E(1)$ with its *complex conjugate*, $E(2)$:

$E(1)$	1	i	-1	$-i$
$E(2)$	1	$-i$	-1	i
$E(1) \otimes E(2)$	1	1	1	1

That is, when working with a separately degenerate representation, one has to elaborate on the statements made in Chapter 10 about direct products. The way

---

† A short, readable, paper which deals with the problems covered in this section is R. L. Carter, 'Representations with Imaginary Characters', *J. Chem. Educ.*, **70** (1993), 17.

to proceed is reasonably straightforward. Thus, the general expression for an overlap integral given in texts on quantum mechanics is

$$S_{ab} = \int \psi_a^* \psi_b \, \delta v$$

the asterisk on ψ_a^* indicating the complex conjugate of ψ_a. If ψ_a and ψ_b are not complex this reduces to the simple form considered in Section 10.1,

$$S_{ab} = \int \psi_a \psi_b \, \delta v$$

However, when ψ_a and ψ_b are both complex the more general form must be used. In such a case complex conjugate irreducible representations must be used when carrying out the associated group theory. An explicit example of this will be met in the next section.

Problem 11.4   Modify the discussion of selection rules in Section 10.4 so that it covers the case where the wavefunctions are complex.

11.4   THE π ORBITALS OF CYCLOBUTADIENE

We now return to the problem of the π-electrons of cyclobutadiene. We know that these π electrons interact with each other—they form π bonds of some sort, and so the first problem is that of finding the π molecular orbitals which they occupy. This will be tackled in two stages. First, the irreducible representations generated by the transformations of the four carbon p_π orbitals is determined and their symmetry-adapted combinations generated. Second, the approximate relative energies of these symmetry-adapted combinations will be determined. It is easy to show that the transformations of the four carbon p_π orbitals of cyclobutadiene in the C_4 point group (Figure 11.1) generates the reducible representation

E	C_4	C_2	C_4^3
4	0	0	0

and that this gives rise to $A + B + E$ irreducible components.†

The determination of the symmetry adapted combinations is straightforward, and follows the projection operator procedure detailed in Chapter 4 very closely. Using the labels shown in Figure 11.1 for the four p_π orbitals and

---

† Reducible representations like this one—in which the number which is the order of the group appears in the identity operation column with all other entries zero—are called 'the regular representation' (of the particular point group). They always span each and every irreducible representation, the number of times an irreducible representation is spanned being given by the number in the identity operation for that particular irreducible representation (i.e. the dimension of the irreducible representation). The regular representation plays a part in the proof of some theorems of group theory.

neglecting overlap between these orbitals, the two E components given in Table 11.1 are used separately to give the linear combinations:

$$\psi(A) = \tfrac{1}{2}(a + b + c + d)$$
$$\psi(B) = \tfrac{1}{2}(a - b + c - d)$$
$$\psi_1(E) = \tfrac{1}{2}(a - ib - c + id)$$
$$\psi_2(E) = \tfrac{1}{2}(a + ib - c - id)$$

Problem 11.5 Use the projection operator technique to obtain the above linear combinations. The normalization of the E functions will be discussed in the text below.

Hint: The derivation is similar to that detailed in Section 4.6.

As indicated in the above problem, the only difficult point in this derivation concerns the two E functions. First, a hidden catch. In using the projection operator technique to generate a function transforming as a component of a separably degenerate representation one has to use its *complex conjugate* in the derivation. Thus, using the characters of the *second* E component in Table 11.1, the function listed above as $\psi_1(E)$ is obtained by the projection operator technique in un-normalized form: $a - ib - c + id$. It is easy to show that this procedure has given the correct answer—that $\psi_1(E)$ transforms as the *first* E component in Table 11.1. As Table 11.1 shows, the effect of a C_4 rotation on a function transforming as the first E component is to multiply it by i. Now this rotation permutes the p_π orbitals thus:

$$
\begin{array}{ccc}
a & \rightarrow & b \\
\uparrow & & \downarrow \\
d & \leftarrow & c
\end{array}
$$

so that it turns $\psi_1(E)$ into

$$b - ic - d + ia$$

which is $i(a - ib - c + id) = i\psi_1(E)$, as expected for the first E component. The next step is to normalize $\psi_1(E)$; that is multiply it by a coefficient such that:

$$\int \psi^* \phi \, \delta v = 1$$

where ϕ^* is the complex conjugate of ϕ (and ϕ is the normalized $\psi_1(E)$). The complex conjugate of a function is obtained by replacing i by $-i$ within it, so the complex conjugate of $\psi_1(E)$ is:

$$a + ib - c - id = \psi_1(E)^*$$

This function has been met before, it is $\psi_2(E)$; $\psi_1(E)$ and $\psi_2(E)$ are complex conjugates of each other.

It follows that the overlap integral of $\psi_1(E)$ with itself has the value

$$\int \psi_1(E)\psi_1^*(E)\psi\,\delta v = \int (a + ib - c - id)(a - ib - c + id)\,\delta v$$
$$= \int aa\,\delta v + \int bb\,\delta v + \int cc\,\delta v + \int dd\,\delta v = 4$$

where, as mentioned earlier, it has been assumed that the functions a, b, c and d do not overlap each other. The fact that a and b, for example, do not overlap each other means that the integral $\int ab\,\delta v$ is equal to zero. Because a and b are separately normalized $\int aa\,\delta v = \int bb\,\delta v = 1$. From the value of the overlap integral obtained above, 4, it follows that the normalization constant for $\psi_1(E)$— and, equally, $\psi_2(E)$—must be $\frac{1}{2}$, the value used in the linear combinations above.

Problem 11.6 Show that $\int \psi_2(E)^*\psi_2(E)\,\delta v = 4$.

11.5 THE ENERGIES OF THE π ORBITALS OF CYCLOBUTADIENE IN THE HÜCKEL APPROXIMATION

The limit at which simple group theory can help the discussion has now been reached. To proceed, chemical knowledge has to be added or, failing that, chemical intuition! In practice, this means that the next step involves using some model which provides a recipe for obtaining relative orbital energies. Such a model was used in earlier chapters of this book when a nodal plane criterion was used to obtain orbital energies—the more nodes that an orbital contains the higher its energy is expected to be. This model was augmented by an overlap criterion—the greater the overlap between two orbitals, the larger the energetic consequences of the interaction between them. The latter part of the discussion of Section 7.2 provides a good example of the augmentation of symmetry arguments by these models.

In the present section the nodal plane argument† will be used in a more mathematical form. The mathematical form to be used is that contained in Hückel theory; this is the simplest of all mathematical models of chemical bonding and one that is particularly appropriate to unsaturated organic molecules.‡ It will be recalled that in Section 10.3 energy integrals were encountered:

$$\int \psi_a \mathcal{H} \psi_b\,\delta v$$

and it is these that are important in Hückel theory. In this application, the orbitals ψ_a and ψ_b are p_π orbitals and so, in cyclobutadiene, they are the

---

† When the functions obtained in the previous section were obtained it was assumed the overlap between p_p orbitals on adjacent carbon atoms in cyclobutadiene is zero. It would therefore scarcely be convincing to use an overlap model at this point!
‡ It is now commonly extended, in a purely numerical form, to inorganic molecules too.

orbitals a, b, c and d of Figure 11.1. The energy of each of these orbitals, before each is involved in any interaction with its partners, is the same. This energy is conventionally designated α. For the orbital a we have, then

$$\int a \mathcal{H} a \, \delta v = \alpha$$

with similar expressions for b, c and d. The energy of interaction between adjacent p_π orbitals is called β. So, the interaction between a and b is

$$\int a \mathcal{H} b \, \delta v = \int b \mathcal{H} a \, \delta v = \beta$$

with similar expressions for the pairs b/c, c/d and d/a. Those p_π orbitals which are not adjacent are assumed not to interact so that, for instance,

$$\int a \mathcal{H} c \, \delta v = \int c \mathcal{H} a \, \delta v = 0$$

and similarly for b/d.

To obtain the energy, within the Hückel model, of the A combination

$$\psi(A) = \tfrac{1}{2}(a + b + c + d)$$

we simply have to evaluate

$$\int \psi(A) \mathcal{H} \psi(A) \, \delta v = \tfrac{1}{4} \int (a + b + c + d) \mathcal{H} (a + b + c + d) \, \delta v$$

Expansion of the right-hand side of this expression and substitution of α, β and 0 as appropriate for the resulting integrals gives the energy of $\psi(A)$ as

$$\mathscr{E}[\psi(A)] = \alpha + 2\beta$$

Problem 11.7   (i) Show that the energy of $\psi(A)$ is $\alpha + 2\beta$. (ii) Show that the energy of $\psi(B)$ is $\alpha - 2\beta$.

As the above problem should have demonstrated it is a simple matter to show that the energy of the $\psi(B)$ orbital is

$$\mathscr{E}[\psi(B)] = \alpha - 2\beta$$

but that of $\psi_1(E)$ is a little more difficult. This is because the form of the energy expression appropriate to complex functions has to be used. This is

$$\int \psi_a^* \mathcal{H} \psi_b \, \delta v$$

In our case we take ψ_b to be $\psi_1(E)$ then ψ_a^* is its complex conjugate, that is, $\psi_2(E)$. It follows that we have to evaluate

$$\mathscr{E}[\psi_1(E)] = \tfrac{1}{4} \int (a + ib - c - id) \mathcal{H} (a - ib - c + id) \, \delta v$$

On expansion of this expression all of the complex quantities disappear.

Problem 11.8   Show that the energy of $\psi(E)$ is α.

The energy of $\psi_2(E)$, which is given by

$$\mathscr{E}[\psi_2(E)] = \tfrac{1}{4}\int (a - ib - c + id)\mathcal{H}(a + ib - c - id)\,\delta v,$$

will be the same as that for $\psi_1(E)$ because the right-hand side of this expression on expansion is identical to that for $\psi_1(E)$. We have, then,

$$\mathscr{E}[\psi_1(E)] = \mathscr{E}[\psi_2(E)] = \alpha$$

Evidently, for the present problem at least, it is entirely reasonable that $\psi_1(E)$ and $\psi_2(E)$ should be called 'degenerate'. Actually, this degeneracy between them is general—the algebraic expressions obtained for their energies were identical and so the degeneracy did not result from the Hückel approximations, which came later.

Because the interaction between two of the p_π orbitals is one that leads to a stabilization—it requires more energy to ionize an electron from a stabilized orbital than from an isolated p_π orbital—the energy β is negative (as too is α, but as its contribution to all of the energy levels is the same its value does not affect the relative order of orbital energies). It is concluded that the relative energies of the π molecular orbitals of cyclobutadiene are those given in Figure 11.3. There are four p_π electrons—one from each carbon atom— located in these orbitals and so we conclude that in the most stable arrangement they will be distributed as shown in Figure 11.3, the degenerate E orbitals containing one electron each; these two electrons will, in the ground state, have parallel spins (the maximum spin multiplicity principle). The total π electron stabilization, compared to four carbon p_π orbitals of energy α, is 4β (2β from each electron in the A orbital).

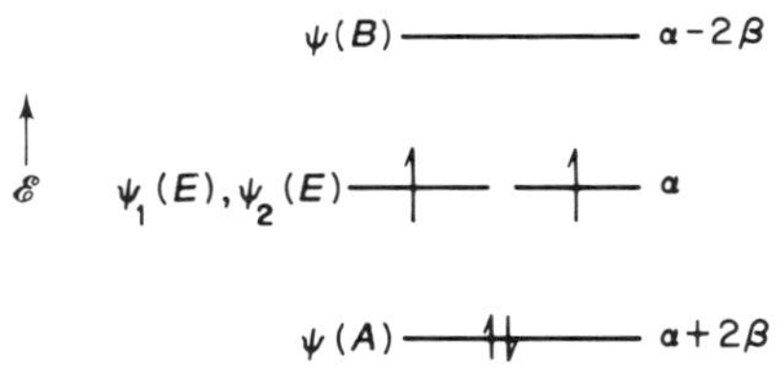

Figure 11.3 Relative energies of the π molecular orbitals of cyclobutadiene in the Hückel approximation.

Suppose that, instead of delocalized π system, we had two localized, non-interacting, π bonds—that is, suppose cyclobutadiene is rectangular, rather than square:

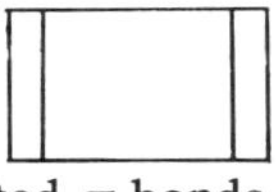

rather than

Each of the two isolated π bonds will have the form derived in Section 4.4 for ethene

$$\frac{1}{\sqrt{2}}(e + f).$$

The energy of this function is given by

$$\tfrac{1}{2}\int(e+f)\mathcal{H}(e+f)\,\delta v$$

which, on expansion and substitution on the Hückel values for the integrals, leads to an energy of

$$\alpha + \beta$$

There would be two such π bonding orbitals, each doubly occupied so that the total π electron stabilization would, again, be 4β. As far as the π electrons are concerned, at this level of approximation, there is nothing to choose between rectangular and square cyclobutadiene. Cyclobutadiene is a very reactive compound—it readily dimerizes (a reaction that can be discussed by the methods of the next section)—but it has been prepared at 35 K in an argon matrix. In fact, in the spin singlet ground state the molecule is rectangular with somewhat localized double bonds; in its spin triplet ground state it is square. For a simple discussion of this point see E. Heilbronner, 'Why Do Some Molecules Have Symmetry Different from that Expected?', *J. Chem. Educ.*, **66** (1989), 471.

Of the π-electron wave functions obtained working in the C_4 point group, two, $\psi(A)$ and $\psi(B)$ are identical in form to two that would have been obtained working in either C_{4v} or D_{4h} (the symmetry labels would have been different, of course). On the other hand, the $\psi_1(E)$ and $\psi_2(E)$ wavefunctions are different. The reason for this can be traced back to the existence of operations in C_{4v} and D_{4h} which have the effect of either mixing or interchanging the two degenerate functions. If the x and y axes, through the carbon atoms as shown in Figure 11.1 had been taken, then the $2\sigma_v$ mirror planes discarded in C_{4v} would have had the effect of interchanging x and y (and so also any functions transforming like them). Hence, they would have transformed *as a pair*. In contrast, there is in C_4 no operation which will interchange or mix $\psi_1(E)$ and $\psi_2(E)$—they are separate functions, although degenerate. The only way that we can, in C_4, obtain those E functions which would have been obtained in C_{4v} is to mix $\psi_1(E)$ and $\psi_2(E)$ although, of course, this is not permissible in the C_4 group itself. Taking the sum and difference of $\psi_1(E)$ and $\psi_2(E)$, the sum gives:

$$\psi_1(E) + \psi_2(E) = \tfrac{1}{2}(2a - 2c)$$

or, renormalizing

$$\psi_1'(E) = \frac{1}{\sqrt{2}}(a - c)$$

and the difference gives:

$$\psi_1(E) - \psi_2(E) = \tfrac{1}{2}(2ib - 2id)$$

or, renormalizing remembering that the complex conjugate of $(ib - id)$ is $(-ib + id)$ one obtains

$$\psi_2'(E) = \frac{1}{\sqrt{2}} (b - d)$$

The functions $\psi_1'(E)$ and $\psi_2'(E)$ are those E functions which would have been obtained working in C_{4v} (or D_{4h}).

Problem 11.9 Work in either the point group D_{4h} or C_{4v} and
(a) obtain the explicit forms of the four p_π molecular orbitals of cyclobutadiene,
(b) check that the doubly degenerate functions obtained have an energy of α within the Hückel approximation.

Before we finally leave the C_4 group there is one further point that should be made. In deriving the C_{2v} character table in Chapter 2 it was asserted that there is no other set of characters other than those considered there which, when substituted for the operations of the C_{2v} group in the group multiplication table would give a table which is arithmetically correct. The possibility of complex characters such as those which occur in the C_4 character table was not explored, although a hint of their existence was given in Problem 2.6. However, it is clear that substitution of a set of characters such as

$$1 \qquad i \qquad -1 \qquad -i$$

in the C_{2v} character table would not lead to a multiplication table which is arithmetically correct (because, for instance, when i multiples i it gives -1 on the leading diagonal rather than 1). It is evident from this discussion, and can be readily checked, that the multiplication tables of C_4 and C_{2v} are not isomorphic. Any operation in C_{2v} carried out twice leads to E, whereas in C_4 the C_4 and C_4^3 operations have to be carried out four times to give E. This, incidentally, explains the appearance of i in the C_4 character table. Because

$$C_2 \times C_2 = E$$

only characters of either 1 or -1 for the C_2 operation are possible for a singly degenerate irreducible representation (because $1 \times 1 = -1 \times -1 = 1$). In the C_4 group we also have that

$$C_4 \times C_4 = C_2$$

In other words, the character for the C_4 operation, squared, must give the character of the C_2. This presents no problems when the character for C_2 is 1, because that for the C_4 can then be either $+1$ or -1 (leading to the A and B irreducible representations of C_4). When the character for C_2 is -1, however, the only possibilities are that the character for the C_4 operation is either i or $-i$ (either of these squared gives -1), leading to the two components of the E irreducible representation.

One example of the application of symmetry to the energy levels of π-electron systems has just been given. There are many others, but, having established the principles and procedures involved, the subject will not be pursued further in detail. Suffice to say that the concept of aromaticity in organic chemistry is closely related to the type of stabilization arguments used when comparing square and rectangular cyclobutadiene. Roughly speaking, aromatic systems are those for which the delocalized system is more stable than any corresponding localized one.

11.6 SYMMETRY AND CHEMICAL REACTIONS

There have been many attempts to apply symmetry concepts to molecular reactions. This is a difficult area; it is necessary to assume some geometry for the key step in the reaction and often the only reasonable symmetry is low and so of little help. Further, large molecular distortions are usually involved in chemical reactions; that is, the molecules involved are vibrationally very excited. This has two consequences. First, the analysis given of vibrations in Chapter 9 evidently needs modification for large amplitude vibrations—when there are several symmetry-related atoms in a molecule the evidence is that one bond breaks before the others, whereas the discussion of Chapter 9 would lead us to expect several bonds to break simultaneously. This is akin to the failure of simple molecular orbital theory at large internuclear distances (it predicts a mixture of dissociation products), a failure which is also of relevance. Second, as was seen at the end of Chapter 9, the actual group which is applicable to the problem usually differs from the point group which, formally, is being used.

By far the most fruitful of the applications of symmetry to molecular reactivity has been the symmetry correlation method introduced by Woodward and Hoffmann and which is applicable to many organic reactions. A simple example of the application of their approach will be given, although the formalism which will be used was introduced by other workers. Consider the possible reaction of two ethene molecules to give cyclobutane, a molecule which, for simplicity, will be assumed to be planar

Written like this, it seems a perfectly feasible reaction, yet it is not one that readily occurs; the question then is 'why does it not occur?' The answer is not difficult to find. Pictorially, place two ethene molecules close together so that they are just about to react. The 'before' and 'after' reaction bonding arrangements are shown in Figure 11.4. The actual symmetry shown in Figure 11.4 is D_{2h}, but it is common to work in C_{2v}, so that the geometrical constraints on the molecular arrangement are not as rigid as required by D_{2h} symmetry. If symmetry constraints arise in C_{2v} (as they do) they are likely to be yet more severe in D_{2h}. Working in C_{2v} and choosing the C_2 axis as shown in

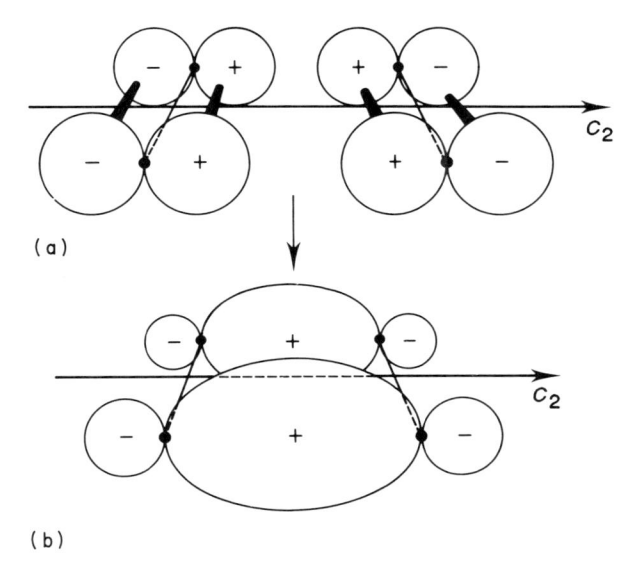

Figure 11.4 (a) Each π-bonding orbital of each C_2H_4 molecule transforms in C_{2v} as A_1. (b) The two new σ-bonding orbitals in C_4H_8 transform together in C_{2v} as $A_1 + B_1$.

Figure 11.4, it is easy to show that the symmetry species subtended by the two π bonding molecular orbitals in the two ethene molecules shown in that figure is $2A_1$. It is these two π orbitals that are involved in the reaction and that are assumed to smoothly become the two new C–C σ bonds as the reaction takes place. These two new C σ bonds give rise to the symmetry species $A_1 + B_1$. This is not the same as those generated by the π orbitals with which the problem started. There is a discontinuity; the π bonds cannot smoothly become the new σ bonds and so a ready reaction is not to be expected.

Let us look at this further by asking whether there exist any B_1 orbitals in the two ethene molecules? The answer is 'yes'. There are two of them and they are derived from the two π antibonding orbitals of the two ethene molecules (Figure 11.5). Correspondingly, the σ antibonding orbitals corresponding to the two newly formed C–C σ bonds in cyclobutane have symmetries $A_1 + B_1$ (Figure 11.5). We are led to the orbital correlation diagram shown in Figure 11.6 which shows the correspondences between the 'before' and 'after' reaction orbital patterns. In this figure the detailed pattern of σ orbital energy levels in cyclobutane has been obtained using the nodal pattern method of determining relative related energy levels met in the early chapters of this book—the more nodal planes, the higher the energy.

Problem 11.10 Use the nodal criterion (used, for example, in Section 4.7) to show that it is reasonable to expect both a_1 levels in Figure 11.6 to be more stable than the corresponding b_1 levels.

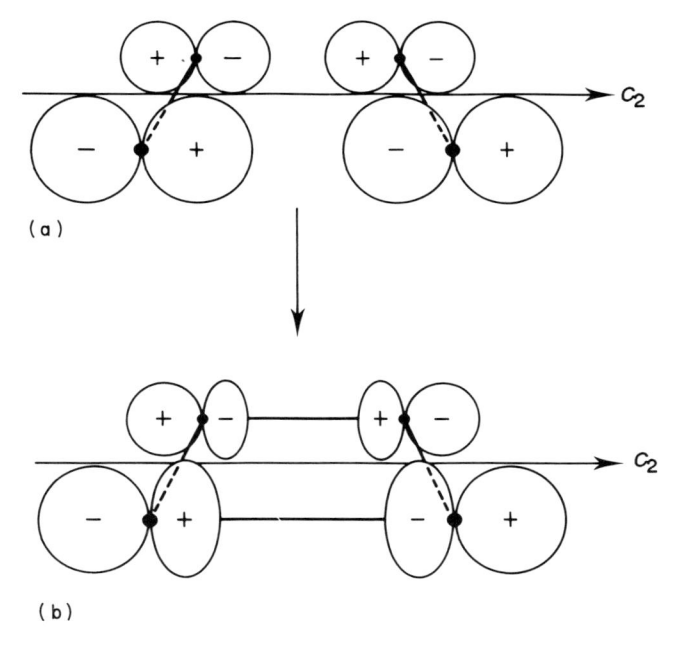

Figure 11.5 (a) Each π-antibonding orbital of each C_2H_4 molecule transforms in C_{2v} as B_1. (b) The two new σ-antibonding orbitals in C_4H_8 transform together in C_{2v} as $A_1 + B_1$. (The diagram shows *two* orbitals; if regarded, however, as a single symmetry-adapted orbital it is the B_1. The A_1 is obtained by changing the phases of all lobes of the σ-antibonding orbital at the 'front' of the diagram.)

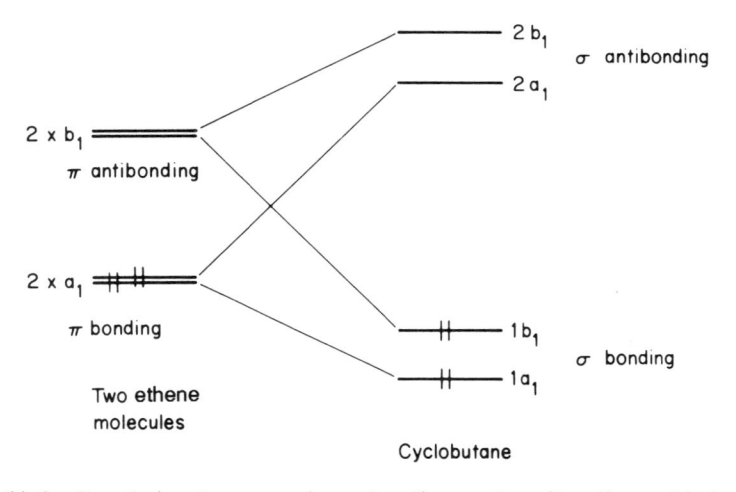

Figure 11.6 Correlation between the π-bonding and antibonding orbitals of two ethene molecules and the σ bonding and antibonding orbitals of cyclobutadiene.

It is clear from Figure 11.6 that there will be no strong bonding driving force to form cyclobutane from two ethene molecules. As the reaction proceeds the lowest energy orbitals would be expected to be filled. Up to the energy level crossing-over point in the middle of Figure 11.6 this means that the two A_1 orbitals will be filled. But while the stability of one increases with decreasing separation between the two ethenes (as a π bond becomes a σ bond) that of the other will decrease rapidly (as a π bond becomes a σ antibonding orbital) so that no reaction is to be expected.

There is an alternative approach to this problem, an approach which is based on states rather than orbitals. The ground state electronic configuration of two ethene molecules is $(a_1)^2(a_1)^2$, a configuration which gives rise to a $^1A_1$ term (we shall be concerned with spin singlet terms throughout the following discussion). A configuration such as $(a_1)^2(b_1)^2$ is an excited state configuration but also gives rise to a $^1A_1$ term (as is readily seen since the quadruple direct product $A_1 \otimes A_1 \otimes B_1 \otimes B_1 = A_1$). In cyclobutane the situation is reversed. The ground state configuration (considering only the newly formed σ orbitals) is $(1a_1)^2(1b_1)^2$. In contrast, $(1a_1)^2(2a_1)^2$ is an excited state configuration. The important thing is that both of these configurations give rise to $^1A_1$ terms. The appropriate *term correlation diagram* is shown in Figure 11.7, where the *non-crossing rule* has been invoked (terms of the same symmetry species only cross in very rare circumstances). Physically, this application of the non-crossing rule in the present example arises because electron repulsion favours electrons being as spatially separated as possible and the energy gained from this separation can contribute the energy apparently required to promote an electron to a higher orbital. Figure 11.7 demonstrates rather more clearly than does Figure 11.6 that ethene should not be expected to spontaneously dimerize to cyclobutane. Even here, our discussion is somewhat simplified but it does correctly indicate that one can sometimes be misled by a simple 'filling of the

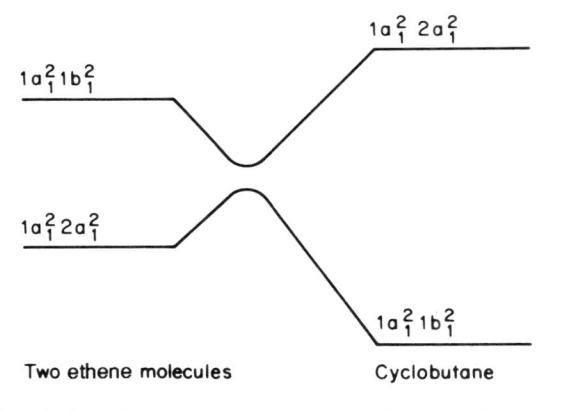

Two ethene molecules   Cyclobutane

Figure 11.7 Correlation between $^1A_1$ terms in the dimerization of two ethene molecules to give cyclobutane.

lowest orbitals' approach to chemical bonding. It might be appropriate to remind the reader that the complicating effects of repulsive forces on simple pictures of chemical bonding were also encountered in the first chapter of this book.

The discussion which has been presented above can readily be extended to photochemically-induced reactions (that is, reactions involving electronically excited molecules). Many very readable accounts of the topic have been written but these tend to use symmetry arguments in a rather less formal manner than the present text. Commonly, particular symmetry operations are selected and orbital behaviour classified as either A (antisymmetric) or S (symmetric) under these operations; these labels are equivalent to the characters -1 and 1 used in this book.

One final cautionary note. In this discussion the concern has been with a single reaction mechanism. Other mechanisms may exist which provide an alternative, and more accessible, route to a particular product. Thus, although ethene does not dimerize to cyclobutane, reaction between the ethene derivatives $CH_3(H)C=C(H)OC_2H_5$ and $(CN)_2C=C(CN)_2$ proceeds smoothly at room temperature to give the corresponding cyclobutane derivatives. In this case there is evidence that a zwitterion intermediate, $(CN)_2C^{\ominus}-C(CN)_2-CH(CH_3)-{}^{\oplus}CH(OC_2H_5)$ is formed. Symmetry arguments are powerful, but nature may be yet more cunning and have unexpected tricks!

11.7 SUMMARY

Discussion of the π orbitals of cyclobutadiene has provided a relatively simple example of the use of a group containing complex quantities in its character table (p. 230). It is necessary in such cases to work with complex conjugate basis functions and an example was provided in deriving the Hückel energies of cyclobutadiene (pp. 233, 235). The fact that the C_4 group contains no S_n operations enabled a discussion of optical activity (p. 231). Molecules having symmetries without such axes are, in principle, optically active (p. 231). Finally, it was shown that symmetry correlations can give insight into some chemical reactions (p. 240). Correlations between molecular terms may be preferable to simple orbital correlations (p. 243).

12
Space Groups

12.1 THE CRYSTAL SYSTEMS

Crystals and crystal structure determinations are well known and important areas of chemistry and in this chapter it will be assumed that the reader has some familiarity with them. Although the solid state is becoming an increasingly important part of chemistry, solid state symmetry is a much neglected area, leading to conceptual gaps which make the subject much more difficult than need be the case. To meet this problem, in the present chapter some subject areas will be explored which are normally given a very brief treatment; the relationships between the seven crystal systems, the 14 Bravais lattices, the 32 crystallographic point groups and the 230 crystallographic space groups. An attempt will be made to answer the fundamental question 'why are there 230 space groups and not some other number'?

A convenient starting point for the discussion is the concept of an empty lattice. In talking about crystal structures one expects to be concerned with the arrangement of atoms in space and the way that they fill unit cells. Despite this very reasonable expectation, our starting point will be quite different; no atoms, no unit cells, just the fiction of an empty lattice. The reason for this approach is that it will enable the introduction of atoms into the lattice as a step which is quite distinct from anything to do with the lattice itself. As will be seen, the step of introducing atoms adds possibilities which do not exist for the bare lattices. What, then, is a 'bare lattice' (and why is it a fiction)? For our purposes it is convenient to regard a lattice as arising from a three-dimensional network of vectors, vectors that connect equivalent points in an empty space. Strictly, it is the points that form the lattice, although group theoretically the vector set is more important (the members of this set correspond to the operations of the group of translations which turns the lattice into itself). In empty space all points are equivalent, they form a continuum, each point being equivalent to all of the (infinite number) of neighbours as well as being equivalent to all more distant points—and this is why the concept with which we shall work is basically a fiction, for all our equivalent points are separated from each other by some finite distance. The vectors connecting these separated and equivalent points fill all of three-dimensional space in the manner indicated in Figure 12.1(a); they look rather like sets of parallel fishing nets, the corners of the mesh (and it is these corners that are the points of the lattice) in any one net being linked by additional pieces of string sideways to the corresponding

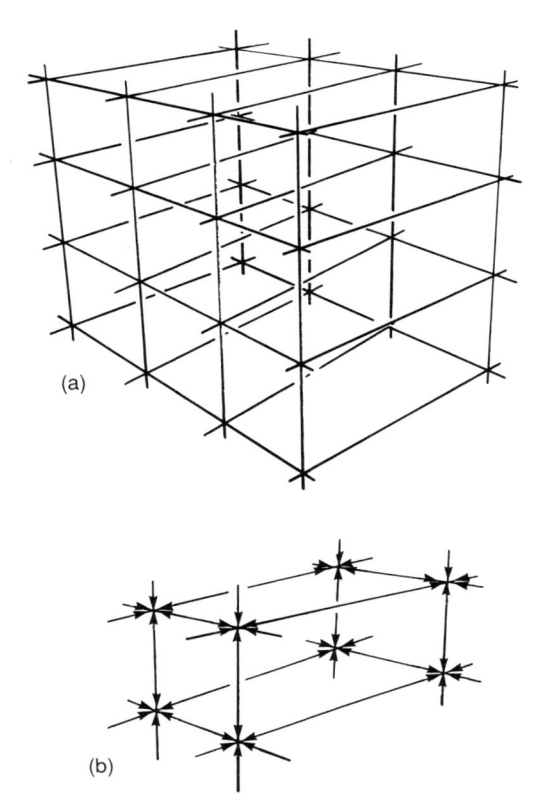

Figure 12.1   (a) A lattice typical of a crystal. Equivalent points in space are linked by sets of equally spaced parallel lines extending to infinity. For simplicity of construction, in this figure all lines are drawn as intersecting at right angles but this is not a general requirement.
(b) A single segment of the lattice of (a). Each line in (a) is really two superimposed vectors, one of the negative of the other, represented here by an arrow-head at either end of each line.

adjacent corners of the nets on either side of it. It is as if there were a three-dimensional array of string-edged boxes filling all space. It is tempting to call these boxes 'unit cells', and, indeed, this is what they would normally be called. However, this temptation will be resisted—as has been said earlier, no atoms, no unit cells. Rather, an arrowhead will be attached at each end of every piece of string to demonstrate that sets of *translation vectors* are under consideration; this has been done in Figure 12.1(b). Each vector has a negative, which is why arrowheads are needed at each end of each segment of string, at each equivalent lattice point. Our concern is to enumerate all the possible different symmetries that can be spanned by such sets of lattice points, and of course the translation vectors. This will lead us to the seven crystal systems.

The argument that follows is directed at, and limited to, finding the symmetries of all those lattices conventionally called 'primitive', for there is one for each crystal system.

> **Problem 12.1**   The subject matter of the following paragraphs is unlikely to be new to the reader, although it is equally unlikely that it will have been well understood. Write down, briefly, that which you know about the seven crystal systems. This should help focus attention on any problem areas that exist.

It is natural to start with a set of translation vectors in which the vectors are all of the same length and all mutually perpendicular; this is the most symmetrical arrangement possible. A cubic lattice results (Figure 12.2).† The

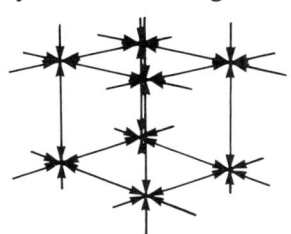

Figure 12.2   A segment of a cubic lattice drawn in the manner of Figure 12.1(b). This segment has O_h symmetry; this is the symmetry of the point at the centre of the cube shown (although this point is not itself indicated) and also the symmetry of each point at which the vectors meet.

fact that there is an infinite set of vectors means that it is sensible to pause before passing from the statement that 'it is a cubic lattice' to the statement that 'the lattice has O_h symmetry'. It is true that in this lattice each lattice point has O_h symmetry. That is, the operations of this O_h point group turn a given vector either into itself or into an equivalent vector. However, if one were to choose another point in space other than a lattice point, the symmetry would usually be different, although the lattice would still be turned into itself by the relevant symmetry operations. For example, points 'along' the vectors have C_{4v} symmetry, except the mid-points of such vectors, which have D_{4h} symmetry. However, once these problems inherent in the statement have been recognized one can say that 'the lattice has O_h symmetry'. At the end of the present chapter a particular choice of unit cell will be described which makes the lattice symmetry unambiguously clear.

Having established the symmetry of the most symmetrical of lattices, it seems reasonable to hope to obtain all the other possible lattices by reducing the symmetry. Unlike their string counterparts, vectors can be continuously deformed—stretched, contracted and re-orientated. The most natural way of proceeding is to focus on the O_h symmetry of the cubic lattice and to lower that symmetry by considering all the subgroups of O_h. This is the path that will be followed but first a basic and important fact. In Figure 12.1(b) each vector was

---

† The reader can be forgiven for relating this arrangement to the properties of the primitive cubic unit cell; note, however, one danger in this association—it leaves unanswered the question of just how the body-centred and face-centred cubic unit cells arise.

drawn as two-headed because both a vector and its negative interrelate equivalent lattice points. Such a statement holds for each and every vector and so for each and every sum of vectors, no matter how twisted and convoluted the path traced in three-dimensional space by such a sum—the resultant vector and its negative interrelate equivalent lattice points. There is a much more succinct way of stating this. It is that the lattice is centrosymmetric. This statement is true of all the translation vector sets, no matter how low their symmetry otherwise. *All lattices are centrosymmetric.* It is certainly not true to say that all unit cells are centrosymmetric and so a good reason for excluding the phrase 'unit cell' from the discussion is evident. The fact that the packing of atoms within a lattice can destroy the inherent centrosymmetry of that lattice will arise naturally when atoms are added. This is the reason that consideration of atoms is temporarily excluded from the discussion. It also simplifies the problem of finding the lattices which can be obtained from the cubic by a reduction in symmetry. Each acceptable symmetry will be that of a point group, a subgroup of O_h. The inherent lattice centrosymmetry means that all subgroups of O_h which lack a centre of symmetry can immediately be excluded. Further, of the groups which remain, only those which are consistent with a lattice are possibilities. What does this mean? Suppose the lattice has a C_n rotation axis, $n > 2$ (C_2 are special and necessitate a separate discussion). There cannot be just a single C_n axis, there must be an infinite set of them, regularly arranged in space—and all parallel. The regular arrangement arises because associated with each axis there will be a set of translation operations,

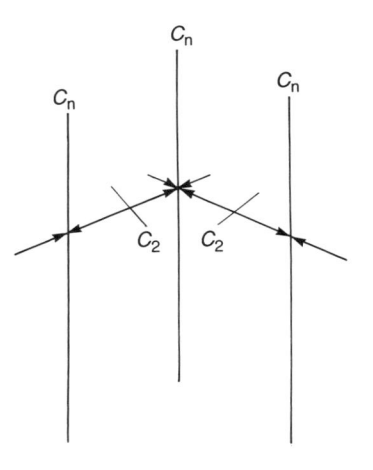

Figure 12.3 Perpendicular to a set of C_n axes there will be sets of translation vectors, as shown in this figure. In a plane perpendicular to the C_n axes and bisecting the translation vectors shown, will be sets of C_2 axes. For simplicity, and to emphasize the fact that translation vectors do not *have* to coincide with rotation axes, arrow-heads are only shown on the translation vectors, not on the rotation axes.

some along each C_n axis and others perpendicular to it—the double-headed arrows of the discussion above. The double-headed nature of these arrows along the C_n directions means that perpendicular to the C_n axes there will be C_2 axes and that these C_2 axes are interrelated by the C_n (Figure 12.3). This means that subgroups of O_h without such sets of axes must be excluded (whilst the C_n axes persist). Which groups are left? Relatively few. D_{4h} is the highest containing a fourfold rotation; D_{3d} is the highest with a threefold (we shall see that it is these highest that we need to concern ourselves with; others come in later). What of the n = 2 case, is it the same? The key difference between C_n, n > 2, and the n = 2 case is that for the latter, a perpendicular C_2 axis is rotated onto *itself* by the 'original' C_2, whereas for C_n, n > 2, it is rotated into a different C_2 axis. In the latter cases, therefore, the mutual compatibility of all the C_2 axes is assured. This is not so for the case where n = 2. They can be compatible—and this is the situation for the D_{2h} group, which conforms to the pattern we have so far required of acceptable subgroups, containing three mutually perpendicular twofold rotation axes. Alternatively, the perpendicular twofold axes can be mutually incompatible and thus destroy each other. This mutual destruction occurs whenever the angle between the two sets of additional 'twofold' axes is not 90°. The point is illustrated in Figure 12.4 which shows a rectangular grid, a two-dimensional lattice. In it the 'original' C_2 axes carry this simple label whereas those perpendicular to the first set, assumed to arise because of the double-headed arrow pattern (although which 'come first' is, of course, entirely arbitrary) are labelled C_2 (perp). Now attempt to add a third set of twofold axes perpendicular to those labelled C_2. If they are

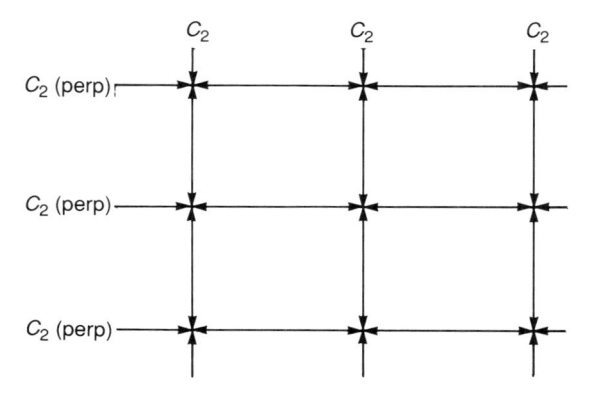

Figure 12.4 A layer of mutually perpendicular—and mutually compatible—twofold rotation axes, labelled C_2 and C_2(perp). To extend the structure into three dimensions, a third set of C_2 axes is added, perpendicular to the 'original' C_2 and through the points of intersection of C_2 and C_2(perp). Unless these 'new' axes are perpendicular to both C_2 and C_2(perp) then, although the 'new' are compatible with the 'original' C_2, they are not compatible with the C_2(perp) and mutual self-destruction occurs, leaving only the 'original' C_2.

added through the points of intersection of the already-existing twofold axes and perpendicular to the plane of the paper then the D_{2h} case is obtained. If they are not added perpendicular to the plane of the paper although perpendicular to the 'original' C_2, then their addition leads both to self-destruction and to the destruction of the C_2(perp). The highest group that can be obtained in this situation is C_{2h}, which is centrosymmetric and contains a single twofold rotation operation. Destruction of that twofold axis by relaxing the requirement that additional axes be perpendicular to the first (an inescapable requirement for C_n axes, $n > 2$) leads to the last subgroup of O_h that has to be considered, C_i, which contains no operation except inversion in a centre of symmetry, the lowest symmetry that a lattice can have. Provided that all lattices can be obtained by reduction in symmetry from O_h, all have been generated. Alas, not all lattices can be obtained by such a reduction in symmetry; since O_h contains no sixfold rotation operation, a point group containing one can never be obtained starting with it. Yet there is a perfectly good lattice that is based on a point group that contains such operations. If this is so, it at once prompts the question 'how many others have been missed?'. Really, this is a question about whether there are any other point groups with symmetry operations not present in O_h which can give rise to acceptable lattices. Clear candidates are groups with fivefold and sevenfold rotation operations. Can they be excluded (and any others that may well occur to the reader)? The answer is 'yes', as is now demonstrated.

Consider Figure 12.5, which shows a two-dimensional lattice. Clearly, any constraints that apply to this lattice must apply to a three-dimensional lattice also, for any three-dimensional lattice may be regarded as linked networks of two-dimensional lattices, just as the fishing nets were linked at the beginning of this section (remember, the head-to-head linking of vectors in the manner of Figure 12.1 leads to straight line arrays, never to kinks or bends). As we are looking for all rotation axes that are compatible with translation symmetry, we simply require that the lattice points of Figure 12.5 be interrelated both by pure translations and by pure rotations and enquire into the compatibility of these two requirements. With no loss of generality for $n > 2$, assume that the C_n axes are perpendicular to the page and that one passes through each of the points indicated. The points A and B in the top row are separated, in general, by an

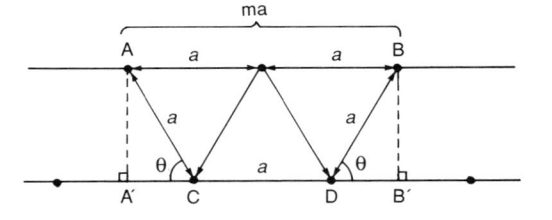

Figure 12.5 A fragment of a two-dimensional lattice, used in deducing which C_n axes through the dots and perpendicular to the lattice are compatible with the lattice. In this diagram the case $m = 2$ is shown but in principle m can assume any integer value.

integral number, m, of translation steps, a, and so by a distance of ma (the points C and D are separated by a single translation step, a, and so m will be a small number. For simplicity, in the diagram the case with $m = 2$ is shown). For purposes of the argument, the row of points below the top may be regarded as free to slide around, subject only to the requirement that the surrounding points are interrelated by the C_n axes through the points of the second row. This is indicated by the arrowheads emanating from the points in the second row, the angle of rotation of the C_n axis being taken as θ so that, from Figure 12.5:

$$AB = A'B' = a + 2a \cos \theta = ma$$

so that
$$2a \cos \theta = (m - 1)a$$

and
$$2 \cos \theta = (m - 1)$$

$$\cos \theta = \frac{(m - 1)}{2}$$

Now, m is an integer and $\cos \theta$ can only have values from 1 to -1. The possible solutions are therefore rather limited and are detailed in Table 12.1 below.

Table 12.1

Value of m	Value of $\cos \theta$ $\left(\text{this equals } \dfrac{(m-1)}{2}\right)$	Comments
4	1.5	no solution possible
3	1	$\theta = 0$ or $360°$ corresponding to a C_1 axis
2	0.5	$\theta = 60°$ corresponding to a C_6 axis
1	0	$\theta = 90°$ corresponding to a C_4 axis
0	-0.5	$\theta = 120°$ corresponding to a C_3 axis
-1	-1	$\theta = 180°$ corresponding to a C_2 axis
-2	-1.5	no solution possible

Problem 12.2 Work through the above argument but instead of choosing $m = 2$, as in Figure 12.5, take $m = 3$.

Clearly, the only non-trivial rotation operations that are consistent with a lattice are C_2, C_3, C_4 and C_6. This means that the only case not covered by a reduction in symmetry from O_h is that of the C_6 rotations. The highest centrosymmetric point group containing a C_6 axis which also satisfies the other requirements discussed above is required. This is D_{6h} and it completes the list of acceptable lattices because there is no lattice that can be obtained from it which satisfies all the requirements and which has not already been generated.†

The seven point groups that have been obtained, together with their associated lattices, characterize the seven *crystal systems*. For convenience, these are listed in Table 12.2. One point about this table is to be noted. This is that the characteristics of each *lattice* have been detailed. Such compilations of characteristics are found in many texts but frequently under a different heading, that of 'unit cells'. In such texts the unit cell is chosen as the smallest parallelepiped defined by the translation vectors. There are two reasons that this precedent is not followed here, quite apart from the fact that the name 'unit cell' has been avoided. These are that, first, the translation vectors are the fundamental quantities and it is better to recognize them as such. Second, as will be seen, there is no unique choice of unit cell for any crystal structure—

Table 12.2

Crystal (lattice) system	Characteristic point group	Lattice vector characteristics
Cubic	O_h	$a = b = c$; $\alpha = \beta = \gamma = 90°$
Tetragonal	D_{4h}	$a = b = \neq c$; $\alpha = \beta = \gamma = 90°$
Orthorhombic	D_{2h}	$a \neq b \neq c$; $\alpha = \beta = \gamma = 90°$
Monoclinic	C_{2h}	$a \neq b \neq c$; $\alpha = \gamma = 90°$, $\beta > 90°$
Triclinic	C_i	$a \neq b \neq c$; $\alpha \neq \beta \neq \gamma$
Hexagonal	D_{6h}	$a = b \neq c$; $\alpha = \beta = 90°$, $\gamma = 120°$
Trigonal	D_{3d}	$a = b = c$; $\alpha = \beta = \gamma \neq 90°$

The quantities a, b and c are the absolute magnitudes of the three primitive translation vectors that define the lattice. The angles complement the axes; thus α is the angle between the vectors associated with b and c, β that between the vectors associated with c and a (it helps avoid hidden problems if one is consistent in the order in which axes are listed by being cyclic: $a \rightarrow b \rightarrow c \rightarrow a$). Note that in the monoclinic system there is conventionally a departure from the system followed for point groups where the axis of highest symmetry is chosen as z (the choice of β as the unique angle means that the y axis is chosen as the C_2 in monoclinic systems). The sixfold axis is not immediately evident in the vectors defining the hexagonal system but when they are used to define a lattice—and, so, an extended lattice generated—the sixfold axis pattern becomes evident (it is along the vector associated with c). However, the threefold axis is not along any of the vectors used to define the trigonal lattice—the vectors associated with a, b and c are interrelated by the threefold axis.

---

† Although it comes equally well from O_h, it is often convenient to think of D_{3d} as a subgroup of D_{6h}.

just some choices that are more convenient for some purposes than others (the choice that is convenient for X-ray crystallography may be inconvenient for other purposes).

All of the seven three-dimensional lattices (or, rather, remembering the context in which they were sought, the seven crystal systems) which arise from translationally-related repeat units have now been obtained. The relevance to crystal structures is clear. However, we should not close our eyes to other possibilities. For instance, suppose the repeat layers were not related one to another by simple translation operations. What if they were arranged in a spherical layer fashion, much as the layers of an onion? Perhaps such crystals could be formed if a nucleating unit led to crystal growth by successive spherical shells being added. As the shape of the popular form of a soccer ball shows, such a crystal could have fivefold rotation axes (six of them, all passing through the point which was the original crystal nucleus, giving a crystal with icosahedral symmetry), in complete contrast to the discussion above, where it was found that C_5 axes are not possible for a regular array of translationally-related units.

12.2 THE BRAVAIS LATTICES

In the preceding section the concept of a lattice was explored. It was found that all three-dimensional lattices have to conform to one of seven symmetry types, each characterized by a unique centrosymmetric point group, and that these are normally spoken of as 'the seven crystal systems'. However, there is more to say on the topic of lattices, even the topic of empty lattices. The question is this: 'is it possible, for any of the seven lattices, that a second, identical, lattice be taken and interpenetrated into the first to give an arrangement which retains the symmetry of the first lattice?' The question may be phrased in a rather less accurate but more colourful way. Suppose the first lattice were made of red string. Is it possible to construct within it a displaced but otherwise identical lattice made of blue string which does not destroy the symmetry pattern of the red? Such questions are inadmissible because any given crystal structure can only have a single lattice† (and so talk about 'interpenetrating lattices', while sometimes useful for teaching purposes, cannot accurately describe reality). However, having recognized the error, let us continue to sin—for teaching purposes! The question is illustrated in Figure 12.6 for the cubic lattice. The lattice generated in Section 12.1 is shown in Figure 12.6(a); it is shown again together with an identical interpenetrating lattice (showed dotted rather than

---

† This statement originates in the fact that there is a single translational subgroup of the space group. In discussions it is often very convenient to talk of the lattice composed of one set of atoms and to relate it to the lattice composed of another set. Such language is convenient rather than strictly accurate.

coloured) in Figures 12.6(b), (c) and (d). In Figure 12.6(b) the additional lattice is placed in an arbitrary position. Not surprisingly, the symmetry of the first lattice is destroyed as, too, simultaneously, is that of the second—the combined lattice is probably of no higher symmetry than C_i—it is difficult to be sure without a more detailed specification of the positioning of the two lattices in Figure 12.6(b). In contrast, in Figure 12.6(c), were it not for the fact that one is shown dashed and the other with solid lines, one would not know which lattice is the 'original' and which the 'added'. The two lattices are arranged in a mutually compatible manner but have no points in common. The combined arrangement is of O_h symmetry. Because of the relationship between the two sets of translation vectors, this new lattice is called 'the *body-centred cubic* lattice' (a diagonal of the 'solid-line' lattice is shown dotted; the 'dashed' lattice has a point lying at the mid-point of this diagonal). However, as has been indicated above, it is incorrect to think of the body-centred cubic as defined by two sets of translation vectors. It is a single lattice, defined by a single set of translation vectors. Clearly, this set is neither of those used in the construction above but, in some manner, contains both. Discussion of the set will be deferred until later. Note, however, that the basic vectors of this single set cannot be mutually perpendicular—if they were, the pattern of the original—primitive cubic—lattice would be regenerated. In Figure 12.6(d) is shown another example of sets of interpenetrating cubic lattices which are mutually compatible but this time involving four such lattice sets (a solution to the question originally set which goes beyond the assumption in that question—that only a single additional lattice need be considered). These sets, together, define the *face-centred cubic* lattice. It is possible to define this lattice by a single vector set, but, again, the vectors are not mutually perpendicular.

All of the three possible variants of the cubic lattice, the primitive, the body-centred and the face-centred have now been generated. To see why there are no more, consider again the way the additional lattices were generated. In the body-centred lattice, the second set of lattice points were placed at what, for the 'original' lattice were positions of O_h symmetry. Clearly, the 'second' lattice was compatible with the 'first'. The case of the face-centred cubic was different. The lattice points of the 'first' lattice that were occupied by the 'second' were only of D_{4h} symmetry. Had just a single 'second' lattice been interpenetrated with the 'first' then the symmetry of the 'first' would have been destroyed and lattice points that were originally of O_h symmetry would have been reduced to D_{4h}. This problem was overcome by, effectively, adding a C_3 axis to this reduced D_{4h} symmetry and thus bring it back to O_h. This was achieved by adding three points around each 'original' lattice point, a step which required that three additional lattices, not just one, be added. It is less immediately obvious that this same step serves to turn each of the 'second' lattice points, originally of D_{4h} symmetry, into points of O_h symmetry. However, the fact that the combined lattice 'looks the same' whichever lattice

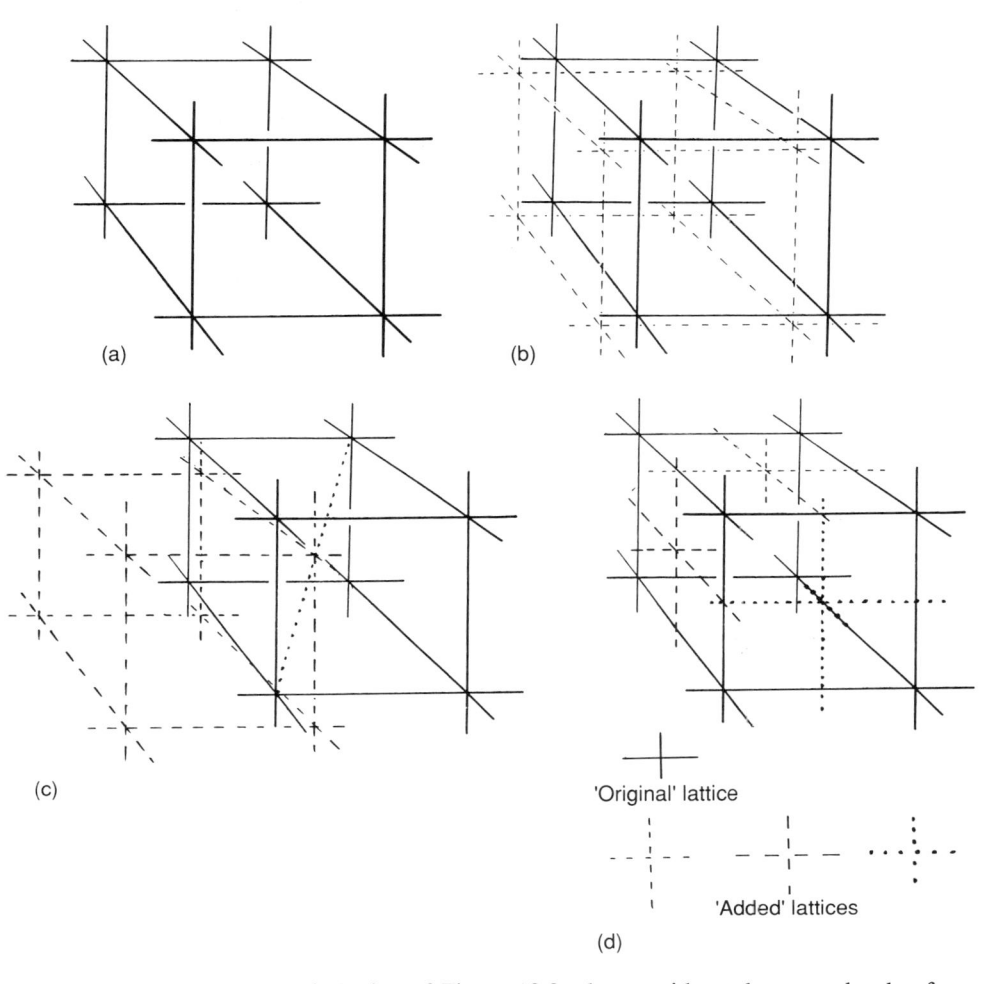

(a)

(b)

(c)

'Original' lattice

'Added' lattices

(d)

Figure 12.6   (a) The cubic lattice of Figure 12.2, shown without the arrow-heads of that figure.

(b) The lattice of (a) with an identical lattice (shown dashed) displaced from the 'first' by arbitrary translations (which means that the vectors associated with the two vector sets remain parallel). Except in very special cases such a 'second' lattice will destroy much of the symmetry of the 'first'. So, in this figure, the fourfold rotation axes of the 'first' lattice are destroyed by the presence of the 'second' and vice versa.

(c) One of the special cases of (b) occurs when the 'corners' of the 'second' lattice coincide with the centre of the cube defined by the first, because both have O_h symmetry. A diagonal of the 'first' lattice is shown dotted; the second lattice intersects the mid-point of this diagonal.

(d) A second special case occurs when *three* 'additional' equivalent lattices are added to the 'first'. In this figure the 'additional' lattices are shown small-dashed, large-dashed and dotted.

is called 'first' establishes the point. The question then arises as to whether the same sort of trick can be played a different way. Is it possible, for example, to add a 'second' lattice at points of D_{3d} symmetry of the 'first' and, by adding four of them in all, thus regenerate an O_h lattice? The answer is 'no'; the reason is that although there are unique points of D_{4h} symmetry in the 'original' lattice there are none of D_{3d}. All of the possible cubic lattice patterns have been generated.†

The type of argument developed in the preceding paragraphs can also be applied to the six other different possible lattice symmetries. When all of the lattice patterns consistent with each various lattice symmetry type have been obtained, there are 14 in all. They are known as the 14 *Bravais lattices*, named after the Frenchman who first recognized their existence. Note three things. First that our lattices are still empty. Second, that each of the 14 Bravais lattices is associated with a unique pattern of vectors—it is this vector pattern that distinguishes the various lattices. So, there are three qualitatively different vector sets that define cubic lattices. As has been mentioned above and will be seen later, the vectors defining a cubic lattice do not *have* to be mutually perpendicular. This point is an important one; despite it, all the Bravais lattices have the symmetry of the crystal system and so all three cubic Bravais lattices *are* cubic, and of O_h symmetry. However, unit cells constructed with the basic translation vectors of the Bravais lattice used to define their edges do not have the crystal symmetry unless the lattice is one of those conventionally labelled 'primitive'. This point is actually the same as the final one, one that anticipates our discussion somewhat but is included here because it will be rare for there to be a reader who has not encountered the content of this chapter elsewhere, at least superficially. The final point is to emphasize that *all* the fourteen Bravais lattices are primitive (and so the unit cells just mentioned also primitive). It is necessary to make this point with some considerable strength because of the names conventionally given to the Bravais lattices, which could be rather misleading—only half of them are actually called 'primitive'. For completeness it is convenient at this point to list all 14 Bravais lattices and this is done in Table 12.3.

Problem 12.3 Unconventionally, the name 'unit cell' has been avoided so far in this text. Make a short(!) list of the arguments given so far in support of this avoidance (weightier arguments will be given later).

One of the most evident things about Table 12.3 is its rather patchwork character—so, some crystal systems have body-centred lattices, others do not.

---

† The way that these have been generated demonstrates a limitation in the way that the seven crystal systems were derived earlier in the text. *All* of the cubic lattices have O_h symmetry—this is why the derivation of the seven crystal systems had to be confined to 'those lattices conventionally called primitive'. Strictly, the derivation of the seven crystal systems should have been concerned only with groups and subgroups but it was felt that a less abstract discussion would be more easily followed.

Table 12.3

Crystal system (point group)	Bravais lattice	Number of 'primitive' lattices needed in the construction
Cubic (O_h)	Primitive	1
	Body-centred	2
	(All) Face-centred	4
Tetragonal (D_{4h})	Primitive	1
	Body-centred	2
Orthorhombic (D_{2h})	Primitive	1
	Body-centred	2
	One-face-centred	2
	(All) face-centred	4
Monoclinic (C_{2h})[a]	Primitive	1
	(One) face-centred	2
Triclinic (C_i)	Primitive	1
Hexagonal (D_{6h})[b]	Primitive	1
Trigonal (D_{3d})[c]	Primitive	1

[a] In the monoclinic one face-centred lattice, the face which is centred is parallel to, not perpendicular to, the twofold axis.
[b] The hexagonal lattice is sometimes drawn showing a unit cell with a hexagon as a face, this face being centred by a lattice point. This is not a primitive unit cell (it is actually three times the volume of the primitive).
[c] One variety of trigonal lattice is referred to as 'rhombohedral'. This name arises from the shape of the corresponding unit cell as it is usually drawn. A rhombohedron contains eight edges, all of the same length and each face is diamond-shaped (essentially, the shape is that of a cube stretched or compressed symmetrically by pulling outwards or pushing inwards on a pair of opposite corners).

Some have one-faced-centred lattices, others do not. It has already been seen why there cannot be a one-face-centred cubic—it would not be cubic (actually, it would be a primitive tetragonal lattice). The reason for the non-listing of the other apparently 'missing' lattices is similar—they are each equivalent to one already listed in Table 12.3. In Table 12.4 is given a list of the 'missing' lattices together with their equivalents—which are present.

Earlier in this section the point was made that each and every one of the 14 Bravais lattices is both single (not interpenetrating, notwithstanding our derivation) and primitive, despite the fact that only seven are actually labelled as primitive. Figure 12.7 shows the vector set that gives rise to each of these lattices, together with an indication of the way that this vector set is related to the unit cells commonly used to picture the seven 'centred' Bravais lattices. Because in such cases the primitive translation vectors relate a given lattice point to its nearest neighbours, for the 'centred' lattices the members of this vector set relate a point in the 'first' lattice (in our derivation) to points in the added, interpenetrating (in our derivative), lattice(s).

This section has contained material which differs significantly in emphasis

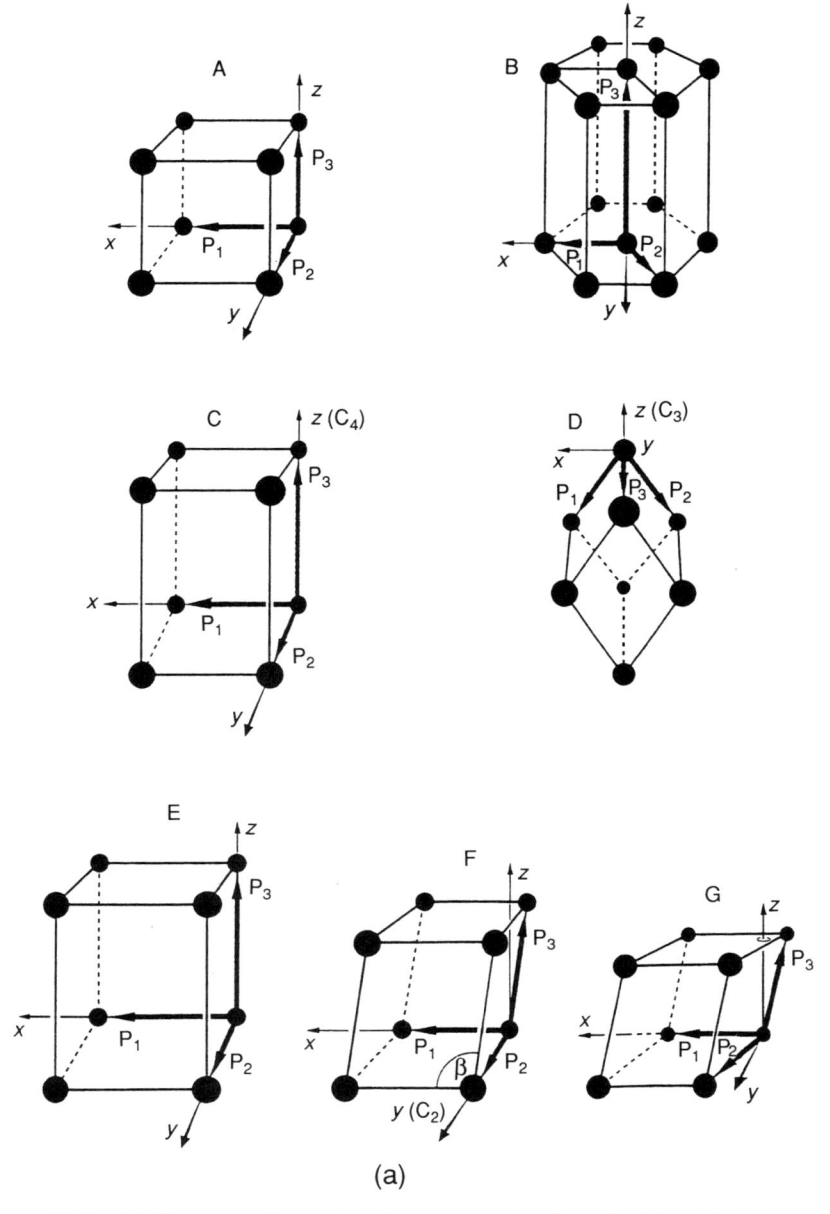

(a)

Figure 12.7 (a) The seven Bravais lattices conventionally called 'primitive', together with the associated primitive translation vectors (shown bold). Notice that only in three cases are the translation vectors in directions parallel to the Cartesian axes of the crystal. A, cubic, B, hexagonal; C, tetragonal; D, trigonal (rhombohedral); E, orthorhombic; F, monoclinic; G, triclinic. Reproduced by permission of the Journal of Chemical Education from Kettle and Norby, *J. Chem. Educ.*, **70** (1993) 959.

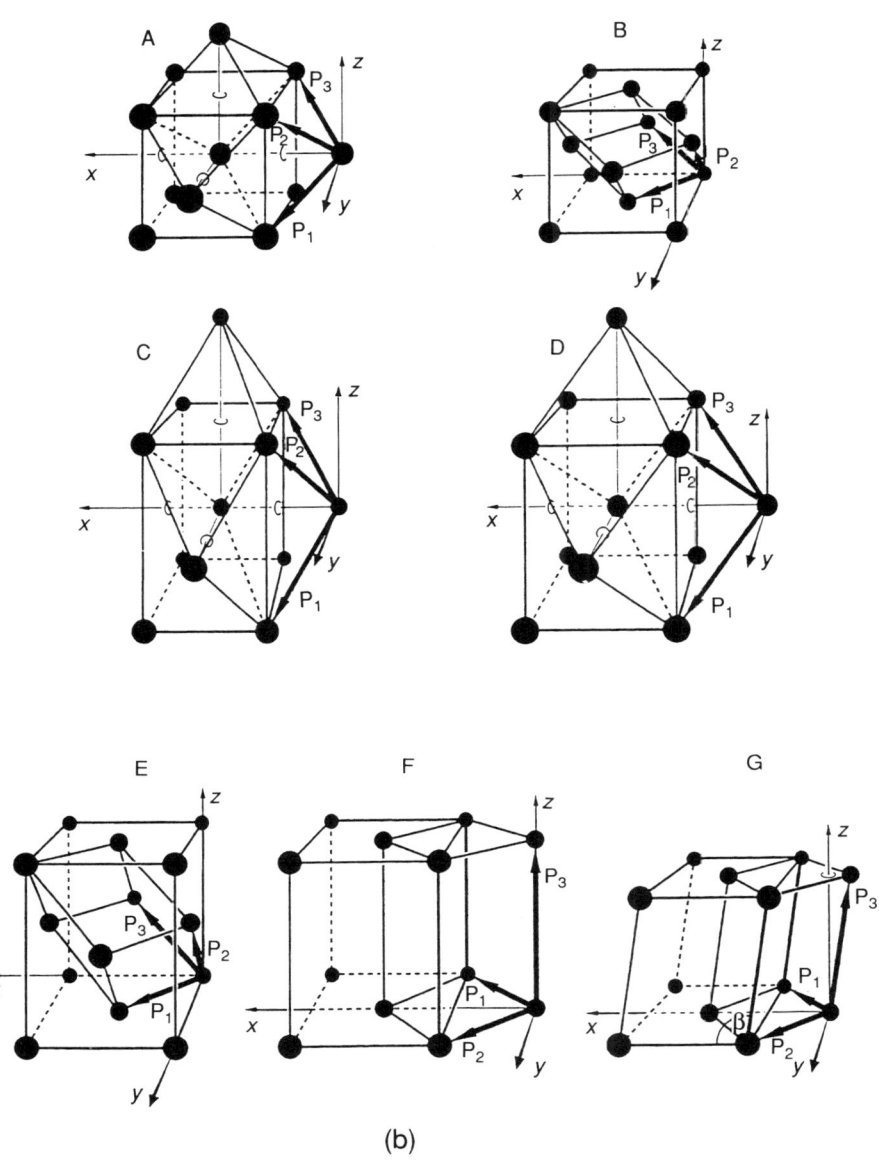

(b)

(b) The seven Bravais lattices conventionally called 'centred', together with the associated primitive translation vectors (shown bold). Notice that the translation vectors are generally in directions rather different from those of the Cartesian axes of the crystal. A, body-centred cubic; B, face-centred cubic; C, body-centred tetragonal; D, body-centred orthorhombic; E, all-face-centred orthorhombic; F, one-face centred orthorhombic; G, face-centred monoclinic. Reproduced by permission of the Journal of Chemical Education from Kettle and Norby, *J. Chem. Educ.*, **70** (1993) 959.

Table 12.4

Crystal system	'Missing' lattice	Equivalent lattice actually listed
Cubic	One-face-centred	Primitive tetragonal
Tetragonal	One-face centred	Primitive tetragonal
	All-face-centred	Body-centred tetragonal
Monoclinic	Body-centred	One-face-centred monoclinic
	All-face-centred	One-face-centred monoclinic
Triclinic	Any centring	Primitive triclinic
Hexagonal	Body-centred	One-face-centred monoclinic
	Unique-face-centred	Primitive orthorhombic
	All-face-centred	One-face-centred monoclinic
Trigonal	Body-centred	Primitive trigonal
(rhombohedral)	Unique-face-centred	Triclinic
	All-face-centred	Primitive trigonal

from that contained in many introductory texts and a reiteration of those points which will be needed later would perhaps be helpful. Only for the Bravais lattices explicitly labelled as 'primitive' is the symmetry of the (primitive) vector sets shown in Figure 12.7 that shown in the 'crystal system' column of Table 12.2. All Bravais lattices (and the *complete* vector sets) show the symmetry of the corresponding crystal system (Table 12.2). It is the desire that all cubic lattices, for instance, have *basis* vector sets of O_h symmetry that leads crystallographers to prefer to work with centred unit cells. This is fine for crystallographers, for their work is not sensitive to the choice of translation vector set and so they are free to choose that which is the most convenient for their purposes. However, in most other work on crystalline materials— spectroscopic and theoretical work provide two important examples—it is vital that the vector set be correctly chosen (the reason for this will become evident in the next chapter) and this is the reason for the approach adopted in the present section.

12.3   THE 32 CRYSTALLOGRAPHIC POINT GROUPS

The point has now been reached at which atoms must be introduced into the lattices although it will continue to be convenient to adopt the fiction that a lattice exists before atoms are introduced into it. Some consequences of the introduction of atoms are quite evident. Thus, although up to this point all lattices have been centrosymmetric, the introduction of atoms means crystal structures, which may or may not retain this centrosymmetry. Clearly, the actual arrangement of atoms, molecules or ions in the lattice is of key importance. Equally clearly there are considerable limitations on the admissible arrangements. For instance, to place atoms randomly in a cubic lattice would immediately destroy the multitude of rotation axes and other symmetry elements essential to a cubic lattice. The arrangement of atoms (the word 'atoms' will be used for simplicity; ions and molecules are not excluded) in a lattice must be consistent with the symmetry of that lattice for the lattice symmetry to be evident in the space group. This prompts the question 'what are the acceptable symmetries for a given crystal system?' As the argument develops it will be seen that this question is correctly put—the different Bravais lattices falling into one crystal system do not have to be distinguished.

Because the symmetry makes it particularly easy to visualize, the cubic case will be detailed. Rather than deal with the full O_h symmetry of the lattice it is easiest to focus on one aspect of this symmetry. This is that in cubic symmetry the x, y and z axes are all equivalent. All acceptable ways of introducing atoms into a cubic lattice must respect this equivalence. While, no doubt the number of acceptable ways of introducing atoms into a cubic lattice is infinite—it is this fact that serves to distinguish one cubic crystal structure from another—the number of different *symmetries* that these arrangements can have is rather limited. The symmetry of any acceptable arrangement has to be one in which x, y and z equivalence is retained. Relatively few point groups satisfy this condition, they are:

$$I_h, I, O_h, O, T_h, T_d \text{ and } T$$

Of these, the first two, of icosahedral symmetry, can be excluded because they require the existence of fivefold axes, and these are inconsistent with a cubic lattice. Were an atomic arrangement with I_h symmetry, for instance, be put into a cubic lattice then the lattice (i.e. the arrangement of other I_h groupings in space) would destroy the fivefold axes, the highest possible effective symmetry of the atomic arrangement would be T, this being the only subgroup of I_h (and I) in the above list. It is concluded, then, that the only symmetries of atomic arrangements consistent with a cubic lattice are:

$$O_h, O, T_h, T_d \text{ and } T$$

Three things are to be noted. First, that only two of these groups (O_h and T_h) contain inversion in a centre of symmetry as an operation. The absence of this operation in the other groups means that the corresponding atomic arrangements will be such as to destroy the centrosymmetry originating in the lattice (remember, the translation operations of the lattice will only move the atomic arrangements, not turn them round in the way needed if the lattice centrosymmetry were to be preserved). Second, groupings of atoms with these symmetries can only form cubic lattices if the symmetry axes of the atomic grouping coincide with the corresponding axes of the lattice. So, an alkali metal salt, $K[MX_6]$, say, has a spherical cation and, probably, an octahedral anion. Both cation and anion, separately, are consistent with a cubic lattice.

However, it does not automatically follow that the salt will crystallize in a cubic space group. Third, there has been nothing in the above discussion which confines it to the case of the primitive cubic lattice. It must therefore be concluded that it applies equally to all the cubic Bravais lattices. This is an important point, the relevance of which will become apparent when all possible space groups are counted.

> **Problem 12.4**   Sketch two possible crystal structures for the salt $K[MX_6]$, one of which is cubic and the other which is not.

The pattern for the other crystal symmetries will follow that set by the discussion of the cubic case. Consider the tetragonal lattices, for instance. Suppose we have a molecule which, itself, is of a cubic symmetry—an octahedral molecule, ML_6, for instance—but which crystallizes with the molecules so arranged that they form a tetragonal lattice. The crystal can be of no higher symmetry than tetragonal. Conversely, if a molecule is of low symmetry, for it to crystallize in a tetragonal lattice then the molecules have to be arranged in groups of four, symmetry related. For an array of atoms to have a symmetry consistent with a tetragonal lattice it must, at least, satisfy the basic requirement of having a fourfold axis of some sort or other. Point groups satisfying this condition (in addition to the cubic, which have been covered) are:

$$D_{4h}, D_4, C_{4h}, C_{4v}, C_4, S_4 \text{ and } D_{2d}$$

The last two of these may cause some surprise because they do not contain C_4 rotation operations— but they do have S_4 rotation–reflection operations and so an essential tetragonality—and this is what is needed for compatibility with a tetragonal lattice. Having made this point, what *is* surprising is that the group D_{4d} is missing—this, at first sight, seems to have the essential requirement of C_4 rotation operations. Indeed it has, but it also has S_8 rotation–reflection operations and this is something that no lattice, cubic included, possesses. It follows that if a group of atoms of D_{4d} symmetry were to be placed in a tetragonal lattice its symmetry would be reduced (D_4 or C_{4v} are the highest symmetries that could result). Finally, note that of the seven point groups just listed, only two, D_{4h} and C_{4h}, contain the operation of inversion in a centre of symmetry. It would not be difficult to apply similar arguments to all the other crystal systems, but no new principle would emerge. We therefore content ourselves with listing, in Table 12.5, the correspondences between the crystal systems and acceptable point groups spanned by the atomic arrangements that may be described by them.

> **Problem 12.5**   Explain in detail (use of Appendix 3 may be needed) why a molecule of D_{4d} symmetry could, at best, give rise to an arrangement of either D_4 and C_{4v} crystal symmetry.

Table 12.5

Crystal system	Acceptable point groups
Cubic	O_h O T_h T_d T
Tetragonal	D_{4h} D_4 C_{4h} C_{4v} C_4 D_{2d} S_4
Orthorhombic	D_{2h} D_2 C_{2v}
Monoclinic	C_{2h} C_2 C_s
Triclinic	C_i C_1
Hexagonal	D_{6h} D_6 D_{3h} C_{6h} C_{6v} C_6 C_{3h}
Trigonal (rhombohedral)	D_{3d} D_3 C_{3v} C_3 S_6

Note the absence of D_{6d} in the hexagonal listing—it has S_{12} rotation–reflection operations, operations not possessed by D_{6h}, the parent of the hexagonal system. In contrast, note the presence of D_{3h}; this seems 'wrong' but is readily explained. The important point is that D_{6h}, the parent group of the hexagonal system, has a σ_h mirror plane reflection operation whereas D_{3d}, the parent of the trigonal system, has no such mirror plane reflection. D_{3h} has this mirror plane and so cannot be associated with a trigonal lattice. The reason that S_6, which looks as if it should be in the hexagonal system, but is in the trigonal, is given in the text.

These point groups number 32 in total and are usually referred to as 'the 32 crystallographic point groups'. In the next two sections it will be seen that they play a key role in determining the number of distinct crystallographic space groups.

Problem 12.6   The following symmetries, although acceptable as those of atomic arrangements, cannot persist in a crystal. In each case give the highest symmetry crystallographic point group arrangement that could result.

$$D_{5h}, D_{5d}, C_{5v}, D_{7h}, D_{7d}, C_{7v}$$

In retrospect, it is possible to see a very simple way of relating a crystal system with the acceptable crystallographic point groups associated with it, a method that has been hinted at more than once in the above section. This relationship depends on the simultaneous satisfaction of two conditions. The first is that acceptable point groups are invariably subgroups of the symmetry of the crystal system. This condition is very important. For instance, it immediately shows that D_{3h} is associated with the hexagonal system and not the trigonal because D_{3h} is a subgroup of D_{6h} but not of D_{3d}. The second condition is that the subgroup is not also the subgroup of the parent group of a lower (i.e. fewer symmetry operations) crystal system. It is the crystal system of lowest symmetry that is relevant. So, D_2 is a subgroup of O_h, D_{4h}, D_{6h} and D_{2h}. Of these, D_{2h} has the smallest number of symmetry operations and so D_2 is associated with the orthorhombic crystal system. It is also for this reason that S_6 is a trigonal crystallographic point group and not a hexagonal.

12.4   THE SYMMORPHIC SPACE GROUPS

The discussion so far is sufficient to enable the first space groups to be obtained but before doing so it is convenient first to review the present position. The seven crystal systems were first obtained as the seven different symmetries of translation vectors that can exist in three-dimensional space. We then found that in several cases there exist more than one distinct set of such vectors, all of the same symmetry—that of the parent crystal system. These subtend the 14 Bravais lattices. It was at this point that atoms were introduced into the discussion. It was found that for each Bravais lattice there exist several symmetry-distinct ways of introducing atoms which are compatible with the symmetry of the Bravais lattice. Distinct space groups will differ *either* in their lattices *or* in the symmetry of their atom arrangement in space—or both. Space groups can be generated by combining each Bravais lattice with each of the corresponding crystallographic point groups. Each group that results will be unique in lattice, point group or both, as required. Effectively, from this point on the following approximate equation will be used to obtain space groups (the question of the points at which this equation is not quite correct will be at the heart of the following discussion):

$$\text{(A Bravais lattice)} + \text{(a corresponding point group)} = \text{(A space group)}$$

How many space groups can be obtained in this way? The answer to this question is detailed in Table 12.6, which the reader should now study.

In the extreme right-hand column of Table 12.6 the actual number of space groups that exist of the sort that have been under discussion is given in parentheses. In some cases the correct number has been obtained, but not in all.

Table 12.6

Crystal system	Number of Bravais lattices (B)	Number of crystallographic point groups (N)	The product (BN)
Cubic	3	5	15 (15)
Tetragonal	2	7	14 (16)
Orthorhombic	4	3	12 (13)
Monoclinic	2	3	6 (6)
Triclinic	1	2	2 (2)
Hexagonal	1	7	7 (8)
Trigonal	1	5	5 (13)
Totals	14	32	61 (73)

So, although perhaps not much is missing, something has to be added to the approach. In particular, the answer to the trigonal case is seriously wrong—and this will necessitate a serious discussion! The other errors are readily dealt with. In Table 12.7, the cases are detailed for which they occur. In this table individual crystallographic point groups and Bravais lattices are given. The table shows the number of space groups that arise from each combination. The argument developed above leads to the expectation that the answer will be '1' in each and every case. It is where the number '2' appears that there is a problem!

The obvious thing about Table 12.7 is that most numbers *are* 1. Those that are 2 do not occur for the highest symmetry crystallographic point groups of a crystal system. This is a relevant point, as study of the C_{2v}, one-face-centred orthorhombic, example shows. In this example, although the lattice is D_{2h}, the crystallographic point group, the filling of the lattice with atoms, destroys all but a single set of parallel twofold axes (but the primitive translation vectors remain mutually perpendicular, which is why C_{2v} is an orthorhombic and not a monoclinic crystallographic point group). Is the lattice face that is centred in the one-face-centred case parallel to or perpendicular to the twofold axes? The answer is that *both* are possible and so two space groups are obtained and not the expected one. The duality arises from a degree of freedom between the lattice and its relationship to the crystallographic point group that was ignored

Table 12.7

System	Crystallographic point group	Bravais lattice and number of symmorphic space groups			
		p;	b.c;	o.f.c;	a.f.c.
Tetragonal	D_{4h}	1	1		
	D_4	1	1		
	C_{4h}	1	1		
	C_{4v}	1	1		
	C_4	1	1		
	D_{2d}	2	2		
	S_4	1	1		
Orthorhombic	D_{2h}	1	1	1	1
	D_2	1	1	1	1
	C_{2v}	1	1	2	1
Hexagonal	D_{6h}	1			
	D_6	1			
	D_{3h}	2			
	C_{6h}	1			
	C_{6v}	1			
	C_6	1			
	C_{3h}	1			

p = primitive, b.c. = body-centred, o.f.c. = one-face-centred, a.f.c. = all-face-centred.

in the analysis of the previous sections. Similarly, in the D_{3h}, hexagonal, case, in the group defining the parent lattice, D_{6h}, there are two distinct sets of mirror plane reflections perpendicular to the sixfold axis. In D_{3h}, only one of the sets is retained. In the parent D_{6h} lattice one of the associated sets of symmetry elements contains the translation vectors, the other bisects the angle between them. Which set is retained in D_{3h}? The answer is that either is possible—but two different space groups are obtained as a result.

Problem 12.7 By sketching a one face-centred orthorhombic lattice and placing atoms in it in two different ways, illustrate the two different space groups which were the subject of the above discussion.

Having thus seen how the relatively small errors in Table 12.6 arise, what of the problem of the apparent gross error in prediction for the trigonal case? The extension of Table 12.7 to cover this crystal system is given below.

Table 12.7 (*continued*)

System	Crystallographic point group	Bravais lattice and number of symmorphic space groups
		p.
Trigonal	D_{3d}	3
	D_3	3
	C_{3v}	3
	C_3	2
	S_6	2

This extension is most strange when compared with the earlier part of Table 12.7 because again the number 1 is expected, if not everywhere, at least to be predominant. It does not appear! Matters would perhaps be improved if there were two primitive trigonal Bravais lattices, not just one, but even then there would be a problem—the 'additional' errors occur for the higher symmetry point groups, not the lower—which is where they were found in the first part of this table. To deal with the former problem first. In fact, there has been a long-standing argument about the number of trigonal Bravais lattices. There have been those who have argued that there are 15, not 14, Bravais lattices, and that two of them are trigonal. If, indeed, there are two trigonal lattices, then they would both have to be primitive—and this sounds like a contradiction in terms and not surprisingly, this is where the argument has arisen!

The primitive trigonal lattice was introduced as one of those obtained when the symmetry of a primitive cubic lattice is reduced. The rhombohedral cell was pictured as obtained when opposite corners of a cube are either symmetrically compressed or stretched. Later, as something of an aside, it was commented that 'although it comes equally well from O_h, it is often convenient to think of D_{3h} as a subgroup of D_{6h}'. The question now arises of whether the primitive trigonal lattice and the rhombohedral lattice are the same or whether they are different. The answer is that they are different. The rhombohedral lattice has just been described as being a distorted primitive cubic; it is typified by three primitive translation vectors that are interrelated by threefold rotation operations. They are shown in Figure 12.8. The other trigonal lattice is most accurately regarded as originating in a hexagonal lattice. Suppose there is a genuinely hexagonal lattice but it is filled with atoms arranged in the point group symmetries characteristic of trigonal systems—D_{3d}, D_3 C_{3v} C_3 and S_6. Of course, trigonal space groups are obtained. Now, should this pattern be regarded as resulting from a hexagonal lattice or a, separate, trigonal? The accepted convention is to work with the final result. So, suppose there were a crystal with a low-symmetry atomic filling but for which it happened that $a = b = c$; $\alpha = \beta = \gamma = 90°$. This would be regarded as a lattice compatible with the atomic filling, and so *not* a cubic. It follows that there is no need to include the lattice vector characteristics given in Table 12.2 for the hexagonal lattice a second time under 'trigonal'. The resulting space groups are called 'primitive'

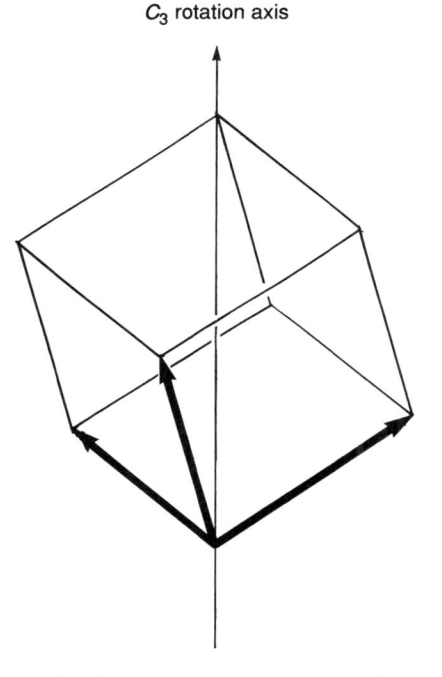

C_3 rotation axis

Figure 12.8 The three primitive translation vectors of the rhombohedral lattice are interrelated by threefold rotation operations. There are no restrictions on the angles between the vectors, although at certain angles special lattices are generated (90° gives the primitive cubic, for instance).

and so distinguished from the trigonals with a rhombohedral lattice (which are called 'rhombohedral'). For simplicity, in some of the tables earlier in this chapter some licence was used in almost equating 'trigonal' with 'rhombohedral'. This was because a discussion such as that just developed would have been quite out of place associated with those tables. It is perhaps worthwhile to look more closely at the difference between rhombohedral and trigonal lattices. The primitive translation vectors of the latter, as has been said, are interrelated by threefold rotation operations. What happens when the angle between each of these vectors becomes 120°? They become coplanar and define a two-dimensional, not a three-dimensional, array. This two-dimensional array, when drawn, is immediately seen to look like a honeycomb; it is a net of hexagons such as is associated with the hexagonal lattice. To obtain a three-dimensional lattice one has to have a primitive translation vector perpendicular to the hexagon net and it is this lattice pattern that gives rise to the trigonal space groups. It is dependent on the fact that two coplanar vectors at 120° suffice to define a planar hexagonal array (the 'missing' vector can be written as a linear combination of the other two when the system is planar). All three trigonally-related vectors are needed—and are sufficient to define a lattice—when they are non-planar. So, there is a clear distinction between the two basic lattices of the trigonal crystal system.

One problem has now been solved; each and every entry in Table 12.7 (contd) should be '2'—but this still leaves unanswered the problem posed by the fact that three are '3'. Why? In fact, the answer has already been given. It was met when discussing the fact that in the hexagonal crystal system there were two different ways of introducing a D_{3h} arrangement of atoms. Either the vertical mirror planes of this group were *coincident* with the directions of two of the primitive translation vectors defining the hexagonal lattice or the mirror planes *interleaved* the vector directions. It is essentially these two possibilities which give rise to the additional trigonal space groups, except that for the D_3 case there are no mirror planes and it is the corresponding alternative orientations of the twofold axes which is relevant. For groups without these symmetry elements this ambiguity does not arise and so these have only the now-expected '2' in Table 12.7 (contd). Of course, the axis orientation ambiguity only occurs for the hexagonal-axes derived trigonal space groups, not for the rhombohedral.

Problem 12.8 Explain why the number 1 does not appear in Table 12.7 (*contd*).

The end of this section has almost been reached but before concluding it there are two questions demanding answers. First, the section was headed 'The Symmorphic Space Groups'—yet the word 'symmorphic' has only been mentioned in a table heading. What is it all about? In this section our concern has been with those space groups that arise from the combination of translation operations with point group operations. In making these combinations, life was made simple by the fact that these two types of operations were quite distinct. The space groups that result are called the *symmorphic* space groups. The name itself is of little significance until we work with the other space groups, the *non-symmorphic* space groups. These will be the concern of the next section. The final question arises because in Table 12.7 the number of space groups of a particular type that exists has been given. This implies that there is some source book containing all such information. Indeed there is, and much study in the field is impossible without a copy to hand. The book is called *International Tables for Crystallography*, the most recent edition appearing in 1983. Details are given in 'Further Reading' at the end of the book. An earlier edition, dated 1952, is preferred by some—it is only half the size of the more recent. As its name makes clear, the book is written for crystallographers. For the present purposes this has one major disadvantage. It means that it is written using the nomenclature of crystallographers, a nomenclature that is rather different from that used so far in this book. However, given the unique position of the book, there is no alternative to working in the crystallographers' notation. So, this section is concluded with a brief introduction to it.

Whereas the concern of the group theoretician is with symmetry operations, the concern of the crystallographer is more with symmetry elements—the symmetry elements associated with a crystal structure influence both the pattern of diffracted beams in an X-ray structure determination and also which occur and which are 'missing' (systematic absences). The reader may have noticed that a few paragraphs above, symmetry elements, rather than operations, were mentioned—preparing the way for the change of emphasis that has to come. It is in the next section that we will be forced to use the crystallographers' notation—because it will enable us to describe new types of operations—but it seems sensible to introduce this notation here, where the ground is familiar. Crystallographers use the Hermann–Mauguin notation (that used in the vast majority of applications of symmetry to chemistry is the Schönflies). As far as point groups are concerned, the Hermann–Mauguin notation is most simply thought of as offering an approximate answer to the question 'for a given point group which are the symmetry elements needed to define it uniquely?' The idea here is that, as group multiplication tables show, combinations of symmetry operations generate other symmetry operations (and much the same can be said of symmetry elements). Thus, a group which contains a twofold axis and a centre of symmetry must also contain a mirror plane perpendicular to the twofold axis (it is C_{2h}). So, what is the smallest list of elements which uniquely specifies a particular group? While the answer to this question is not provided by the Hermann–Mauguin notation, this notation certainly approximates it. Before illustrating this, a point of symbolism. In place of C_n, the notation uses n. So, C_4 is replaced by 4, C_3 by 3 and a group which contains distinct fourfold and threefold axes would be denoted 43 (remember, we are dealing with crystallographic point groups so there is no confusion with a 43-fold axis, one can never exist). It so happens that these two axes serve to uniquely identify the point group. This group is the one that, in Schönflies notation, is called O. Actually, in Hermann–Maugin notation, this group is called 432, but this merely serves to illustrate the point that the notation sometimes only approximates to the minimal defining symmetry element set. Mirror planes containing a C_n axis are denoted m. So, C_{2v} becomes $2mm$ (a minimal set would be $2m$ or mm). When the mirror plane is perpendicular to a C_n axis the notation is n/m. So, the point group $+C_{2h}$ is written $2/m$, a minimal set (it is pronounced 'two over m'). A centre of symmetry is never explicitly indicated unless it is the only non-trivial symmetry element (the group C_i), when it is written $\bar{1}$; centres of symmetry are indicated by the bar—so, a threefold rotation/inversion axis is denoted $\bar{3}$. The short Hermann–Mauguin symbols are most likely to be encountered but usually longer, more explicit, ones exist. Thus, the normal, short, notation which is equivalent to O_h is $4\bar{3}2$, the more explicit version of which is $4/m\bar{3}2/m$ ($4/m$ means that perpendicular to each 4, C_4, axis is a mirror plane). In most cases the long symbols describe the symmetry characteristics of the x, y and z coordinate axes in this sequence; the short symbols are a contraction in which this sequence may well not appear. For cubic space groups, where all three coordinate axes are equivalent, symmetry-distinct axes are detailed (for instance, a fourfold and a threefold). A complete set of Hermann–Mauguin and Schönflies equivalents for the 32 crystallographic point groups is given in Appendix 6, an appendix which also contains more explanations of the notation. A few minutes spent with this appendix (perhaps by covering up the Schönflies and working out the Schönflies equivalents of the Hermann–Mauguin) would be a most helpful preparation for the next section, a section which demands some familiarity with the Hermann–Mauguin notation.

12.5   THE NON-SYMMORPHIC SPACE GROUPS

At the beginning of this chapter it was stated that there are 230 space groups but in the previous section only 73 symmorphic space groups were encountered. It follows that there are 157 *non-symmorphic space groups*, whatever the name means. The vast majority of non-symmorphic space groups are distinguished by the fact that, while the lattices are the Bravais lattices already met, one or more of the point group symmetry elements that combine with them to give complete space groups contain a translation component. This combination of a point group operation with a non-primitive translation is a characteristic of non-symmorphic space groups.† So, a typical situation is

---

† But it does not completely define them—as will be seen, it is possible to obtain the same effect by moving the position of a set of rotation axes in space and the complete definition has to take account of this.

one in which a twofold rotation operation has a translation component added to it. Any such translation component cannot be a primitive translation for all of these have already been included in the translation vector set. Double counting is not allowed! What resolves the problem is the group algebra. A theme of the present section will be the way that the algebra associated with point groups carries over into space groups. The same theme was equally relevant to the last section but its formal inclusion there would have given no new insights. In the present section it is a life-line! A C_2 rotation carried out twice is equal to the identity, leave alone, operation. If a translation is to be included along with the C_2 and the operation carried out twice gives the identity then the associated translation has to be of one half of a primitive translation in the direction of the twofold axis. Carrying out the operation twice would then give a single translation step. But as has already been said, this is in the translation group, not the point group. So, our identity *is* the identity (at least, from the point of view of the point group). The point generated is equivalent to the identity point group operation (together with a primitive translation). The way that operations originally associated with a point group can apparently be transferred to a translation group clearly merits detailed discussion. At the present point it is sufficient to recognize that only well-defined non-primitive translations can be associated with point group operations. The operation in which a non-primitive translation is associated with a rotation operation is called a *screw rotation* and the axis is a *screw axis*. In Hermann–Mauguin notation the screw axis just discussed is denoted a 2_1 axis (pronounced 'two one axis'). Here, the 2 means just what it did at the end of the last section. The $_1$ means that associated with the 2 is a non-primitive translation of an amount equal to one of the two steps needed to give a pure translation. It therefore corresponds to one half of a primitive translation in the direction of the 2 axis. In a similar way, a 3_1 axis involves one third of a primitive translation in the direction of the threefold axis and 4_1 one quarter of a primitive translation. These last two examples show more clearly than the first why the axes are called 'screw' axes. The act of putting a screw into a piece of wood involves a simultaneous rotation and translation of the screw. So here, we have a combination of a rotation with a translation. However, these last two examples also point to a problem. Most screws are right-hand but some are left-hand (many a would-be mechanic has ruined a mechanism because of an unexpected left-handed screw!). Which do we have here? The answer is met by a convention. 3_1 and 4_1 refer to right-handed screws but 3_2 and 4_3 refer to left-handed (these latter two might equally well be written as 3_{-1} and 4_{-1} but this is never done). Right-handed screws go into the wood when rotated clockwise, viewed from the screwdriver end. It is not just rotation operations that can be combined with non-primitive translations. So, too, can mirror plane reflection operations. The identity cannot and the operation of inversion in a centre of symmetry is not. For inversion in a centre of symmetry there is a choice. Either it could be combined with a non-primitive translation (which would have to be one-half of a primitive) or, because inversion is an operation which operates about a unique point, the point can simply be moved to a new position which is displaced from the 'original' by one quarter of the corresponding primitive translation—the overall effect is the same (Figure 12.9). By adopting the latter choice the need to formally specify the translation involved is avoided and this makes life easier. So, the diagrams in *International Tables*, which show centres of symmetry as points, have this latter choice built-in. A similar (sideways) translation of rotation axes occurs when the non-primitive translation concerned is perpendicular to the rotation axis. Mirror plane reflections combined with non-primitive translations (and these are always halves of primitive translations) are called *glide planes* and denoted by the direction in which the translation associated with the glide occurs. So, in an 'a glide' the translation is one half of a primitive translation in the x direction, in a 'b glide' the translation is in the y direction, and so on. Unfortunately, as far as the Hermann–Mauguin notation is concerned, that is all. In this chapter it will be necessary to specify them in more detail. Because a mirror plane is always perpendicular to some axis or other, for economy of labels the mirror plane associated with a glide will be labelled by the perpendicular axis. So, $c(x)$ and

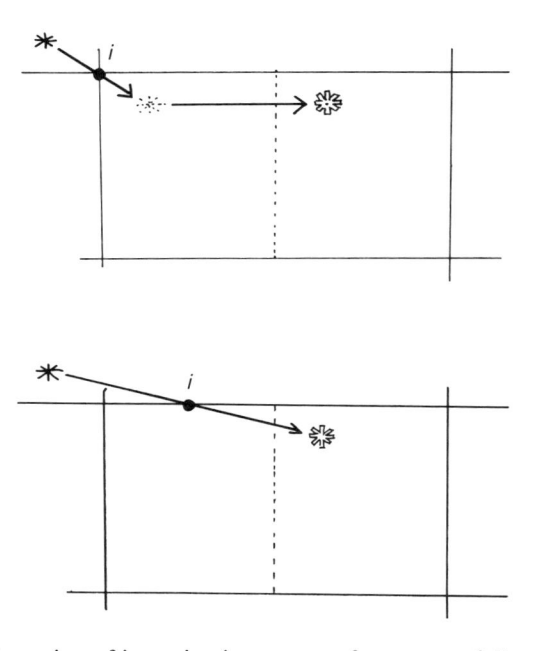

Figure 12.9 The action of inversion in a centre of symmetry followed by a translation of one half of a primitive translation (upper) is equivalent to inverting in a centre of symmetry which has been moved by one quarter of the primitive translation.

b(x) are glide planes, the mirror reflection component of which is perpendicular to the crystallographic x axis (and for which the translation components are along z and y, respectively). It would perhaps be more natural for the x to be written as a suffix but this would introduce problems with the notation for screw axes. The requirement on the non-primitive translation associated with a glide plane is that it lies in a plane parallel to that of the mirror plane. It can, therefore, be composed of half primitive translations along more than one axis. Such glides involving two half primitive translations are denoted by the letter n (for 'net', because they span the diagonals of a two-dimensional net); those involving three by the letter d (for 'diamond', because they occur in the diamond lattice).

Problem 12.9 Draw separate diagrams to illustrate 2_1, 3_1, 4_1, 3_2 and 4_2 screw axes and others to illustrate a, b, c, n and d glides.

12.6 NON-SYMMORPHIC RELATIVES OF THE POINT GROUP D_2

Some considerable space has been spent above on notation. The time has come to put it to good use. There will be two stages in the development. First, non-primitive translations will be introduced into the crystallographic point groups. Second, the modified point groups will be combined with the Bravais lattices to give non-symmorphic space groups. Clearly, it will be impossible to consider every point group, every Bravais lattice and every space group that results from their combination. Examples will be given, hopefully chosen so that the reader becomes confident that the origin of each and every space group could be understood. As a first example consider the orthorhombic crystallographic point group D_2. This group has several advantages. In Table 12.7 it appeared without complications. Having three symmetry-distinct C_2 axes (and Cartesian axes), it will enable us to examine the interplay between different non-primitive translations. The group multiplication table of the D_2 group is small and so easily manageable. Finally, the orthorhombic system has more Bravais lattices than any other crystal system, giving an opportunity to explore all likely problem areas.

A diagram of axes and operations of the D_2 point group is given in Figure 12.10. The corresponding group multiplication table is given as Table 12.8a. A feeling for this table will be essential in the development that follows. In particular, its implications for the way that the point $[x, y, z]$† is converted by the symmetry operations to the four other coordinate sets—and the way that

---

† A word of caution; the same symbols are being used to indicate an axis, as in 2(y), and a general point, as in x, y, z. This has been done because of the familiarity of this usage; with this word of caution no confusion should result.

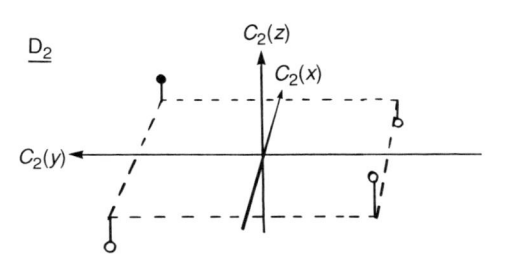

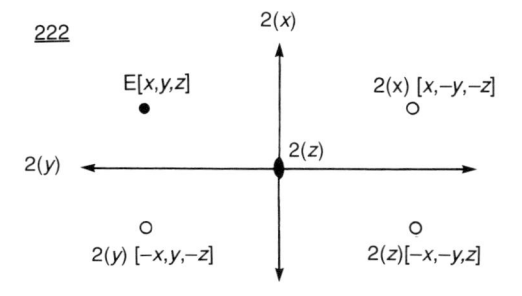

Figure 12.10 The symmetry axes of the point group D_2 (Schönflies notation), 222 (Hermann–Mauguin notation) using the symbolism appropriate to the particular notation. While a perspective view has been adopted for the Schönflies diagram (lower), that given for the Hermann–Mauguin is that which will be adapted in the following figures. Note the representation of twofold rotation axes, particularly that viewed 'end-on' (that along z). In this diagram, as in the following figures, a general point in space is denoted by a solid circle and those into which it is converted by empty circles. For each, the relevant operation and coordinates are given. In the following figures it will be important to follow these carefully; it will often be helpful to compare the entries in them with those given here.

the operations of D_2 convert these points into each other—need to be well understood. The reader would be well advised to stop at this stage and relate each entry in Table 12.8a to the corresponding coordinate transformations in Figure 12.10. To skip this step now may well be to invite problems later!

Problem 12.10 Relate each entry in Table 12.8a to the corresponding coordinate transformations in Figure 12.10.

Table 12.8a

D_2	E	$C_2(x)$	$C_2(y)$	$C_2(z)$
E	E	$C_2(x)$	$C_2(y)$	$C_2(z)$
$C_2(x)$	$C_2(x)$	E	$C_2(z)$	$C_2(y)$
$C_2(y)$	$C_2(y)$	$C_2(z)$	E	$C_2(x)$
$C_2(z)$	$C_2(z)$	$C_2(y)$	$C_2(x)$	E

To help in understanding the symbolism Table 12.8a is repeated as Table 12.8b using the Hermann–Mauguin notation. This notation uses the symbol 1 for the identity but to avoid any ambiguity E will continue to be used.

Table 12.8b

D_2	E	$2(x)$	$2(y)$	$2(z)$
E	E	$2(x)$	$2(y)$	$2(z)$
$2(x)$	$2(x)$	E	$2(z)$	$2(y)$
$2(y)$	$2(y)$	$2(z)$	E	$2(x)$
$2(z)$	$2(z)$	$2(y)$	$2(x)$	E

The table will now be repeated yet again but replacing some or all of the 2 rotation axes (this should be read 'twofold rotation axes') by 2_1 screw axes ('two-one screw axes'). One, two or all three of the 2 can be replaced by a 2_1 — and, because no coordinate axis has any unique properties compared with any other, they can be replaced in any order. The choice to be adopted is given in Table 12.9.†

Table 12.9

D_2	E	$2(x)$	$2(y)$	$2(z)$
	E	$2(x)$	$2(y)$	$2(z)$
Choice 1	E	$2_1(x)$	$2(y)$	$2(z)$
Choice 2	E	$2_1(x)$	$2_1(y)$	$2(z)$
Choice 3	E	$2_1(x)$	$2_1(y)$	$2_1(z)$

The first row in Table 12.9 corresponds to the elements in the multiplication table, Table 12.8b. To obtain a multiplication table corresponding to 'Choice 1', it seems sensible to substitute $2_1(x)$ for $2(x)$ in Table 12.8b and then to ask whether this is a correct step to take. This substitution has been made in Table 12.10.

Table 12.10

Choice 1	E	$2_1(x)$	$2(y)$	$2(z)$
E	E	$2_1(x)$	$2(y)$	$2(z)$
$2_1(x)$	$2_1(x)$	E	$2(z)$	$2(y)$
$2(y)$	$2(y)$	$2(z)$	E	$2_1(x)$
$2(z)$	$2(z)$	$2(y)$	$2_1(x)$	E

† Although the choices of substitutions that follows has an evident logic, it is not that used in *International Tables* — in Choice 1, for instance, the 2_1 is there taken to be along z, not x.

At first sight this table may appear fine, but in fact all of the entries in bold typeface present problems. All involve the $2_1(x)$ operation in some way or other. The first is the identity element resulting from the combination of two $2_1(x)$ operations. Although the entry in Table 12.10 is E, the actual outcome of combining these two operations is a primitive translation along x. It will be necessary to return to this problem later. In the same row as this E are $2(z)$ and $2(y)$ entries, resulting from the combination of $2_1(x)$ with $2(y)$ and $2(z)$, respectively. Where has the non-primitive component of the $2_1(x)$ operation gone? Two other entries in bold typeface are $2_1(x)$ operations which, apparently, have to result from the combination of two operations which do not have any translation component! How can this be—if it can be? The answer is that it can, indeed, be. The reason is the existence of a flexibility noted earlier which must now be developed further.

Consider a space group derived from the C_{2h} point group in which the 2 is replaced by a 2_1. The combination of reflection in the mirror plane with inversion in the centre of symmetry of C_{2h} must now equal 2_1, rather than 2. The translation component in the combination of operations arises because this is an example of the centre of symmetry being displaced by a quarter of a primitive lattice translation. In this particular case, not surprisingly, the displacement of the centre of symmetry is along the 2_1 axis. It is important to note that the displacement takes the centre of symmetry out of the mirror plane. As a result, there is no longer a point through which all of the symmetry elements pass. We are no longer talking about a point group. Indeed, it is by no means clear that we are talking about a group at all, and this, again, is something to which it will be necessary to return. Is it an inevitable consequence of having non-primitive translations associated with point group operations that some of the corresponding symmetry elements no longer pass through a point? The answer is 'yes', so that the problems of the sort under discussion are common to all of the crystallographic point groups associated with non-symmorphic space groups. Can such non-coincidences resolve the problems associated with Table 12.10? Not surprisingly, the answer is 'yes'.

There are two ways in which the argument could be developed. Either the problem of the non-coincidence of symmetry elements could be treated as one requiring an answer to the question 'what is the displacement required?' or the answer to the question could simply be presented. In the latter case all would become clear in solving the problem of what is meant by 'a group' in the present context. In fact, the two approaches are linked as something of a circular argument is involved. Figure 12.11 gives the answer. $2(y)$ and $2(z)$ are displaced relative to each other along x, although each still cuts the $2_1(x)$ axis. Also shown in Figure 12.11 are the effects of the three non-trivial symmetry operations on the point $[x, y, z]$. It was in preparation for diagrams like this that practice with Figure 12.10 was strongly recommended! In Figure 12.11 the

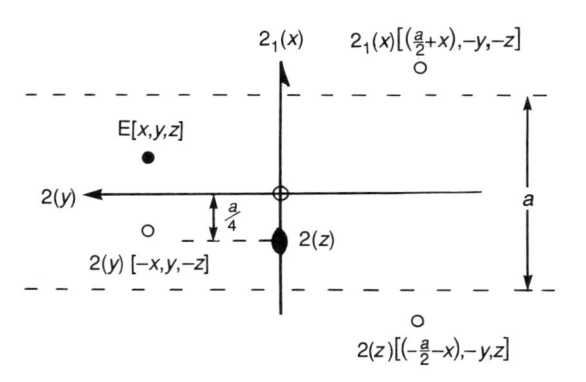

Figure 12.11 The relative arrangements of the symmetry elements of Choice 1 (those of the space group P222$_1$; note that in this label, which in the text has been treated as P2$_1$22, as in the following figures, some liberty has been taken with such things as conventions for the directions of axes, always in the interests of simplifying the discussion). Note the standard convention for showing a 2$_1$ axis in the plane of the paper—a half-headed arrow.

origin has been taken as the intersection of $2_1(x)$ and $2(y)$; a primitive displacement along the x axis, a, has been indicated and set symmetrically about the origin. The $2(z)$ axis is displaced from the origin by $a/4$. This displacement could be either 'up' or 'down'; the latter has been chosen. A table which is the equivalent of Table 12.10 will now be generated but, instead of giving symmetry operations, it gives the coordinates of the points generated. This is Table 12.11. If Figure 12.10 was thoroughly studied then the compilation of Table 12.11 should not prove unduly difficult—certainly, all the y and z entries are those appropriate for Figure 12.10; in any event, the reader should stop and check the entries in Table 12.11 (the use of Figure 12.11 is essential). There is an important point. In contrast to Table 12.10, which was obtained by simple substitution in Table 12.9, the entries in Table 12.11 depend on whether the operations in the left-hand column operate first and are followed by in the first row, or vice versa. Table 12.11 has been compiled with the left-hand

Table 12.11

Choice 1	E	$2_1(x)$	$2(y)$	$2(z)$
E	x, y, z	$(a/2 + x), -y, -z$	$-x, y, -z$	$(-a/2 - x), -y, z$
$2_1(x)$	$(a/2 + x), -y, -z$	$(a + x), y, z$	$(-a/2 - x), -y, z$	$(-a - x), y, -z$
$2(y)$	$-x, y, -z$	$(a/2 - x), -y, z$	x, y, z	$(-a/2 + x), -y, -z$
$2(z)$	$(-a/2 - x), -y, z$	$-x, y, -z$	$(a/2 + x), -y, -z$	x, y, z

column entries operating first (this makes the table easier to read). So, for instance,

$$2_1(x).2(y)[x, y, z] = 2_1(x).[-x, y, z]$$

$$= \left[\frac{a}{2} + (-x), -(y), -(z) \right]$$

$$= \left[\left(\frac{a}{2} - x \right), -y, z \right]$$

Problem 12.11   Check Table 12.11.

The only problem area of Table 12.11 should have been in the bottom right-hand corner, in a square of 3×3 entries. The way that these entries differ from the coordinates in Figure 12.11 appropriate to the corresponding individual operation (so, for example, comparing the effect of $2_1(x)$ followed with $2_1(x)$ with that of E, the operation which is expected from Table 12.10) is shown in the box below:

a	0	a
a	0	$-a$
0	0	0

All the differences are integer multiples of a. They are all members of the translation group. Because of their presence, the set of operations E, $2_1(x)$, $2(y)$ and $2(z)$ do not form a group—to form a group, a set of elements has to map onto itself when combined according to the rules of multiplication of that group—the 'closure' requirement, discussed in detail in Appendix 1. The problem is that members of another group, the translation group, have been obtained and their presence prevents closure. Can the rules of multiplication of the would-be-group be modified so that this problem is avoided and the would-be-group turned into a real group? The answer is 'yes'. The multiplication is made 'moduli primitive translations'. That is, it is made impossible to obtain primitive translations by the simple expedient of defining it to be so! More pictorially, all entries such as those in the box above are thrown away. More physically, and in the context of unit cells as usually defined, whatever 'moves out' of one face of the unit cell 'comes back in again' through the opposite face. This is perhaps the point for the author to confess that Figure 12.11 has been made more apparently difficult than need be the case. Had the point $[x, y, z]$ been chosen to lie in the lower left-hand quadrant then all of the generated points would have been within the limits of the a displacement indicated. To have made this choice would have temporarily concealed the fact that in compiling Table 12.11 points lying outside the boundaries of the a displacement could have been produced and might have made the problem appear of a less fundamental nature than is in fact the case.

Significant progress has now been made and it is perhaps timely to review the position reached. It has been seen that when non-primitive translations are combined with point group operations the combination does not lead to a group unless an additional restriction is placed on the group multiplication rules. This restriction only makes any sense in the context of crystals and so it is only in this context that these groups have meaning. Further, it was seen that whereas there is only one D_{2h} point group, it is possible to create three more groups by adding non-primitive translations (these are those listed in Table 12.9). So, it is reasonable to expect many more non-symmorphic space groups than symmorphic. While this expectation is justified, an important restriction working in the opposite direction must be noted. This is that by invoking non-primitive translations in the definition of point group-derived operations, the corresponding primitive translation has automatically been defined. Yet, as has been seen in Figure 12.7, the definition of the non-primitive Bravais lattices of a particular crystal system requires a different choice of primitive translation vectors to that appropriate to the primitive lattice. This variation of choice is in potential conflict with the rigidity imposed by combining non-primitive translations with point group operations. In particular, for none of the so-called non-primitive Bravais lattices do the primitive translation vectors form an orthogonal (mutually perpendicular) set. For there to be an association of non-primitive translations with point group operations perpendicular axes tend to be required (more on this later). So, the majority of non-symmorphic space groups are associated with primitive lattices. Of the 157, 113 are primitive and only 44 non-primitive, notwithstanding that there are equal numbers of primitive and non-primitive Bravais lattices.

As an example of the consequences of there being two point group operations associated with non-primitive translations consider Choice 2 of Table 12.9. Following the pattern set by the previous example, in Figure 12.12 a diagram is shown which indicates how an original point is transformed when it is operated upon by the set of operations of Choice 2. The screw axes have been taken as along x and y and their intersection has been chosen as origin. Perhaps predictably, $2(z)$ is displaced from this origin by translations along two axes, by $a/4$ and $b/4$ (in the previous example where there was just a single screw axis it was displaced by translation along a single axis). Again, as in the previous example, it could have been arranged that all transformed points fall within the bounds set by the primitive translation units shown in Figure 12.12, a and b. This could have been achieved by placing both the identity point, $[x, y, z]$, and $2(z)$ in the lower left-hand quadrant. A less comfortable arrangement has, in fact, been chosen in which all generated points (indicated by empty circles) fall outside the a, b bounds, safe in the

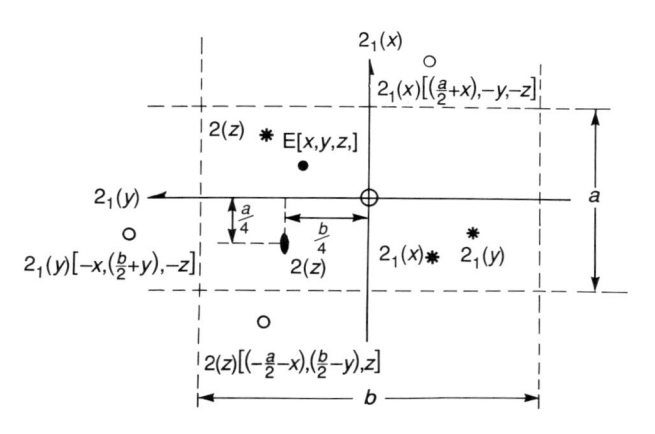

Figure 12.12   Transformations of the point $[x, y, z]$ produced by the operations $2_1(x)$, $2_1(y)$ and $2(z)$ of Choice 2 (corresponding to the space group $P22_12_1$). In this figure, as in Figure 12.11, the transformed points are indicated by empty circles. However, in addition, the corresponding points moduli primitive translations are shown (as stars). In each case they are labelled with the same operation as that containing the primitive translations.

knowledge that the 'moduli primitive translations' requirement means that there are equivalent points within these bounds; the latter have been indicated by stars in Figure 12.12. The actual coordinate changes associated with Choice 2 are shown in Table 12.12, which, like Table 12.11, is compiled with the operation in the left-hand column operating on the coordinates implied by the entry in the first row.

Table 12.12

Choice 2	E	$2_1(x)$	$2_1(y)$	$2(z)$
E	x, y, z	$(a/2 + x), -y, -z$	$-x, (b/2 + y), -z$	$-(a/2 + x), (b/2 - y), z$
$2_1(x)$	$(a/2 + x), -y, -z$	$(a + x), y, z$	$(a/2 - x), -(b/2 + y), z$	$(-x, -(b/2 - y), -z$
$2_1(y)$	$-x, (b/2 + y), -z$	$-(a/2 + x), (b/2 - y), z$	$x, (b + y), z$	$(a/2 + x), (b - y), -z$
$2(z)$	$-(a/2 + x), (b/2 - y), z$	$-(a + x), (b/2 + y), -z$	$-(a/2 - x), -y, -z$	x, y, z

Problem 12.12   Check Table 12.12.

The entries in Table 12.12 may be compared with the coordinates of Figure 12.12; again, it is only the nine entries in the bottom right-hand square that are of interest. The differences are shown below:

$$
\begin{array}{ccc}
a & a - b & -b \\
0 & b & b \\
-a & -a & 0
\end{array}
$$

Again, all the differences are an integer number of primitive translations so that, again, if multiplication is made moduli primitive translations a group is obtained. This exercise could be repeated for Choice 3 of Table 12.11—but the general pattern is clear. Figure 12.13 gives a figure appropriate to Choice 3. No

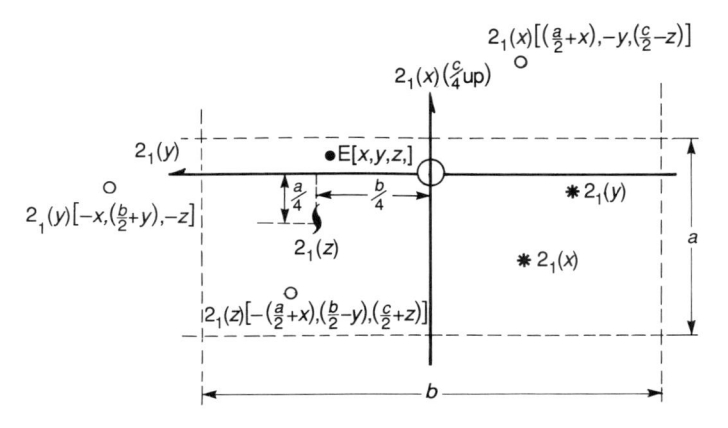

Figure 12.13   Transformations of the point $[x, y, z]$ produced by the operations $2_1(x)$, $2_1(y)$ and $2_1(z)$ of Choice 3 (corresponding to the space group $P2_12_12_1$). Note the way that a 2_1 axis is conventionally shown when viewed end on [the $2_1(z)$]. In this figure, to emphasize that a primitive translation does not have to be chosen to be symmetrically placed with respect to symmetry elements, one has been chosen asymmetrically placed.

screw axes intersect and, really, the diagram should have an indication of the out-of-paper translation distance, c. In a table of coordinate transformations and, derived from it, a box of differences akin to those earlier, entries in units of a, b and c would be found—the moduli primitive translation requirement to obtain a group would still hold.

Problem 12.13   Carry out the above exercise for Choice 3 of Table 12.9.

The next step is to explore the compatibility relationships between the four choices of Table 12.9 and the orthorhombic Bravais lattices. Table 12.3 shows that there are four of these lattices, the primitive, the one-face-centred, the body-centred and the all-face-centred. Clearly, the primitive lattice presents no compatibility problem. In all of the three additional choices that have been considered, the axes chosen (in some cases there was no choice) were all mutually perpendicular. So, too, are the translation vectors which characterize the primitive orthorhombic lattice and so it is concluded that the following three space groups all exist:

$$P222_1 \qquad P2_12_12 \qquad P2_12_12_1$$

A word about notation is needed. The initial 'P' indicates a Primitive lattice. Following are three symbols which indicate, in order, the rotation axes associated with the x, y and z axes. The reason that the 2_1 comes last in the first symbol (and the 2 last in the second) is the convention that the z axis is unique, when a choice exists. For completeness, the corresponding symmorphic space group is P222 (it was not given earlier to avoid the simultaneous introduction of space group notation and Hermann–Mauguin). Although it is the Hermann–Mauguin notation which is almost invariably used to define space groups, there is a Schönflies alternative. The above three groups, for instance, are D_2^2, D_2^3 and D_2^4 respectively (P222 is D_2^1) where the superscript is simply a running number.

Next, the one-face-centred Bravais lattice. Following the convention that z is unique, the face that is chosen to be centred is that perpendicular to this axis. It is denoted C, as opposed to A and B which would characterize the other possibilities. So, the question to be answered is 'which of the following exist?':

$$C222 \qquad C222_1 \qquad C2_12_12 \qquad C2_12_12_1$$

Clearly, the first presents no problems—it was included when counting the symmorphic space groups (Table 12.7). What of the others? Here the guiding principle is that lattice and 'point group' (quotes are used to indicate that this is a convenient, but not quite correct, term) must be mutually compatible. They will be incompatible if, for example, one requires that the primitive translation vectors all be perpendicular when the other requires that one pair are not perpendicular. Now, as Figure 12.7 shows, the translation vectors that define the one-face-centred orthorhombic lattice are not all mutually perpendicular (p_3 is perpendicular to p_1 and p_2 but this pair are not mutually perpendicular—if they were, a tetragonal lattice would result). So, while p_3 could be associated with a 2_1 axis, the other two could not. That is, a space group $C222_1$ is expected but not $C2_12_12$ or $C2_12_12_1$—these last two require that two and three, respectively, of the primitive translation vectors defining the lattice be oriented along twofold axes of the Bravais lattice, but, as has been seen, only one is. Similarly, as Figure 12.7 also shows, none of the primitive translation vectors defining the all-face-centred orthorhombic Bravais lattice (denoted F) is directed along twofold axes (these latter are in the directions of x, y and z in Figure 12.7). It follows that none can be associated with 2_1 screw axes and so the three following space groups have an incompatibility between the 'point group' and the Bravais lattice and do not exist:

$$F222_1 \qquad F2_12_12 \qquad F2_12_12_1$$

Of course, the space group F222 exists for here there is no requirement that the twofold axes and primitive translation vectors coincide. Parallel arguments apply to the body centred-orthorhombic lattice, also shown in Figure 12.7. This lattice is conventionally labelled I; it is concluded that the following space groups do not exist:

$$I222_1 \qquad I2_12_12 \qquad I2_12_12_1$$

As expected, the first two do not exist; unexpectedly however, *International Tables* list $I2_12_12_1$! The reason is that this is the name conventionally given to one of the space groups mentioned earlier in a footnote (page 270).† It is a space group derived from D_2 in which the three twofold axes are retained but the non-symmorphic nature of the group is manifest in that none of them intersect; the translational components are manifest in that the twofold axes are all displaced by $\frac{1}{4}$ translations perpendicular to the direction of the twofold axis. Each twofold axis is associated with a non-primitive translation but it is not such as to generate a screw axis (despite the label given to the group).‡ There is only one other space group of this type—the cubic group conventionally labelled $I2_13$.

In this section, it has been shown how to obtain all of the space groups that are derived from the combination of the elements of the point group D_2 (and its derivatives) with the relevant Bravais lattices. The principles used are general and so the reader should be in a position to apply them to any such combination and thus to understand the origin of each and every space group. No doubt, there will be problems of detail which will be encountered but the broad principles are in place. The major omission is that none of the examples considered contains a glide plane. This omission is remedied in the next section, where the most commonly encountered space group, $P2_1/c$, is studied.

12.7 THE SPACE GROUP $P2_1/c$ (C_{2h}^5)

The first and most obvious question when one encounters the symbol $P2_1/c$ (pronounced 'pee two one over see') is 'what does it mean?'. The P and 2_1 should present no problems—a primitive lattice in some as yet uncertain crystal system but which certainly contains a twofold rotation axis which, in this space group, has become a 2_1 screw. This group is derived from the C_{2h} point group, which means that there must be a 1:1 correspondence between the operations of C_{2h} and the point-group derived operations of $P2_1/c$, just as exemplified earlier for the case of the D_2 point group. In $P2_1/c$ the C_2 (2) of C_{2h} is replaced by a 2_1 and the σ mirror plane is replaced by a glide. At the end of Section 12.4 and in Appendix 6 it was seen that the C_{2h} point group in the Hermann–Mauguin notation is represented as $2/m$ (the symbol $/m$ indicating

---

† It can be argued that the use of the label $I2_12_12_1$ is not ideal.
‡ The diagram given in *International Tables* shows 2_1 screw axes but these arise from the doubling of the group (as indicated by the I symbol).

that the mirror plane is perpendicular to the 2 axis). At the beginning of Section 12.5 it was recognized that a, b and c glides exist, indicating half primitive translations in the direction of the x, y and z axes respectively. At that point there was no discussion of how the existence of these glides is incorporated into the symbolism. The answer is that the letters a, b or c replace the m in $2/m$ so that one talks of $2/a$, $2/b$ and $2/c$ (all of which are groups provided the moduli primitive translation requirement is applied to the group multiplication). When combined with a primitive lattice these point groups give the space groups $P2_1/a$, $P2_1/b$ and $P2_1/c$, where, following convention, the symbol is written on a single line, and not in bold type. These three are not different space groups; the only way they differ is in the choice of axis labels—indeed, there is fourth equivalent choice which is quite often used in crystallography, it is $P2_1/n$, where the 'n' indicates a translation involving half-translations along two coordinate axes. In this case it is more than a choice of axis labels which is involved, it is a change of axes (so that, for example, an axis is chosen which lies between the 'original' x and y). This may seem rather obtuse but, in fact, such a choice can be crystallographically convenient. Care has to be taken, however. In the symbol which corresponds to the crystallographer's 'standard' setting, $P2_1/c$, the 2_1 axis is the y axis, not the z which would be conventional in C_{2h}. One other point, in that the 'parent' point group is C_{2h}, there must be a centre of symmetry somewhere, because C_{2h} contains one. In the non-symmorphic space groups $P2_1/a$, $P2_1/b$ and $P2_1/c$ this centre of symmetry is displaced out of the (former) mirror plane by one quarter of a primitive translation. This may seem rather complicated but Figure 12.14 should clarify the situation; the space group shown there is $P2_1/c$. The displacement of the centre of symmetry is related to, but a bit more convoluted than, that in the example discussed at the end of Section 12.5.

In Figure 12.14 the coordinates generated by the symmetry operations of the $2_1/c$ group (moduli primitive translations, of course) are shown. The reader will find it very useful practice to work through the generation of these. In Figure 12.15 is shown the diagram that appears in *International Tables* for the $P2_1/c$ space group; its connection with Figure 12.14 should be evident. There may appear to be too many screw axes, glides and centres of symmetry in Figure 12.15—after all, the correspondence with C_{2h} discussed above requires one of each. Any one of each sort of symmetry element may be selected from Figure 12.15—all of the others are then such that the corresponding operation is equivalent to the selected one plus a translation (always a primitive translation although possibly a sum of them). One final point; there is no space group $C2_1/c$. The reason should be obvious from the earlier discussion and Figure 12.7.

Problem 12.14 Make a copy of Figure 12.14 but excluding the coordinates of each point. Without reference to the original until the task is completed, add coordinates to your copy of Figure 12.14.

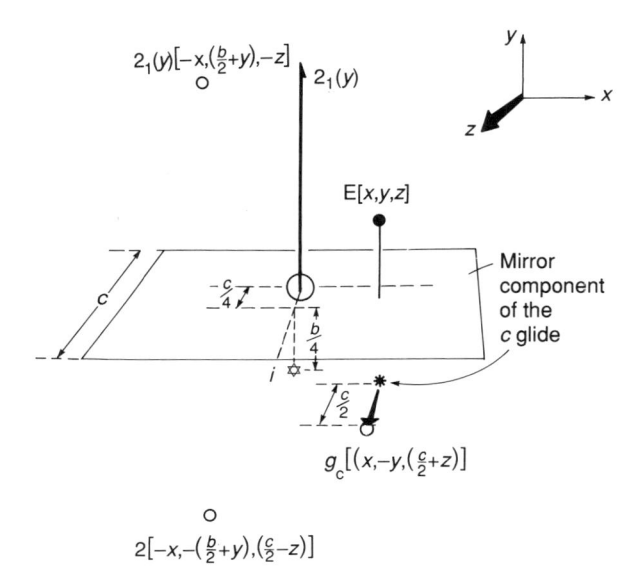

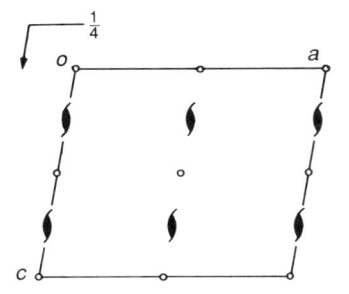

Figure 12.14 Pictorial perspective representation of the symmetry elements of the space group P2₁/c. The centre of symmetry is here represented by the open six-point star; note its displacement from both the 2₁ and the glide plane. The glide operation is shown divided into two components; the act of reflection leads to the solid star; the end product of the complete operation is shown, as usual, by an empty circle.

Figure 12.15 The diagram given for the space group P2₁/c in *International Tables* (viewed down the y axis). Note the way that the glide plane is $b/4$ above the plane containing the centre of symmetry is represented (top left-hand corner). Centres of symmetry are conventionally represented by circles, as here—beware confusion with a different use for this symbol in Figures 12.10–12.14. In this diagram nine such centres of symmetry are shown but they differ only in the way that they add translations to the basic operation of inversion in any one of them. In this diagram the origin is taken as at the top left-hand corner.

12.8 UNIT CELLS

In the early part of this chapter care was taken to avoid use of the term 'unit cell'. Later, and particularly when space groups were discussed, it crept in, although its use was kept to a minimum. The name is so simple and useful that it cannot long be avoided. Why then should it be so assiduously avoided in the present text? The reason is that the concept of a unit cell is more complicated than one might suppose and it is preferable to avoid basing arguments on an unknowingly simplified concept. Where, then, lies the problem? The answer is that for no crystal structure is there a unique unit cell. Indeed, quite the opposite. For every crystal structure there is an infinite number of acceptable unit cells, *all* primitive (of course, crystallographers will prefer non-primitive unit cells for some structures). This infinite choice is important—it is the reason for including this section—but it should be contrasted with the common use of the expression 'the unit cell is ...' in research papers and textbooks when referring to individual crystal structures, a statement that by the use of the 'the' implies that only a single choice exists for a unit cell. Of course, the use is justified in that for one reason or another there is often a single *convenient* choice, but it is important not to overlook the possibility that for a different purpose a different choice might be preferable.

In Figure 12.6 a two-dimensional grid is shown that might be one layer in an orthorhombic lattice. The figure shows several alternative choices of two-dimensional unit cells, all of the same area but differing in shape. Clearly, a similar set of constructions is possible in the third dimension of an orthorhombic (or any other) lattice. For each of these constructions an infinite number of variants exist, at least for an infinite lattice (and on the atomic length-scale the lattice of a real crystal is, effectively, infinite). It is easy to see that there is an infinity of choices of unit cell. But this discussion has been too restrictive! Yet more choice exists. First, it has unquestioningly been accepted that unit cells should be bounded by plane faces. Again, this is a choice of convention and convenience; it is not a requirement. The faces of a unit cell can be curved, dimpled, re-entrant, whatever. There is one requirement on a

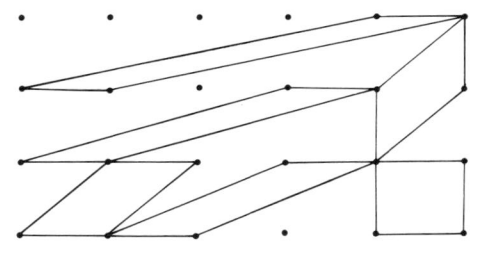

Figure 12.16 Six different equally acceptable (although not necessarily equally convenient) choices of two-dimensional unit cell for the two dimensional lattice shown. All of the choices have identical areas. Clearly, there is an infinite number of acceptable choices. Similar considerations apply to the three-dimensional case.

unit cell; that when repeatedly operated on by the primitive translation vectors it generates the entire crystal (or crystal lattice, depending on the context). One context where it might often be sensible to choose a curved unit cell is in the presentation of the results of a crystal structure determination. It is all too common to find in diagrams of a unit cell of a crystal structure that part of a molecule is in one unit cell and part in an adjacent one. In such cases a single molecule appears as two or more fragments, with no chemical bonds between the fragments shown—because they are parts of different molecules. In such a case, if unit cells are needed, it would be sensible to define their boundaries in such a way that they envelop a complete molecule; however, such boundaries would have to be curved and, at present, this is not computationally convenient.

Problem 12.15   Draw a diagram similar to that in Figure 12.16 but with all unit cell edges curved.

The second reason that our statement that there is an infinite variety of choice of unit cell was too restrictive lies in the fact that all the unit cells considered had one thing in common; they were each bounded by three pairs of parallel faces. Even if we restrict ourselves to plane faces, a unit cell can have many more than three sets of parallel plane faces. Indeed, if one were to select one choice of unit cell as being preferred to all others then it would generally be one with many facets, many faces. After all that has just been said about unit cells this is a strange message, that one choice of unit cell is to be preferred, made the more so by the fact that it so obviously is at variance with that which is familiar to chemists—the unit cells of crystallographers.

Problem 12.16   Repeat Problem 12.3—the answer should be longer this time!

12.9   WIGNER–SEITZ UNIT CELLS

Wigner–Seitz unit cells are essential to a full understanding of the solid state. This is because Brillouin zones, which are at the heart of solid state physics, are a sort of Wigner–Seitz unit cell.† The construction of Wigner–Seitz unit cells is perhaps best explained in a somewhat unreal, anthropomorphic, way. Suppose the reader is reduced to the dimensions of an atom and is standing at a lattice site (alternatively, that the lattice is so enlarged that a person can stand inside it). From the chosen lattice site draw lines to all other (equivalent, of course) lattice sites. The chosen lattice site bristles with lines, rather like a curled up hedgehog/porcupine. Now, exactly halfway along each line, construct a plane

---

† The development of Brillouin zone theory is not given in the present text; the author has written about it elsewhere in a manner which is entirely compatible with the present discussion (see S. F. A. Kettle, *Physical Inorganic Chemistry,* Spektrum (Freeman), Oxford, 1995), Chapter 17.

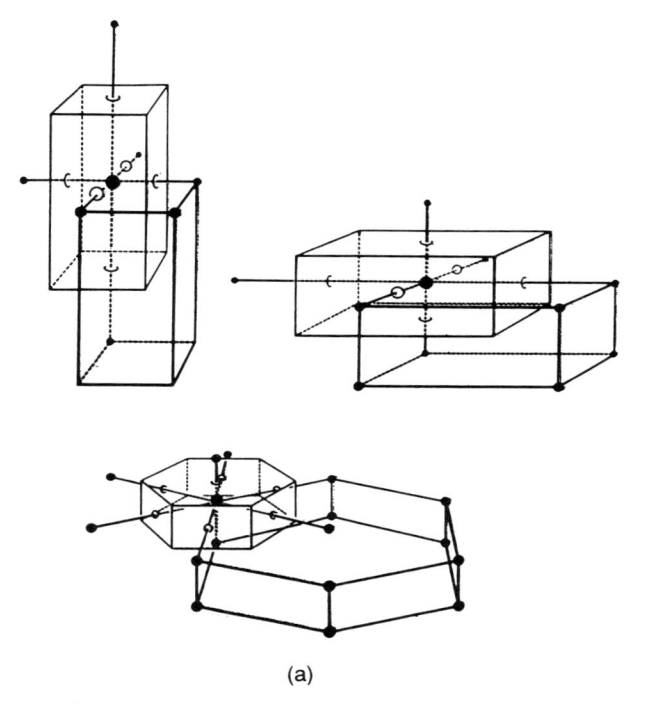

(a)

Figure 12.17   (a) The Wigner–Seitz unit cells of three lattices that are conventionally denoted 'primitive'. As shown by the upper two (tetragonal and orthorhombic), the Wigner–Seitz unit cells 'look the same' as the conventional unit cells, although they differ in that a lattice point occurs at the centre, not at the corners. For the hexagonal lattice (bottom) there is a different shape because the Wigner–Seitz unit cell has eight, not six, faces. The hexagonal lattice shown is primitive, despite first appearances; the conventional unit cell does not show the hexagonal symmetry in an immediately apparent way (although it has sides of equal length and angles of 120°). Reproduced by permission of the Journal of Chemical Education from Kettle and Norby, *J. Chem. Educ.*, **71** (1994) 1003.

perpendicular to the line; it is perhaps helpful to think of these planes as being rather solid. Standing at the chosen lattice site, and forgetting the lines used in their construction, the reader will be surrounded by the planes originating from the shortest lines, those running to the nearest lattice points. These planes will intersect and, although the planes themselves run to infinity, all that will be seen is the box formed by their intersection immediately around the chosen lattice point. On the real, atomic, scale, this box is the Wigner–Seitz unit cell of the lattice. In Figure 12.17 examples of Wigner–Seitz unit cells are given. Note several things. First, a Wigner–Seitz unit cell, by its very construction, contains only a single lattice point; all Wigner–Seitz unit cells are primitive. Third, the number of faces of a Wigner–Seitz unit cell is determined by the number of neighbours and their disposition in space. As we will see, it is not unusual for a Wigner–Seitz unit cell to have a dozen or so faces—and this is not unreasonable.

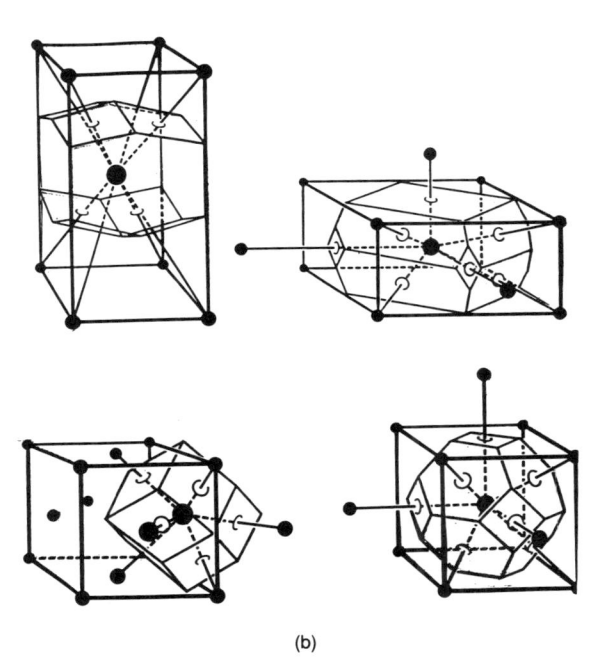

(b)

Figure 12.17   (b) The Wigner–Seitz unit cells of some lattices that are conventionally denoted 'centred'. The top two are both Wigner–Seitz unit cells of the body-centred tetragonal lattice and illustrate the way that the qualitative shape of the cell can depend on an axial ratio. The bottom two are (left) of the face-centred cubic and of the body-centred cubic (right). Note that all of the unit cells of (b) show the symmetry of the corresponding lattice (D_{4h} or O_h). Reproduced by permission of the Journal of Chemical Education from Kettle and Norby, *J. Chem. Educ.*, **71** (1994) 1003.

After all, each sphere in an array of close-packed spheres has twelve nearest neighbours. Fourth, each Wigner–Seitz unit cell has the symmetry of its crystal system. That is, whereas the primitive unit cells of centred lattices shown in Figure 12.7 all had symmetries lower than those of their crystal systems—these are listed in Table 12.2—and is the reason that crystallographers prefer to work with centred unit cells, all of the cubic Wigner–Seitz unit cells are of O_h symmetry, for example.

What makes the Wigner–Seitz unit cell unique? Two things. First, it is the only choice of primitive cell which invariably has the point group symmetry of its crystal system. Second, because it has a property shared by no other choice of unit cell, a property that is evident from its construction. This is that it contains all (general, not lattice) points in space that are closer to the chosen lattice point than to any other lattice point. This is an important property. Suppose, for instance, that some spectroscopic property of a crystal is studied; the particular form of spectroscopy is unimportant. It is possible that the spectrum obtained would show evidence of interaction between individual atoms/bonds/molecules (depending on the particular spectroscopy) and their environment. In order better to understand the spectrum, one might attempt to calculate the interaction between a given atom/bond/molecule and every other in the crystal. Even with the fastest and most powerful of modern computers, this is a near-impossible task, one that would exhaust any research budget. As a compromise, it might be decided to carry out calculations of the interactions between the given atom/bond/molecule and all those others with which it + interacts more strongly than does any other equivalent atom/bond/molecule (because these are likely to be the most important interactions). Which, then, are the atoms that have to be considered? The answer is simple. All those contained within the Wigner–Seitz unit cell which has the atom/bond/ molecule at its centre. The only assumption contained within this statement is that the magnitude of the relevant individual interactions decreases with increase in separation between the interacting centres, as do all interactions of recognized chemical importance. There are more than 14 Wigner–Seitz unit cells because although in principle there is one for each different Bravais lattice, the actual shape of a Wigner–Seitz unit cell depends on an axial ratio. In a primitive tetragonal lattice, for instance, does the unique translational vector have a magnitude which is greater or less than the magnitude of the other two translational vectors? The symmetry of the Wigner–Seitz unit cell is D_{4h} in both cases but the cells look qualitatively different. In total, there are 24 different-looking Wigner–Seitz unit cells.

12.10   SUMMARY†

Although there are only seven crystal systems there are 14 associated Bravais lattices, all of which are centrosymmetric (pp. 248, 256). Corresponding to each crystal system there is a set of point groups which may describe the symmetry-distinct arrangements of molecules in space compatible with the crystal system (p. 261). The symmorphic space groups are obtained as the combination of all Bravais lattices and all crystallographic point groups of each crystal system, due allowance being made for alternative orientation arrangements (p. 264). The non-symmorphic space groups are similarly obtained but for each there is a non-primitive translation associated with one or more point group operations and/or with the relative disposition in space of the corresponding symmetry elements (such a movement of a symmetry element automatically changes the translation component contained in the associated operation) (p. 270). For any crystal there is only one choice of lattice but an infinite number of choices of unit cell. Of these, the Wigner–Seitz unit cell invariably shows the symmetry of the crystal system (287).

---

† Two of the topics in this chapter are dealt with in more detail in 'Really, your lattices are all primitive, Mr Bravais', S. F. A. Kettle and L. J. Norby, *J. Chem. Educ.*, **70** (1993) 959, and 'The Wigner–Seitz unit cell', S. F. A. Kettle and L. J. Norby, *J. Chem. Educ.*, **71** (1994) 1003.

13
Spectroscopic Studies of Crystals

13.1 TRANSLATIONAL INVARIANCE

The discussion in Chapter 12 contained two contributory components—that coming from the lattice and that coming from the crystallographic point group. It was the admissible combinations of these components which are manifest in the 230 space groups. However, the discussion was rather different from that in other chapters of this book in that there was no mention of a character table, only of symmetry elements and operations. In progressing towards the introduction of character tables into the discussion there are, again, two distinct approaches which may be adopted, that through the lattice and that through the crystallographic point group. The former is appropriate to solid state physics, leading to the development of band theory; relevant to the electronic energy level patterns in solids. The latter is more appropriate to spectroscopic studies of solids and since spectroscopy has been a theme of earlier chapters it is the one which will be followed here.† A major distinction between the two approaches arises because there is, effectively, an infinite number of translation operations in the translation group of the lattice but the crystallographic point group is finite. The full space group contains, in some way, both the translation group and the point group. Whichever approach is followed, one should really be working with the full space group and its character table and so the relationship between this group and the group actually used becomes of importance. There are two aspects to this relationship—the mathematical and the physical. Although both are important it is arguable that the latter has the greater importance. Unless a mathematical relationship has some physical significance it remains nothing more than an elegant irrelevance. First then, it is necessary to look at spectroscopic measurements on crystals in the large, in order to discover those aspects which enable mathematical simplifications.

The most general and relevant aspect of spectroscopic measurements on crystals is that of scale. A typical translation vector in a crystal relating adjacent equivalent points has a magnitude of a few ångstroms (this is a quantity which would normally be quoted as the length of a unit cell edge). In contrast, the wavelength of visible radiation is of the order of a few thousand ångstroms. In the infrared the wavelength is much longer and even in the vacuum ultraviolet

---

† The author has given an account of the translation group approach elsewhere, in *Physical Inorganic Chemistry* (Spektrum (Freeman), Oxford, 1995), Chapter 17.

it is a few hundred ångstroms. Only for X-rays does the wavelength of the radiation become comparable to or smaller than the length of a typical primitive translation vector in a crystal. This is relevant to the classical explanation of the interaction of light with matter given in Section 10.4. There, typically, the electric vector of an incident light wave was seen as inducing a transient charge displacement in a molecule. This charge displacement changes sign with the oscillations of the electric vector; when these oscillations coincide with a natural frequency of the molecule then resonance occurs, typically with transfer of energy to the molecule. This picture carries over, unmodified, into the spectroscopy of crystals. However, it can be enlarged by recognition of the inequality between the wavelength of the incident radiation and the magnitude of a typical primitive translation vector. In a crystal, molecules which are close to each other and related by pure translation operations will experience essentially the same incident electric vector originating in the light wave. The fact that different molecules are related by a pure translation ensures that they have precisely identical orientations with respect to the incident light wave (ignoring any small imperfections in the crystal). To a first approximation, then, for all common spectroscopies the interaction of light with a crystal is a translation-independent process. Translation-related molecules behave in the same way; such translations can therefore be ignored. If a proper account can be given of the behaviour of a single molecule then that of all of its translation-related counterparts follows. The same cannot be said of molecules that are related by point group operations (or composite operations with a point group component). A 2 (C_2) rotation, for example, 'turns a molecule over' so that two molecules interrelated by a 2 (C_2) would experience precisely opposite transient dipoles induced by the electric vector of a light wave. For two molecules interrelated by a 2_1 screw operation, the 2 would ensure that the induced dipoles are out-of-phase; the non-primitive translation component of the operation would be irrelevant for the reasons given above. The conclusion is that 'the only important aspect of solid state symmetry which is relevant to spectroscopy is that contained in the crystallographic point group'. Even if the actual 'point' group is a derivative of a real point group (if the actual 'point' group contains non-primitive translation operations, as discussed in Chapter 12), then it still will be possible to work in one of the 32 crystallographic point groups. This, then, is the physical picture. Can it be given a more formal, mathematical, justification? The answer is 'yes'; indeed, more than one such formal justification exists. The different justifications are those associated with different models—in particular the unit cell group model and the factor group model. Fortunately, these models invariably lead to identical predictions although the latter is perhaps the closer related to the detailed discussion above. The next section is devoted to these models. One final word about the physical picture. The content of the present chapter is based on the assumption that there is an interaction between the molecules in the crystal under study. If they behave as isolated individuals then there are no spectroscopic complications arising from the fact that the solid state is involved. Indeed, quite the opposite. The molecules may behave as if they were in the gas phase (where there certainly would be no molecule–molecule interactions) but they are, however, fixed in a crystal and this means that, in contrast to the gas phase, they are fixed in their orientations. This is a topic discussed at the end of Chapter 10, where it was pointed out that this can mean a change in the spectral bands excited as the orientation of the crystal is changed (provided that oriented, polarized, radiation is used). This, so-called *oriented gas model* will not be discussed in detail although it should be emphasized that it can well happen that it is applicable to some spectral bands arising from a crystal—but that one of the models to be covered in the next section has to be used for others. This is because, for instance, some vibrational modes of a molecule may be well insulated from those of the surrounding molecules but other vibrational modes of the same molecule are not. A vibration which changes the dipole of a molecule is more likely to be coupled with the same vibrations of other molecules than is a vibration which is, say, quadrupole active because dipole–dipole coupling attenuates less rapidly with distance than does quadrupole–quadrupole. Another possibility is that a molecule is sensitive to its general environment but, none the less, is insulated from specific interactions with other molecules. It is sensitive only to the symmetry of the site in the crystal at which the molecule is situated. Almost always, this site is of lower symmetry than that of the isolated molecule and so the *site symmetry* model is characterized by the relief of degeneracies and by the increased strength (and, perhaps, the appearance) of transitions forbidden in the isolated molecule. The first task of any analysis is to determine which model is appropriate for each spectral feature. An example of the application of the oriented gas and site symmetry models will be given later in this chapter.

13.2 THE FACTOR GROUP AND UNIT CELL GROUP MODELS

In this section, as in the previous, it is convenient to talk in terms of 'molecules', although in the appropriate context the discussion could equally well apply to atoms or to ions. The first approach to be considered is the *factor group* model. Clearly, the first task is to define what is meant by 'factor group'. In principle any group could have one or more associated factor groups. The character tables of factor groups are invariably simpler than those of the parent group to which they relate†—an attractive feature; for the case of space groups, the character tables of the corresponding factor groups are enormously

---

† This statement is true whenever a factor group is non-trivial. A few groups have only factor groups which are trivial, being identical to the parent group itself. In such cases the parent group is itself very simple, having no non-trivial invariant subgroup.

simpler. The concept of a factor group is closely linked to that of an invariant subgroup, a topic mentioned in a footnote in Section 4.3 and dealt with more fully in Section 8.1. In the latter section it was shown that when a group is the direct product of two invariant subgroups then its character table is the direct product of the character tables of those of the two invariant subgroups. The particular case considered was a demonstration that the point group C_{2v} is the direct product of the groups C_2 and C_s. Let us consider this case again; Table 8.3 is particularly relevant. We have that:

$$C_2 \otimes C_s = C_{2v}$$

(remember, the symbol $\otimes$ is used to indicate a direct product). In full:

$$\{E \quad C_2\} \otimes \{E \quad \sigma\} = \{E \quad C_2 \quad \sigma_v \quad \sigma'_v\}$$

The left-hand side of this expression can be written differently:

$$\{E\{E \quad \sigma\} \; C_2\{E \quad \sigma\}\}$$

Written in this form, one sees a generality; the 'inner' group $\{E \quad \sigma\}$ could be varied without changing the general form of the expression. Alternatively, in this particular form of the expression, it can be regarded as a constant which multiplies both the E and the C_2. As a constant, it is playing the role of an identity element. The group $\{E\{E \quad \sigma\} \; C_2\{E \quad \sigma\}\}$ is said to be the *factor group* of C_{2v} with respect to the group $\{E \quad \sigma\}$, the group which plays the part of an identity element. It is most important that the character table of the factor group is that (really, is isomorphic to that) of C_2. This is a simple, almost trivial, example of a factor group. Some meaning would be attached to it if it were possible to make some measurement on a molecule of C_{2v} symmetry, the result of which was independent of the σ_v and σ'_v operations (but the author, despite some effort, can only think of highly contrived examples!). The result would depend only on the E and the C_2. In such a case, the factor group above, a group isomorphic to C_2, contains all of the relevant information. One could work in the full group but to do so would be to add nothing new.

The situation is quite different in the solid state. As discussed in the previous section, to a very good approximation, spectroscopic phenomena are independent of the translation operations. Effectively, all measurements concern transitions which transform as the totally symmetric irreducible representation of the translation group. A detailed study of the way that these phenomena transform under the translation operations is therefore not of any value. Further, the translation group is always an invariant subgroup of the full space group. It is therefore possible to form a factor group of the space group with respect to the relevant translation group (the detailed operations of which are therefore not of concern). So important are these factor groups that they are simply referred to as 'the factor group' (of a particular space group). Just as each and every space group is different so, too, are the corresponding factor groups. Sometimes, the difference will lie in the details of the translation group and so not be evident. More evident will be the relationship between the group of the point group-derived operations and the corresponding crystallographic point group. These will adopt the form described in the previous chapter for the D_2 group (Cases 1, 2 and 3) and for C_{2h} (the space group $P2_1/c$). In practice, one works with the character table of the relevant crystallographic point group. It is necessary to make the correct substitutions (of 2_1 for 2, for example) and it is here that *International Tables of Crystallography* prove invaluable,† although they do not contain the character tables themselves. A word of consolation. Those accustomed to working with point groups and not with factor groups may well find the prospect of handling screw rotations and glide planes somewhat daunting. In contrast, those accustomed to working with factor groups welcome the appearance of screws and glides! The reason is that, almost invariably, the character generated under such operations is zero, no matter the problem under discussion.‡ An example of the use of a factor group in a vibrational analysis of a solid is given in the next section.

Apparently quite different from the factor group approach is the *unit cell* model. In this, the problem of the translation operations is dealt with by the simple expedient of ignoring them! The justification for this simplification is that given in the previous section—that the spectroscopic phenomena under consideration are translation-independent. The method consists of considering a unit cell of the crystal and the operations which interrelate the molecules that it contains. These operations are taken to be moduli primitive translation and so of the type considered in the last chapter for the D_2 point group and the $P2_1/c$ space group. Any operation that takes an object out of the unit cell is held to bring it back again through the opposite face (so, if it disappears through the top face, it reappears through the bottom—where the operation is completed). The unit cell method ends by using the same mapping of crystallographic point group operations onto their space group derivatives as does the factor group and so the two methods lead to identical results. Of the two, possibly because of its more evident connection with the results of crystal structure determinations, the unit cell method is perhaps the more popular. It has to be recognized, however, that it suffers from two potential weaknesses. The first is that it gives undue prominence to a particular choice of unit cell. As has been emphasized, there is an infinite number of choices of acceptable unit cell for any crystal. A

---

† But beware the problem of centred unit cells in *International Tables*. Because the size of the unit cell is increased so, too, is the number of point group-derived operations which interrelate points within the unit cell—a doubled unit cell means a doubling in the number of operations and so on. The 'extras' are really translation operations masquerading as point group-derived operations, appearing as glides and screws. In looking for the correct set of point group-derived operations (strictly, a set which multiply correctly, as described in Chapter 12) pure point group operations should always be retained in preference to derivatives containing a non-primitive translation component.

‡ The exceptions to this statement are few. A long polymer chain, aligned along a screw axis, in which one monomer unit is related to the next by the screw operation, is one such exception. For instance, in such a molecule a vibration could map onto itself under the screw operation.

particular choice of cell invites statements such as 'because in the unit cell they are well separated ...', which are strictly unacceptable. Acceptable alternatives are along the lines 'because in the crystal they are well separated ...'. Second, it is usual for the crystallographically determined unit cell to be used in the unit cell model. This poses a problem when the crystallographic unit cell is centred because it is a primitive unit cell which has to be used in the unit cell model—and the method lays down no rules for moving from the centred to the primitive. Errors have appeared in the literature as a result. Authors have been known to work with the centred cell rather than the primitive (and so predict too many spectral features). Others, aware of the problem, have worked with the centred cell and simply divided the predictions by the relevant factor (those at the right-hand side of Table 12.3). Unfortunately, this procedure does not guarantee the correct answer either. An example of the use of the unit cell method is given in the next section.

13.3   EXAMPLES OF USE OF THE FACTOR AND UNIT CELL GROUP MODELS

As indicated above, in practice the factor group and unit cell group models lead to identical predictions. Indeed, despite the fact that they were developed rather differently in the previous section, someone looking over the shoulder of a spectroscopist might well have some difficulty in deciding which of the two was being used. The reason is that the development of the factor group model given above concentrated on how the translation group could be factored out of the problem. This having been agreed, the next step is to turn to the character table of the relevant crystallographic pointgroup—and this is the first step in the unit cell model also. The two methods differ in subtle ways, which relate to the nuances of their different models. In the following account, these differences will be slightly exaggerated. Further, since it was presented as a problem area above, the unit cell model will be applied to a crystal structure which, crystallographically, is treated as having a centred unit cell.

13.3.1   The $\nu(CO)$ spectra of crystalline $(C_6H_6)Cr(CO)_3$

While the effects which are the subject of this chapter may be found in all forms of spectroscopic measurements on crystals, they are more important in some forms than in others. Roughly, the more local the phenomena observed, the less important are the effects. So, in Mössbauer spectroscopy, where the excited states of suitable nuclei are probed, the phenomena are so local that, essentially, only the atoms bonded to the atom under study have any influence. On the other hand, if in a particular form of spectroscopy the spectral bands are very broad, the effects can be masked within the bandwidth. Many measurements made in the visible and ultraviolet regions of the spectrum, where electronic transitions are studied, fall in this category. The solid state effects can be measured and studied but rather special conditions are often needed—low temperatures, single crystals together with polarized radiation and, perhaps, doping of the crystal with an isomorphous diluent. One of the spectroscopic areas in which the phenomena are easily studied is that of vibrational spectroscopy, an area which has the advantage that the reader may well encounter the relevant phenomena in the laboratory. Particularly attractive for study are transition metal carbonyl species. For these, the $\nu(CO)$ modes are the particular concern. They fall in a region of the spectrum which is almost free from other modes, making assignment easy. They are associated with strong spectral bands, making measurement easy. They couple together rather strongly giving symmetry-determined modes, making interpretation easy.† The species which is the subject of this section, $(C_6H_6)Cr(CO)_3$, crystallizes in a space group which is rather simpler than the $P2_1/c$ of Section 12.7—the space

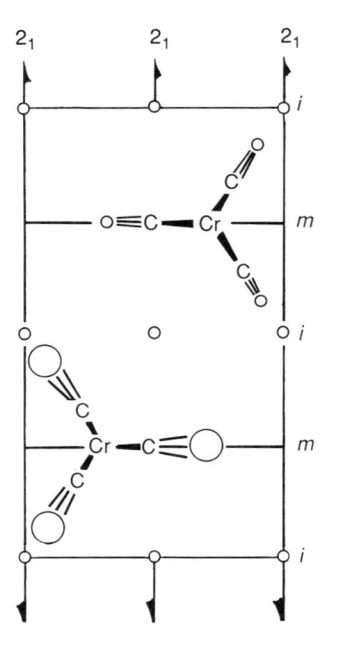

Figure 13.1   The crystal structure of the species $(C_6H_6)Cr(CO)_3$; only the $Cr(CO)_3$ groups are pictured, with their perspective being exaggerated. A primitive unit is shown and consists of two molecules. The space group is $P2_1/m$ (C_{2h}^2) and the molecular site symmetry is C_s.

---

† Easy it may be for simple species but, inevitably, research exploits this to enable the study of species which are complicated and the spectral interpretation no longer easy!

group P2$_1$/m. There are two molecules in the primitive unit ('in the unit cell' would be the usual terminology). As the m in the P2$_1$/m indicates, the space group contains mirror planes and the $(C_6H_6)Cr(CO)_3$ molecules lie on these, as shown in Figure 13.1. This means that the site symmetry is C_s, in contrast to the molecular symmetry which is C_{3v}. The factor group is isomorphic to the crystallographic point group, which is C_{2h} (in P2$_1$/m the C_2 of C_{2h} is 'replaced' by the 2$_1$). In Figure 13.1 and in the following discussion, only the $C \equiv O$ stretching modes will be considered.

Because the molecular symmetry is C_{3v}, the prediction of the symmetries of the $\nu(C \equiv O)$ stretching modes follows the discussion of Section 9.2, $A_1 + E$ modes being obtained. Section 10.4 shows that these modes are both infrared and Raman active. This isolated molecule model is also that of the oriented gas, leading to identical predictions. The oriented gas model differs from that of the isolated molecule because the former would recognize that the molecular C_3 axes are almost exactly aligned along the crystal c (z) axis. This means that if a single crystal were studied then if the incident infrared radiation were polarized along z then only the A_1 mode would appear with any great intensity in the infrared spectrum. When polarized perpendicular to this axis, only the E mode would be seen. With suitable experimental arrangements, a similar separation could also be achieved in the Raman. Without these experimental data, the isolated molecule and oriented gas models both simply predict two bands, coincident in infrared and Raman.†

The site group model is based on the fact that the molecules of $(C_6H_6)Cr(CO)_3$ are symmetrically arranged with respect to the mirror planes of P2$_1$/m. The three $\nu(C \equiv O)$ vibrators of each molecule are therefore, collectively, in an environment of C_s symmetry. The relationship between the groups C_{3v} and C_s was dealt with in Section 8.2 (Tables 8.5 and 8.6), from which it follows that the major effect of site symmetry is to split the degeneracy of the E modes of C_{3v}. Spectral activities remain unchanged so the predictions of the site group model is for three infrared bands (two of which, the components of the split E mode, will probably be close together) and three Raman bands, coincident with those in the infrared.

The factor group model will be dealt with in some detail. At the heart of the model is the fact that the translation operations can be ignored. Only the point group-derived operations isomorphous to those of C_{2h} need be considered. The fact that it is the C_{2h} point group which is relevant is evident in the alternative name given to P2$_1$/m in *International Tables*—C_{2h}^2, a symbol which includes the crystallographic point group. As mentioned in the previous chapter, the superscript 2 has no particular importance, it simply indicates that this space group is the second of the C_{2h}s listed in the *Tables*. The relationship between

---

† A further distinction between isolated and oriented gas models is that there is an environment-induced frequency shift of spectral features from the isolated molecule to the oriented gas model.

the operations of $P2_1/m$ and C_{2h} are:

C_{2h}	E	C_2	i	σ_h
$P2_1/m$	E	2_1	i	m

where, as before, a somewhat mixed nomenclature has been adopted for the operations of the crystallographic point group.† This means that the character table for the C_{2h}^2 ($P2_1/m$) factor group can be derived from that given for C_{2h} in Appendix 3 and is (conventionally, there is no reference to the translation group which has been 'factored out'):

C_{2h}^2	E	2_1	i	m		
A_g	1	1	1	1	R_z	x^2; y^2; z^2; xy
B_g	1	-1	1	-1	R_x; R_y	yz; zx
A_u	1	1	-1	-1	T_z	z
B_u	1	-1	-1	1	T_x; T_y	x; y

The next task is to use the six $\nu(C\equiv O)$ vibrators of two $(C_6H_6)Cr(CO)_3$ molecules (Z, the number of molecules in the primitive unit, is 2) as a basis to generate a reducible representation. In doing this it has to be remembered that, for example, both of the mirror planes in Figure 13.1 are effectively equivalent. The act of reflection of an object in two different mirror planes in this figure will lead to results which differ only in primitive translations—and these have been taken out of the problem by the use of the factor group. As far as the point group component is concerned, the final results are identical. So, the fact that the two molecules shown in Figure 13.1 lie on two apparently different mirror planes is no problem; the mirror planes are treated as one. Alternatively, each $M(C\equiv O)_3$ unit is reflected in the mirror plane on which it lies. From Figure 13.1 the reducible representation generated by the transformation of the $\nu(C\equiv O)$ vibrators is easily shown to be:

E	2_1	i	m
6	0	0	2

which has components $2A_g + B_g + A_u + 2B_u$.‡ As the character table above shows, all of the modes with a g suffix are Raman active and all of those with a u are infrared. The factor group predictions are therefore for three infrared

---

† As noted earlier, in the Hermann–Mauguin notation the identity element is denoted by 1; a centre of symmetry is $\bar{1}$. However, these are symmetry elements and in group theory it is operations that are relevant. Further, characters such as 1 and -1 will be generated in the application of these operations. To avoid possible confusion, the Schönflies symbols E and i are therefore used to denote the operations.

‡ Note, as mentioned earlier, the character of 0 under the operation which contains a non-primitive translation component. This arises because such operations almost invariably interrelate molecules.

bands and three Raman bands, non-coincident with the infrared. The observed spectra are shown in Figure 13.2 and are entirely in accord with these predictions. In both infrared and in Raman, two bands are close together and identified as derived from the split E mode discussed under the site symmetry model above. The sequence of increasing complexity:

Oriented gas model → site symmetry model → factor group model

should not be taken as meaning that the applicability of a more sophisticated model automatically invalidates all of the conclusions derived from a simpler model. So, as here, the site symmetry model can help in the application of the factor group.

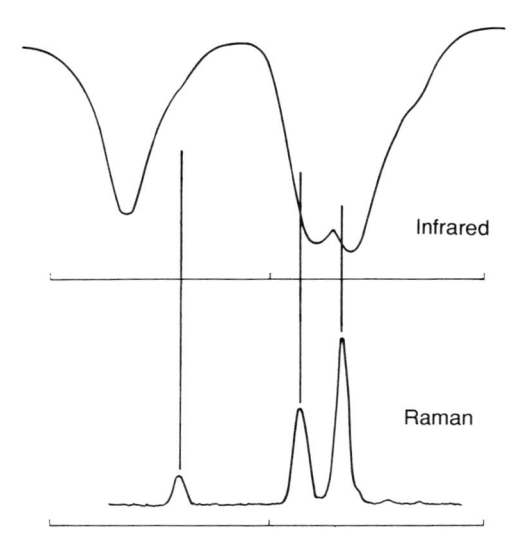

Figure 13.2 A comparison of the infrared and Raman spectra of crystalline $(C_6H_6)Cr(CO)_3$ in the $\nu(C \equiv O)$ region. Either the infrared or Raman on its own could be interpreted as a originating in the $A_1 + E$ modes of the isolated molecule (the E mode of the C_{3v} molecule being split by the lower site symmetry, C_s). However, comparison of the two indicates a general non-coincidence, explicable only in terms of the factor group model.

13.3.2   The vibrational spectrum of a $M(C \equiv O)_3$ species crystallizing in the $C2/c$ (C_{2h}^6) space group using the unit cell model

One of the simplest of the centred space groups is $C2/c$ and this simplicity is why it has been chosen as an example; it provides sufficient generality for more complicated cases subsequently to be treated with some confidence. The diagram that appears in *International Tables* for this space group is given in Figure 13.3. The C in $C2/c$ indicates that it is the face perpendicular to the c (z) axis which is centred, the unique (twofold) axis being b (y) (because, as was mentioned in the previous chapter, this is the convention formonoclinic systems). As the alternative name for the space group, C_{2h}^6, shows, the relevant unit cell group is C_{2h}. The first question that arises is that of the relationship between the operations of the two groups. Figure 13.3 shows an immediate problem, one that has been mentioned previously—but largely in a different context and with a different explanation. Figure 13.3 contains too many symmetry elements. For instance, not only are there twofold rotation axes (shown as the arrows pointing along y) but, interleaving them, 2_1 axes (shown as half-headed arrows pointing in the y direction). As befits C_{2h}, the glide planes are perpendicular to the twofold axes but, again, there are two sorts. Those shown dotted are glides in which the translation component is along c (out of the plane of the paper), as required by the $/c$ in the name of the space group. Glides in which the translation contains both c and a components (one half of a unit cell edge in each case) are shown dot-dashed. There is no mention of these latter glides in the name of the space group. Finally, although less obvious, there are twice as many centres of symmetry as are expected (as careful comparison with Figure 12.15, showing $P2_1/c$, C_{2h}^5, will confirm). These doublings result from the fact that Figure 13.3 contains two primitive units, units which are interrelated by a pure translation operation. If, as is convenient for crystallographers, Figure 13.3 is regarded as containing a *single* unit cell then this pure translation has to be combined with point group operations if the symmetry-relationship between all points is to be recognized. Hence the doubling. The 'extras' are indicated by the fact that they contain extra translation components compared with their genuine counterparts. So, the

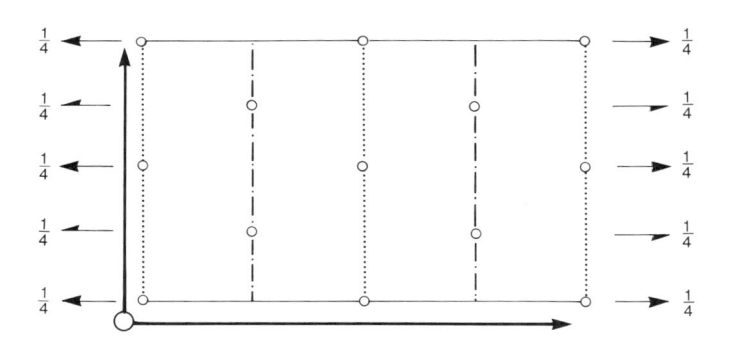

Figure 13.3 The diagram for the space group $C2/c$ that appears in *International Tables of Crystallography*. The translation vectors multiples of which generate the entire crystal from this unit cell have been added (that out-of-the-plane has been shown in symbolic fashion; actually, it is not perpendicular to the other two—the lattice is monoclinic). The meaning of the two different types of dotted lines is given in the text.

2_1 and glide containing a and c translation components are discarded. It is the position in space of the 'extra' centres of symmetry which contains their translation component (this aspect was discussed in Chapter 12; the reason that the twofold rotation axes are at $\frac{1}{4}c$ was also covered there).

A unit cell model requires a unit cell. As has been emphasized many times, there is no unique choice. Two are given in Figures 13.4(a) and 13.4(b) (which should be thought of as cross-sections of three-dimensional unit cells). That in Figure 13.4(a) is perhaps the more obvious, being a rectangular block from the crystallographic unit cell of *International Tables*. That shown in Figure 13.4(b) is the Wigner–Seitz unit cell, which for some purposes has advantages (to show that it is a Wigner–Seitz unit cell it is drawn with a

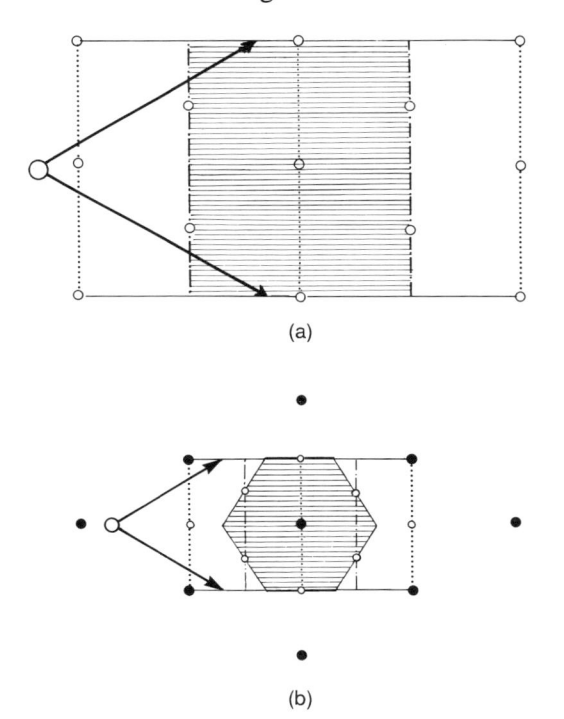

(a)

(b)

Figure 13.4   (a) A primitive unit cell obtained from that in Figure 13.3 by cutting it in half. The (genuine) primitive translation vectors, multiples of which generate the entire crystal from this unit cell have been added (that out-of-the-plane has been shown as in Figure 13.3).
(b) The Wigner–Seitz unit cell corresponding to that in Figure 13.3. The (genuine) primitive translation vectors, multiples of which generate the entire crystal from this unit cell, have been added (that out-of-the-plane has been shown as in Figure 13.3). As required, these are the same as those shown in (a). So that the construction of this unit cell can be followed, equivalent points are shown as black balls and the scale has been reduced compared to that in (a).The origin of the translation vectors is not the same in Figures (a) and (b).

smaller scale so that adjacent equivalent points can be shown). In both parts of Figure 13.4 the primitive translation vectors in the plane of the paper are shown; they are identical, as they have to be, and are non-orthogonal (not at 90°)—in contrast to the crystallographically preferred choice of axes (shown in Figure 13.3).

In Figure 13.5 sets of $M(C \equiv O)_3$ groups in the unit cell of Figure 13.4(a) are shown (there seems to be no actual species with data that enable the discussion to be of a real-life example). The four sets of $M(C \equiv O)_3$ groups are inter-related by the operations of the unit cell group; the ability to carry out these conversions is at the heart of the method. The results are:

$$C2/c \quad \begin{matrix} E & 2 & i & c \\ \alpha & \beta & \gamma & \delta \end{matrix}$$

Of these, only the operation of the c glide presents any difficulty. The c axis is perpendicular to the plane of the paper, so that the c glide operation involves reflection in a mirror plane (the mirror plane of C_{2h}) followed by a translation of $c/2$ perpendicular to the plane of the paper. The result of this $c/2$ translation depends on the choice of direction of translation, down or up. One of these will lead to the generation of a $M(C \equiv O)_3$ group within the unit cell; the other to the generation of a $M(C \equiv O)_3$ group in the adjacent unit cell. In the unit cell group, the closure requirement (Appendix 1) is achieved by the $M(C \equiv O)_3$ group which 'goes out' of the unit cell 'coming back in' through the opposite face (Figure 13.6). That is, the unit cell group is defined so that it does not matter whether the translation of $c/2$ in the c glide is 'up' or 'down', they lead to the same result. A similar situation holds for 2_1 screw axes; for 3_1 and similar screw axes the situation is a little more complicated.† The character

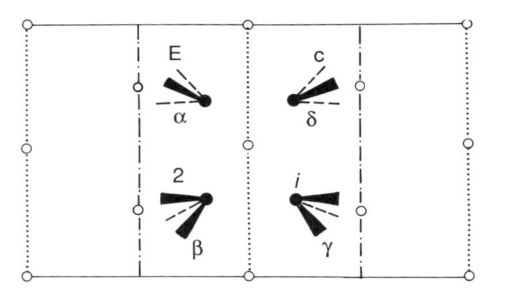

Figure 13.5   The interconversion of four sets of $M(C \equiv O)_3$ groups in the primitive unit cell of Figure 13.4(a). The relevant operations of Figure 13.3 are used (here, i is used to denote inversion in a centre of symmetry and c to indicate the c glide).

---

† If a 3_1 operation 'takes a point out' of the unit cell, then its reappearance through the opposite face means that it is equivalent to $(3_{-1})^2$. This is perhaps most readily seen by analogy with the point group relationship $C_3^+ \equiv (C_3^-)^2$, contained in Table 8.2.

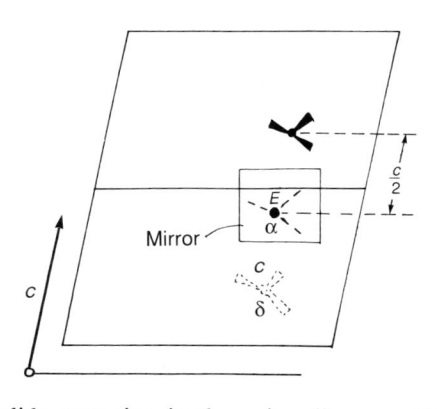

Figure 13.6 The c glide operation in the unit cell group. The 'starting' molecule (centre) is first reflected in a mirror plane (shown as a square around the molecule; this mirror plane is perpendicular to the 2 (C_2) axis, as required in the point group C_{2h}). This reflection is followed by a $c/2$ translation to complete the c glide operation. If the $c/2$ translation is upwards (to give the black molecule) then this molecule 'reappears' in the original unit cell (to give the molecule shown dotted). Had the $c/2$ translation component been taken in the downwards direction, the dotted molecule would have been that generated without the need to 'come back into' the unit cell.

table for the C_{2h}^6 unit cell group is obtained from that for C_{2h} using the correspondences:

C_{2h}	E	C_2	i	σ_h
C2/c(C_{2h}^6)	E	2	i	c

and is:

C_{2h}^6	E	2	i	c		
A_g	1	1	1	1	R_z	x^2; y^2; z^2; xy
B_g	1	-1	1	-1	R_x; R_y	yz; zx
A_u	1	1	-1	-1	T_z	z
B_u	1	-1	-1	1	T_x; T_y	x; y

The transformations of the four $M(C \equiv O)_3$ groups of Figure 13.5 (and so the corresponding $\nu(C \equiv O)$ vibrations) are straightforward and give rise to the reducible representation

E	2	i	c
12	0	0	0

which has $3A_g + 3B_g + 3A_u + 3B_u$ components, leading to a prediction of six infrared active modes and six Raman active, with no coincidences. Several comments are relevant. Because all of the characters (except that for the identity operation) are 0 no error would have resulted had we, in error, worked with the crystallographic (doubled) unit cell and divided the resulting reducible representation by two. However, if the problem had been one in which there were just $M(C \equiv O)$ groups in the primitive unit, each lying on a 2 (C_2) axis, then these would have given a non-zero character under this operation. On the other hand, they would have given a zero character under the 2_1 operations which, apparently, exist in the (doubled) unit cell. The reducible representations generated would have depended on just how this (unreal) dilemma was handled. It is scarcely likely that the correct prediction would have been obtained by dividing any of the possible answers by two!

13.4 SUMMARY

Most spectroscopic measurements on crystals involve phenomena which are translationally invariant (p. 291). As a consequence, great simplification results and knowledge of the crystallographic point group is sufficient to enable spectral predictions (and/or interpretations) to be made (p. 293). The factor and unit cell groups lead to identical predictions but are only relevant when there is coupling between corresponding transitions in different molecules (p. 296).† As the phenomena observed become increasingly localized, the site symmetry and oriented gas models become more applicable (p. 293).

---

† Several alternative names exist which are used to describe the resulting splittings; different names tend to be the province of different areas of spectroscopy. While the names 'factor group' or 'unit cell group' splittings are readily understood because of the content of the present chapter, the alternative names 'correlation field' or 'Davydov' splitting may also be encountered.

Appendix 1
Groups and Classes: Definitions and Examples

A1.1 GROUPS

In Chapter 2 a definition was given of a group which was adequate for the purposes at that point in the text but which was incomplete; all of the requirements were not detailed. The first object of this appendix is to remedy this deficiency and to accurately define the word 'group'. At some points in this appendix it will implicitly be assumed that the group under discussion does not contain an infinite number of elements. This excludes $C_{\infty v}$ and $D_{\infty h}$ — but all of the general statements made can be shown to apply to these two groups also.

Suppose we have a collection of elements (some examples will be given shortly which will help to indicate the breadth of the term 'element'). The set of these elements, $A, B, C, \ldots$ form a group G, written as

$$G = \{A, B, C, \ldots\} \text{ if:}$$

1 There is some *law of combination* which relates the elements one to another. No matter what the precise nature of the operation of combination, it is called *multiplication*. So, the fact that A combines with B to give C would be written:

$$AB = C \tag{A1.1}$$

At the end of this section several different laws of multiplication will be given to illustrate equation (A1.1).

Note: (a) Whenever a group is specified it is, formally, necessary to also specify the law of combination. (b) The order in which elements multiply is important. There is NO general requirement that, for instance,

$$AB = BA$$

so it must be assumed that, in general,

$$AB \neq BA \tag{A1.2}$$

More detailed consideration of this inequality will lead to the concept of *class* later in this appendix.

2 Multiplication is closed (the *closure requirement*).
That is, the product of any two elements within a group is an element within the group.

Note: 'An element' here means a *single* element. Multiplication is single valued; there can never be any ambiguity about the outcome of a multiplication. Thus, it can happen that $AB = C$ and $AB = D$ if and only if

$$C = D$$

3 Multiplication is *associative*.
One might think that once the multiplication of two elements is defined there would be no problem about multiplying any number together. This is not the case. Consider the triple product

$$ABC$$

Because we only know how to multiply pairs of elements we have to select a pair from this trio and multiply them first. There is a choice between

$$(AB)C \quad \text{and} \quad A(BC)$$

But suppose $AB = C$ (as above) and $BC = D$. Then our products are

$$CC \quad \text{and} \quad AD$$

It is by no means evident that these are equal unless this equality is introduced as a requirement. This is just what the statement 'multiplication is associative' does. It means that it must be true that

$$(AB)C = A(BC)$$

for the elements to form a group.
Note: This means that a string of elements can now be multiplied together. Thus,

$$(AB)CD = A(BC)D = AB(CD)$$

4 The group contains a *unit element* (often denoted E or I). This unit element plays a role which in some ways resembles that of the number 1 in ordinary arithmetic. Thus, when it multiplies any other element of the group, A, say, the product is A.

$$\text{i.e.} \quad EA = AE = A \qquad (A1.3)$$

5 For each element in the group there is a unique element which is its *inverse*.
Loosely speaking, the inverse of an element 'undoes' the effect of that element. Thus, C_3^- is the inverse of C_3^+ (Chapter 6).
The inverse of the element A is usually written A^{-1} (in ordinary arithmetic think of multiplying by, say, the number 7. This multiplication can be

cancelled out by multiplying again, this time by the number $7^{-1} = 1/7$). That is,

$$AA^{-1} = A^{-1}A = E \qquad (A1.4)$$

Note: The element which has here been called A^{-1} would normally have another label within the group; it could be B, for instance, or it could be A itself if A were self-inverse. The label A^{-1} is here used in preference to, say, B, because the latter label does not reveal the special relationship between A and A^{-1} given by the equation above.

Problem A1.1 Apply the relationships given above to the elements of the C_{2v} group $(E, C_2, \sigma_v, \sigma_v')$ and thus, formally, show that they comprise a group.

A1.2 SOME EXAMPLES OF GROUPS

1 *Permutation groups* were briefly encountered towards the end of Chapter 9. The groups formed by the operations permuting n objects form a fascinating subject for study. The character table for the so-called 'symmetric group' (the permutation group) with $n = 2$ is isomorphic to that of C_2, that for $n = 3$ is isomorphic to C_{3v} and that for $n = 4$ is isomorphic to T_d. The groups with $n \geqslant 5$ are not isomorphic to any point group. The symmetric groups are of potential importance when identical particles are of interest. In chemistry these particles could be identical nuclei but more frequently they are electrons.

The symmetric group with $n = 3$ has six operations, the three particles being labelled a, b and c. If (a) indicates that a is not permuted, (ab) means 'interchange a and b' and (abc) means cyclically permute a, b and c, then the six operations are:

$(a)(b)(c)$	(abc)	(acb)	$(ab)(c)$	$(ac)(b)$	$(a)(bc)$
E	P_1	P_2	X_1	X_2	X_3

Problem 9.4 gives an example of the combination of permutation operations of this type. Using the shorthand symbols indicated (P—cyclic **P**ermutation; $X = $ e**X**change) the following group multiplication table is obtained.

		First Operation					
		E	P_1	P_2	X_1	X_2	X_3
	E	E	P_1	P_2	X_1	X_2	X_3
	P_1	P_1	P_2	E	X_2	X_3	X_1
Second	P_2	P_2	E	P_1	X_3	X_1	X_2
Operation	X_1	X_1	X_3	X_2	E	P_1	P_2
	X_2	X_2	X_1	X_3	P_2	E	P_1
	X_3	X_3	X_2	X_1	P_1	P_2	E

This multiplication table is isomorphic to that for C_{3v} given in Table 8.2 (substitute systematically C_3^+ for P_1, C_3^- for P_2, X_1 for $\sigma_v(1)$ etc.).

Problem A1.2   Check that the above multiplication table is correct.

2   *Substitution groups* are fun—and have played an important part in the development of group theory—but do not seem to have any general application. Consider the six functions (which, strictly, should be written $E(x)$ etc.):

$$E = x \qquad P = 1/(1-x) \qquad Q = (x-1)/x$$
$$R = 1/x \qquad S = 1-x \qquad T = x/(x-1)$$

These form a group when the law of combination is substitution as function of a function. Thus,

$$SR = S(1/x) = 1 - (1/x) = (x-1)/x = Q$$

and

$$PT = P[x/(x-1)] = \cfrac{1}{1 - \left(\cfrac{x}{x-1}\right)} = 1 - x = S$$

The multiplication table, given below, is also isomorphic to that of the C_{3v} group (Table 8.2). The reader should check this. But he or she should be warned. The table isomorphic to C_{3v} given in the previous example was set out in a way that should have made the isomorphism self-evident. This is not true for that given below, more work will be required to demonstrate the isomorphism.

			First Operation			
		E	P	Q	R	S	T
	E	E	P	Q	R	S	T
	P	P	Q	E	T	R	S
Second	Q	Q	E	P	S	T	R
Operation	R	R	S	T	E	P	Q
	S	S	T	R	Q	E	P
	T	T	R	S	P	Q	E

Problem A1.3   Check that the above multiplication table is correct.

3   An example of a two-colour group has been given in the discussion associated with Figure 2.5. There, the changing of a colour was introduced as a component of a symmetry operation. Colour groups are of some importance in chemistry in the context of space groups, although beyond that discussed in Chapters 12 and 13. Many of the operations of space groups have the effect of relating one molecule in a crystal lattice to another. But what if the molecules are not quite identical? For instance, the molecules could be atomically identical but have opposite magnetic properties (because the electron spins are arranged in opposite ways, for example). In this case the operation—put colloquially—of 'turn the magnet over' is similar to the 'change the colour' operation; it forms a composite with another symmetry operation to relate not-quite identical objects. *Two-colour* space groups are also known as *black and white* groups or *Shubnikov* groups. Polychromatic groups also exist.

4   With the exception of those groups indicated above, all of the point groups discussed in this book relate to our three-dimensional world. However, it is possible to add to this picture that of electron spin. It is not at all obvious why it should be the case, but this has the effect of doubling the number of operations in the group compared to the corresponding point group without spin. Hence, these are called *double groups*.† They have the rather unexpected property that the identity operation corresponds to a rotation of 720°. The double groups are of importance in some areas of chemistry, in particular, in the theory of transition metal ions and of the lanthanides and actinides.

A1.3   THE CLASSES OF A GROUP

When in the previous section the definition of a group was detailed it was found necessary to recognize that the multiplication of any two elements, A and B, of a group could not be assumed to be *commutative*. That is, it is not generally true that

$$AB = BA$$

[when either A or B is the identity, E, the equation is always true—it is equation (A1.3)]. This equation may hold for some pairs of operations within a group but not others (for example, it is true for all pairs of σ_v operations in the C_{3v} point group, but is untrue when a C_3 is combined with a σ_v, see Table 8.2). Groups for which it is true for all pairs of elements are *Abelian* point groups. C_2 (Chapters 2 and 3), D_{2h} (Chapter 4) and C_4 (Chapter 11) are examples of Abelian point groups. In Abelian point groups there are never two elements 'in the same class'. Non-Abelian point groups may have more than one element in each class and so, in giving a more precise meaning to the word 'class', equation (A1.2) is a good starting point since it applies to at least some of the operations of non-Abelian groups:

$$AB \neq BA$$

---

† The author has given a non-mathematical account of them in another book, *Physical Inorganic Chemistry*, Spektrum (Freeman), Oxford, 1995, Section 11.5.

Multiply each side of this equation, on the right, by the operation A^{-1}. This gives:

$$ABA^{-1} \neq BAA^{-1}$$

But $AA^{-1} = E$ [equation (A1.4)] and so $BAA^{-1} = BE = B$ [by equation (A1.3)]. That is,

$$ABA^{-1} \neq B$$

The product ABA^{-1} must be equivalent to a single operation in the group. To be general, let us call this single operation D. That is,

$$ABA^{-1} = D \tag{A1.5}$$

There is a hidden symmetry in equation (A1.5). To see this, multiply on the left of each side of the equation by A^{-1} and on the right of each side by A. The result is:

$$A^{-1}(ABA^{-1})A = A^{-1}(D)A$$

Because multiplication is associative, this can be written:

$$(A^{-1}A)B(A^{-1}A) = A^{-1}DA$$

Which, by equation (A1.4), becomes

$$B = A^{-1}DA \tag{A1.6}$$

which is to be compared with equation (A1.5). Because of this relationship between B and D they are said to be *conjugate* elements of the group. But A was picked at random in the above development—no restrictions were placed on it. Suppose a different element, C say, had been chosen in its place? There is no theorem which would require that because

$$ABA^{-1} = D \tag{A1.5}$$

then

$$CBC^{-1} = D$$

Rather, it must be assumed that CBC^{-1} gives yet another element (even if, sometimes, it does not). Consider the case where it does not give D but another element, F, say. So,

$$CBC^{-1} = F \tag{A1.7}$$

But the arguments leading up to (A1.6) above can be paralleled with a similar development to show from (A1.7) that

$$B = C^{-1}FC \tag{A1.8}$$

That is, B is conjugate with F as well as with D. Not surprisingly, this sequence requires that F and D are also conjugate elements, as may be shown by combining (A1.6) and (A1.8).

$$A^{-1}DA = B = C^{-1}FC$$

Consider the two outer expressions and multiply each on the left by A and on the right by A^{-1}.

$$(AA^{-1})D(AA^{-1}) = (AC^{-1})F(CA^{-1})$$

That is,

$$D = (AC^{-1})F(CA^{-1}) \tag{A1.9}$$

Equation (A1.9) is of a form analogous to (A1.5), (A1.6), (A1.7) and (A1.8) provided that it can be shown that (AC^{-1}) and (CA^{-1}) are inverses of each other. If they are inverses then they satisfy (A1.4) and so they should multiply together to give E. We have:

$$\begin{aligned}
(AC^{-1})&(CA^{-1}) \\
&= AC^{-1}CA^{-1} \\
&= A(C^{-1}C)A^{-1} \\
&= A(E)A^{-1} \\
&= AA^{-1} \\
&= E
\end{aligned}$$

That is, (AC^{-1}) and (CA^{-1}) are, indeed, inverses. Now AC^{-1} must be equal to a single element of the group; call it H. CA^{-1} must then be H^{-1} so that (A1.9) becomes

$$D = HFH^{-1} \tag{A1.10}$$

which is of the form required. We conclude that B, D and F are all conjugate elements and comprise a subset of the set of all the group operations. Each set of conjugate elements in a group forms a *class* of the group.

Formally, then, in order to find all members of a group which are of the same class as B each element of the group in turn (including B) is taken as A in the expression

$$ABA^{-1} \qquad \text{[see (A1.5)]}$$

and all of the products are collected together. They comprise all the elements which fall in the same class as B.

As an example consider the substitution group given in the previous section and use its multiplication table (page 309). First, from the table the inverse of each element is identified:

Element	Inverse
E	E
P	Q
Q	P
R	R
S	S
T	T

To obtain all elements in the same class as P, work down this list forming the products of the form APA^{-1}, where A and its inverse are obtained from the listing above. The results are given below

$$EPE = P$$
$$PPQ = P$$
$$QPP = P$$
$$RPR = Q$$
$$SPS = Q$$
$$TPT = Q$$

It is concluded that P and Q are in the same class (a result which could have been anticipated because they are isomorphous with the C_3^+ and C_3^- operations of C_{3v}).

Problem A1.4   Check the above argument.

As a second example we consider the problem encountered in Chapter 11, that C_4 and C_4^3 are in different classes in the C_4 group. The group multiplication table for the C_4 group is (note its diagonal symmetry):

C_4	E	C_4	C_2	C_4^3
E	E	C_4	C_2	C_4
C_4	C_4	C_2	C_4^3	E
C_2	C_2	C_4	E	C_4
C_4^3	C_4^3	E	C_4	C_2

from which it is evident that the inverses are:

Element	Inverse
E	E
C_4	C_4^3
C_2	C_2
C_4^3	C_4

In the class containing C_4 there will be

$$EC_4E = C_4$$
$$C_4C_4C_4^3 = C_4$$
$$C_2C_4C_2 = C_4$$
$$C_4^3C_4C_4 = C_4$$

That is, the operation C_4 is in a class of its own. It is easy to similarly show that C_4^3 is in a class of its own, as too is C_2. This shows that C_4 is an Abelian group.

Problem A1.5 Demonstrate that C_4^3 is in a class of its own.

Problem A1.6 Show that the C_{2v} group is an Abelian group.

A1.4 CLASS ALGEBRA

When the C_{3v} character table was introduced in Chapter 6 it was done so in the form

C_{3v}	E	$2C_3$	$3\sigma_v$
A_1	1	1	1
A_2	1	1	-1
E	2	-1	0

and this, and the character tables of all other non-Abelian groups are given in this form in Appendix 3. Why? Why put elements which fall in the same class, such as C_3^+ and C_3^-, together as $2C_3$? Why not write this character table as:

C_{3v}	E	C_3^+	C_3^-	$\sigma_v(1)$	$\sigma_v(2)$	$\sigma_v(3)$
A_1	1	1	1	1	1	1
A_2	1	1	1	-1	-1	-1
E	2	-1	-1	0	0	0

After all, this is the form in which, effectively, it was used in the projection operator method (see, for instance, page 129). First, we note that not all of the character table orthonormality relationships (Section 5.3) would remain true if this form of character table were used (some columns in the extended character table are identical). There is, however, another and fundamental reason. This is that there exists a *class algebra*. Take the C_{3v} group as an example. It contains three classes with elements

Class 1		E	
Class 2		C_3^+ C_3^-	
Class 3	$\sigma_v(1)$	$\sigma_v(2)$	$\sigma_v(3)$

Express this mathematically, thus:

$$\mathbb{C}_1 = E$$
$$\mathbb{C}_2 = \tfrac{1}{2}(C_3^+ + C_3^-)$$
$$\mathbb{C}_3 = \tfrac{1}{3}[\sigma_v(1) + \sigma_v(1) + \sigma_v(3)]$$

These classes can be multiplied together. Thus,

$$\mathbb{C}_2\mathbb{C}_2 = \tfrac{1}{4}(C_3^+ + C_3^-)(C_3^+ + C_3^-)$$
$$= \tfrac{1}{4}[C_3^+C_3^+ + C_3^+C_3^- + C_3^-C_3^+ + C_3^-C_3^-]$$

which, from Table 8.2, is equal to

$$= \tfrac{1}{4}[C_3^- + E + E + C_3^+]$$
$$= \tfrac{1}{2}E + \tfrac{1}{4}(C_3^+ + C_3^-)$$
$$= \tfrac{1}{2}(\mathbb{C}_1 + \mathbb{C}_2)$$

A class multiplication table can thus be compiled and which is easily shown to be:

C_{3v}	$\mathbb{C}_1$	$\mathbb{C}_2$	$\mathbb{C}_3$
$\mathbb{C}_1$	$\mathbb{C}_1$	$\mathbb{C}_2$	$\mathbb{C}_3$
$\mathbb{C}_2$	$\mathbb{C}_2$	$\tfrac{1}{2}(\mathbb{C}_1 + \mathbb{C}_2)$	$\mathbb{C}_3$
$\mathbb{C}_3$	$\mathbb{C}_3$	$\mathbb{C}_3$	$\tfrac{1}{3}(\mathbb{C}_1 + 2\mathbb{C}_2)$

Problem A1.7   Check that the class multiplication table shown above is correct.

Problem A1.8   Show that the above classes do *not* form a group under the operation of class multiplication. (*Hint:* Refer to the relationships used to define a group at the beginning of this appendix.)

The classes of Abelian groups form groups under class multiplication but this is trivial because the classes are isomorphic to the elements of the Abelian group itself.

Problem A1.9   Check the truth of the above assertion by reference to the C_{2v} point group.

From the class multiplication table given above it is seen that, in general, the product of multiplying two classes together is of the form

$$\mathbb{C}_j\mathbb{C}_i = \sum_k c_k\mathbb{C}_k$$

where the sum k is over all classes and c_k is a coefficient. We now ask what may appear a rather strange question. Is it possible to obtain a linear sum of the classes of the form

$$\mathscr{E} = \sum_j a_j\mathbb{C}_j$$

which has the property that when multiplied by any class, $\mathbb{C}_i$ say, it satisfies an equation of the form

$$\mathbb{C}_i\mathscr{E} = \lambda\mathscr{E}$$

where λ is a number (possibly complex)?

Those with some knowledge of quantum mechanics will recognize this as an eigenvalue equation. The eigenvalues, λ, when determined, lead directly to the characters in the character table (these characters are not the λ's but are related to them by simple, well defined, numerical coefficients). That is, the characters in a character table are intimately related to the classes. This is the reason that character tables are given in the way that they are.

Clearly, the mathematics given above can be developed to provide a method for the calculation of character tables. This development will not be given here but the interested reader will find a very readable account in a book by G. G. Hall, *Applied Group Theory* (Longmans, 1967).

Appendix 2
The Mathematical Basis of Group Theory

This book contains a non-mathematical treatment of what, in fact, is a mathematical subject. This appendix goes some way towards reinstating the mathematics. However, it cannot claim to be comprehensive—if it were, its length would be very much greater.

A2.1 MATRIX ALGEBRA AND SYMMETRY OPERATIONS

An array of quantities—often numbers—such as those given below is called a *matrix*

$$\begin{bmatrix} 3 & 2 \\ 4 & -1 \\ 0 & 2 \end{bmatrix} \quad \text{and} \quad \begin{bmatrix} 3 & 2 & -2 \\ 4 & -1 & 0 \\ 0 & 2 & 3 \end{bmatrix}$$

Clearly, matrices can be square—contain the same number of rows as they have columns—or they may be rectangular; the number of rows may be greater or less than the number of columns. Each number or other quantity appearing in a matrix is referred to as a *matrix element*. If represented by an algebraic symbol a matrix element is often given suffixes to indicate in which row and which column it lies in the matrix.

Matrices of the same size may be added; this is done by adding together the corresponding entries (elements). We illustrate this by adding two matrices; as an aid to clarity the elements of one matrix are given as letters

$$\begin{bmatrix} 3 & 2 & -2 \\ 4 & -1 & 0 \\ 0 & 2 & 3 \end{bmatrix} + \begin{bmatrix} a & b & c \\ d & e & f \\ g & h & i \end{bmatrix} = \begin{bmatrix} (3+a) & (2+b) & (-2+c) \\ (4+d) & (-1+e) & f \\ g & (2+h) & (3+i) \end{bmatrix}$$

Problem A2.1 Fill in the missing quantities in the following matrix equation

$$\begin{bmatrix} \sin^2\theta & \dfrac{1}{\sqrt{2}} & 3 \\ . & . & . \\ 3 & 1 & -\sin^2\theta \end{bmatrix} + \begin{bmatrix} . & . & 0 \\ 1 & \cos^2\phi & -1 \\ -1 & 5 & . \end{bmatrix} = \begin{bmatrix} 1 & \sqrt{2} & . \\ 1 & 0 & -4 \\ . & . & \cos 2\theta \end{bmatrix}$$

The application of matrix algebra to the theory of groups is relatively limited and we shall have no occasion to add or subtract matrices. Key to our use of them, however, is the multiplication of matrices. Matrix multiplication does *not* parallel matrix addition, one does *not* simply multiply corresponding pairs of elements together. Although pairs of elements are, indeed, multiplied, each element in a complete row is multiplied by the corresponding element in a complete column—so that the row and column have to be of equal length, to contain the same number of elements—and the products are added together. It is this sum of products that is an element in the product matrix. To obtain the entry in the mth row and the nth column of the product matrix, the elements in the mth row of the first matrix are multiplied by those in the nth column of the second.

Consider the two matrices which were added above. Now, let us multiply them. The entry at the top left-hand corner of the product matrix, the one in the first row ($m = 1$) and first column ($n = 1$) is given by:

$$\text{first row} \rightarrow \begin{bmatrix} 3 & 2 & -2 \\ \cdot & \cdot & \cdot \\ \cdot & \cdot & \cdot \end{bmatrix} \times \begin{bmatrix} a & \cdot & \cdot \\ d & \cdot & \cdot \\ g & \cdot & \cdot \end{bmatrix} = \begin{bmatrix} (3a + 2d - 2g) & \cdot & \cdot \\ \cdot & & \cdot \\ \cdot & & \cdot \end{bmatrix}$$

$$\uparrow$$
$$\text{first column}$$

and the reader who is unfamiliar with matrix multiplication should check several of the elements of this product.

Problem A2.2   Fill in the blanks in the following matrix equation:

$$\begin{bmatrix} 3 & -1 \\ 2 & \cdot \end{bmatrix} \times \begin{bmatrix} 2 & -1 \\ \cdot & 3 \end{bmatrix} = \begin{bmatrix} 0 & \cdot \\ 10 & \cdot \end{bmatrix}$$

The multiplication of two matrices may be expressed algebraically. If the product of the matrices A and B (A being on the left) is denoted AB, then

$$(AB)_{mn} = \sum_{t} A_{mt} B_{tn} \tag{A2.1}$$

where m and n carry the meanings given above and t is simply a convenient running label which enables us to distinguish the individual matrix element products which have to be added together to give the element in the mth row and nth column of the product matrix AB.

In the example above, the two matrices which were multiplied together were square but this is not a requirement; the sole restriction is the obvious one stated above—that the number of elements in each row of the matrix on the left of the multiplication sign equals the number of elements in each column of the matrix on the right.

The relevance of this to molecular symmetry can be seen by reference to Figure 3.1. This shows the transformation of the hydrogen 1s orbitals, h_1 and h_2, under the symmetry operations of the C_{2v} point group. It was discussed in Section 3.1. Figure 3.1 shows that under the identity operation, E, h_1 and h_2 remain unchanged. This can be expressed by the matrix product

$$\begin{bmatrix} 1 & 0 \\ 0 & 1 \end{bmatrix} \begin{bmatrix} h_1 \\ h_2 \end{bmatrix} = \begin{bmatrix} h_1 \\ h_2 \end{bmatrix} \tag{A2.2}$$

where, following convention, the multiplication sign between the two matrices multiplied together has been omitted. Writing them side by side in this way is taken as meaning that they are to be multiplied. The reader should check that, arithmetically, equation (A2.2) is correct. It may be correct, but what does it mean? On the left-hand side the hydrogen 1s orbitals are written as the elements of a column matrix. The order in which they are written is, ultimately, unimportant but that used is clearly the more natural. When the matrix multiplication is carried out this column matrix is regenerated, unchanged. That is, multiplication of the matrix $\begin{bmatrix} 1 & 0 \\ 0 & 1 \end{bmatrix}$ has a similar effect on h_1 and h_2 as the identity operation. However, had a different matrix been used to multiply the h_1, h_2 column matrix a different result would have been obtained.

Figure 3.1 shows that the C_2 rotation interchanges h_1 and h_2. The reader can readily show this is expressed by the matrix product

$$\begin{bmatrix} 0 & 1 \\ 1 & 0 \end{bmatrix} \begin{bmatrix} h_1 \\ h_2 \end{bmatrix} = \begin{bmatrix} h_2 \\ h_1 \end{bmatrix}$$

Here, the matrix $\begin{bmatrix} 0 & 1 \\ 1 & 0 \end{bmatrix}$ has a similar effect on $\begin{bmatrix} h_1 \\ h_2 \end{bmatrix}$ as the C_2 operation has on the h_1 and h_2; h_1 and h_2 are interchanged.

It is left as an exercise for the reader to show that effect of the σ_v and σ_v' operations on h_1 and h_2 are paralleled in the matrix products:

$$\sigma_v: \quad \begin{bmatrix} 1 & 0 \\ 0 & 1 \end{bmatrix} \begin{bmatrix} h_1 \\ h_2 \end{bmatrix} = \begin{bmatrix} h_1 \\ h_2 \end{bmatrix} \tag{A2.4}$$

$$\sigma_v': \quad \begin{bmatrix} 0 & 1 \\ 1 & 0 \end{bmatrix} \begin{bmatrix} h_1 \\ h_2 \end{bmatrix} = \begin{bmatrix} h_2 \\ h_1 \end{bmatrix} \tag{A2.5}$$

Problem A2.3 Show, by expansion and comparison with Chapter 3, that equations (A2.4) and (A2.5) correctly describe the action of σ_v and σ_v' on h_1 and h_2.

In Chapter 2 it was shown that sets of numbers such as $1, 1, -1, -1$ multiply in a manner which is isomorphic to the multiplication of the operations of the C_2 point group (see Table 2.3, for example). The important thing about the square matrices in equations (A2.2) → (A2.5) is that when multiplied under the rules of matrix multiplication they, too, multiply isomorphically to the C_2 operations. The multiplication of these 2×2 matrices is given in Table A2.1.

Table A2.1

		E	C_2	σ_v	σ_v'
		\multicolumn{4}{c}{Right-hand matrix in the product}			
		$\begin{bmatrix} 1 & 0 \\ 0 & 1 \end{bmatrix}$	$\begin{bmatrix} 0 & 1 \\ 1 & 0 \end{bmatrix}$	$\begin{bmatrix} 1 & 0 \\ 0 & 1 \end{bmatrix}$	$\begin{bmatrix} 0 & 1 \\ 1 & 0 \end{bmatrix}$
Left-hand matrix in the product	$\begin{bmatrix} 1 & 0 \\ 0 & 1 \end{bmatrix}$	$\begin{bmatrix} 1 & 0 \\ 0 & 1 \end{bmatrix}$	$\begin{bmatrix} 0 & 1 \\ 1 & 0 \end{bmatrix}$	$\begin{bmatrix} 1 & 0 \\ 0 & 1 \end{bmatrix}$	$\begin{bmatrix} 0 & 1 \\ 1 & 0 \end{bmatrix}$
	$\begin{bmatrix} 0 & 1 \\ 1 & 0 \end{bmatrix}$	$\begin{bmatrix} 0 & 1 \\ 1 & 0 \end{bmatrix}$	$\begin{bmatrix} 1 & 0 \\ 0 & 1 \end{bmatrix}$	$\begin{bmatrix} 0 & 1 \\ 1 & 0 \end{bmatrix}$	$\begin{bmatrix} 1 & 0 \\ 0 & 1 \end{bmatrix}$
	$\begin{bmatrix} 1 & 0 \\ 0 & 1 \end{bmatrix}$	$\begin{bmatrix} 1 & 0 \\ 0 & 1 \end{bmatrix}$	$\begin{bmatrix} 0 & 1 \\ 1 & 0 \end{bmatrix}$	$\begin{bmatrix} 1 & 0 \\ 0 & 1 \end{bmatrix}$	$\begin{bmatrix} 0 & 1 \\ 1 & 0 \end{bmatrix}$
	$\begin{bmatrix} 0 & 1 \\ 1 & 0 \end{bmatrix}$	$\begin{bmatrix} 0 & 1 \\ 1 & 0 \end{bmatrix}$	$\begin{bmatrix} 1 & 0 \\ 0 & 1 \end{bmatrix}$	$\begin{bmatrix} 0 & 1 \\ 1 & 0 \end{bmatrix}$	$\begin{bmatrix} 1 & 0 \\ 0 & 1 \end{bmatrix}$

Problem A2.4   Check that Table A2.1 is correct.

Table A2.1 should be compared with Table 2.1. Each matrix in Table A2.1 will be found to transform isomorphically to the operation associated with it. Is this property limited to 2×2 matrices? No, provided that they are square matrices, matrices of any order can be found which multiply isomorphically to the operations of the C_{2v} point group. Indeed, the numbers which behaved like this in Chapter 2 may be regarded as 1×1 matrices! As an example of this, the following four matrices describe the transformations of the hydrogen atoms

$$\begin{bmatrix} H_a \\ H_b \\ H_c \\ H_d \end{bmatrix}$$

of Figure 2.16.

$$E: \begin{bmatrix} 1 & 0 & 0 & 0 \\ 0 & 1 & 0 & 0 \\ 0 & 0 & 1 & 0 \\ 0 & 0 & 0 & 1 \end{bmatrix}$$

$$C_2: \begin{bmatrix} 0 & 0 & 1 & 0 \\ 0 & 0 & 0 & 1 \\ 1 & 0 & 0 & 0 \\ 0 & 1 & 0 & 0 \end{bmatrix}$$

$$\sigma_v: \begin{bmatrix} 0 & 1 & 0 & 0 \\ 1 & 0 & 0 & 0 \\ 0 & 0 & 0 & 1 \\ 0 & 0 & 1 & 0 \end{bmatrix}$$

$$\sigma_v': \begin{bmatrix} 0 & 0 & 0 & 1 \\ 0 & 0 & 1 & 0 \\ 0 & 1 & 0 & 0 \\ 1 & 0 & 0 & 0 \end{bmatrix}$$

Further, the multiplication of these matrices is isomorphic to that of the corresponding operations of the C_{2v} point group.

Problem A2.5 Show that the above matrices do, indeed, describe the transformations of the hydrogen atoms of Figure 2.16.

Problem A2.6 Show that the multiplication of the above matrices is isomorphous to that of the operations of the C_{2v} point group (it may be helpful to use Figure 2.17 as a check).

Matrix multiplication, then, provides a method of describing in detail the transformation of several objects under the operations of a point group. But in the text—in Section 3.2—something similar has been described. It was in Section 3.2 that we first used the transformation of several objects under the operations of a point group to obtain reducible representations. Not surprisingly, the two methods, the transformations of several objects and the matrix, are connected. In Section 3.2 what was described was a method of obtaining the characters of reducible representations. The bridge between this and the matrix formalism appears when it is recognized that 'character' is the name given to the arithmetic sum of all of the elements on the leading diagonal (top left to bottom right) of a square matrix. So, application of this definition to the four matrices given immediately above gives their characters as:

	Matrix associated with			
	E	C_2	σ_v	σ_v'
Character of the matrix	4	0	0	0

This set of characters is just that obtained for the reducible representation generated by the transformations of the four hydrogen atoms in Figure 2.16.

Problem A2.7 (cf. Problem 3.2)   Check that the transformations of the hydrogen atoms of Figure 2.16 leads to the above representation.

Note: The representation which has the number which is equal to the order of the group (here, 4) under the E operation and zeros elsewhere is called *the regular representation*. It is of importance because it is used in the proof of some group theoretical theorems (but none which are included in this book). It is generated by a basis set which is not associated with any symmetry element. Thus, here, the hydrogen atoms are in general positions—they do not lie on a mirror plane or symmetry axis and so the regular representation is generated.

The rules for the generation of characters given in boxes in Section 3.2 are now seen as arising from the definition of the character of a matrix and the fact that it is only when its transformation is described by an entry on the leading diagonal that an object remains unmoved under a symmetry operation. (A word of caution; this last statement will need some modification shortly when fractions will appear on the diagonal.)

Just as one distinguishes between reducible and irreducible representations, so one may distinguish reducible from irreducible matrix representations. Irreducible matrix representations will be met later in this section and the connection between reducible and irreducible is covered in Section A2.4. Both the 2×2 and 4×4 matrices given above are sets of reducible matrices.

Those functions whose transformations are described by matrices in the way just described are called *basis functions*. Those basis functions given at the right-hand side of character tables (Appendix 3) are ultimately related to the transformation of the x, y and z coordinate axes. It is therefore important to consider the transformation of a set of coordinate axes under typical group symmetry operations. This is not a difficult problem. For example, the inversion operation, i, is described by

$$i: \quad \begin{bmatrix} -1 & 0 & 0 \\ 0 & -1 & 0 \\ 0 & 0 & -1 \end{bmatrix} \begin{bmatrix} x \\ y \\ z \end{bmatrix} = \begin{bmatrix} -x \\ -y \\ -z \end{bmatrix}$$

Reflection in a mirror plane (let us choose the yz plane as the mirror plane) is:

$$\sigma(yz): \quad \begin{bmatrix} -1 & 0 & 0 \\ 0 & 1 & 0 \\ 0 & 0 & 1 \end{bmatrix} \begin{bmatrix} x \\ y \\ z \end{bmatrix} = \begin{bmatrix} -x \\ y \\ z \end{bmatrix}$$

The problem of the rotation of axes was discussed in Chapter 6 (see Figure 6.4 and the related discussion). When the axes x, y are rotated by an angle α around the z axis and are then relabelled x', y', then as Figure 6.4 shows

$$x' = x \cos \alpha + y \sin \alpha$$

Similarly,

$$y' = -x \sin \alpha + y \cos \alpha$$

We can add, trivially,

$$z' = z$$

It follows that the matrix describing the effect of a rotation, $R_z(\alpha)$, of an angle α around the z axis is the 3×3 matrix in the middle of equation (A2.6).

$$R_s(\alpha) \begin{bmatrix} x \\ y \\ z \end{bmatrix} = \begin{bmatrix} \cos \alpha & \sin \alpha & 0 \\ -\sin \alpha & \cos \alpha & 0 \\ 0 & 0 & 1 \end{bmatrix} \begin{bmatrix} x \\ y \\ z \end{bmatrix} = \begin{bmatrix} x' \\ y' \\ z' \end{bmatrix} \qquad (A2.6)$$

A study of the elements on the leading diagonal of this matrix—those that contribute to the character—will show the basis for the rule given at the end of Section 6.1:

When an axis is rotated by an angle α its contribution to the character for that operation is $\cos \alpha$.

This relationship enables a more detailed study of the rotation of x and y axes by 45° shown in Figures 4.5 and 5.5 as well as the rotation to give the more general set in Figure 5.6; the following discussion is based on these figures and the reader will have to refer back to them.

It is evident that the character generated by the x and y axes under a C_4 operation is identical for either choice of x and y axes shown in Figures 5.4 and 5.5 (the character is 0 because x and y directions are transposed by the operation). It should also be evident that the same character under this operation is obtained for the more general x and y axes of Figure 5.6 (if it is not evident, use equation (A2.6) suitably adapted to the problem). Less evident is the fact that the general axis set gives the same character as the other sets under improper rotations. Consider the operation of reflection in the $\sigma_v(2)$ mirror plane of Figure 5.3, a mirror plane which is the xz plane in Figure 5.4. All three axis sets give a character of 1 for the z axis. For the (x, y) axis sets of Figure 5.4 and 5.5 characters of 0 are obtained for this reflection operation, but what of the axis set of Figure 5.6? If the angle between y' and the adjacent Br–F bond axis contained in the $\sigma_v(2)$ mirror plane is denoted θ then the relationship between x', y' and their images x'', y'' is found to be (Figure A2.1).

$$x'' = -x' \cos 2\theta - y' \sin 2\theta$$
$$y'' = -x' \sin 2\theta + y' \cos 2\theta$$

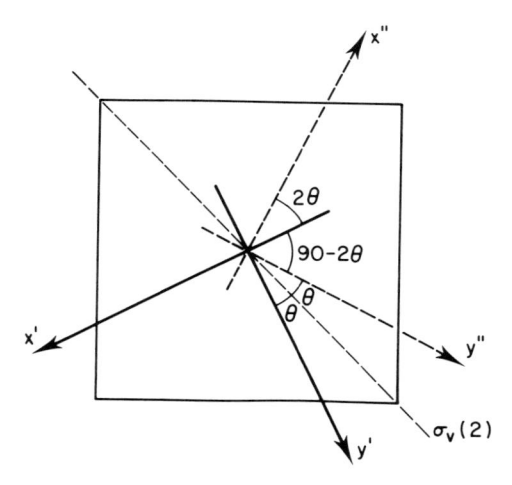

Figure A2.1   The effect of a mirror plane reflection ($\sigma_v(2)$) on x' and y' of Figure 5.6.

That is,

$$\begin{bmatrix} -\cos 2\theta & -\sin 2\theta \\ -\sin 2\theta & \cos 2\theta \end{bmatrix} \begin{bmatrix} x' \\ y' \end{bmatrix} = \begin{bmatrix} x'' \\ y'' \end{bmatrix}$$

Clearly, the character of the 2×2 transformation matrix is 0, just as was the case for the axis choice of Figures 5.4 and 5.5.

So far, in all of the axis transformations that have been considered the z axis has remained unmoved. If it, too varies then the problem becomes that of describing the relationship between two generally orientated sets of axes. It is easy to see that rotation by three independent angles about coordinate axes is necessary to describe the relationship between two sets of arbitrarily orientated Cartesian axes. If the first rotation is about z, then x' and y' must remain in the original xy plane. If the second rotation is about x' then this axis will remain in the original xy plane; it does not assume a general position. Two rotations are not sufficient to place all three original axes in general positions, a third is needed.

The general transformation is shown in Figure A2.2. A rotation by ϕ about z is followed by one of θ about x'. Under this latter rotation z becomes z' and y' becomes y''. The final rotation is one of φ about z', whereupon x' becomes x'' and y'' becomes y'''. Mathematically, equation (A2.6) is applied to each of these transformations in succession. The final result is given in equation (A2.7).

$$\begin{bmatrix} (\cos\psi\cos\phi - \cos\theta\sin\phi\sin\psi) & (\cos\psi\sin\phi + \cos\theta\cos\phi\sin\psi) & (\sin\psi\sin\theta) \\ (-\sin\psi\cos\phi - \cos\theta\sin\phi\sin\psi) & (-\sin\psi\sin\phi + \cos\theta\cos\phi\cos\psi) & (\cos\psi\sin\theta) \\ (\sin\theta\sin\phi) & (-\sin\theta\cos\phi) & \cos\theta \end{bmatrix} \begin{bmatrix} x' \\ y' \\ z' \end{bmatrix} = \begin{bmatrix} x'' \\ y''' \\ z' \end{bmatrix}$$

$$(A2.7)$$

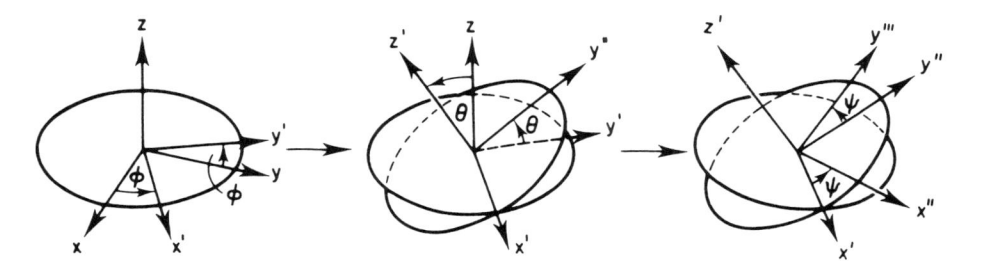

Figure A2.2   The interconversion of two sets of axes, (x, y, z) and (x'', y''', z') which are in a general relationship the one set to the other.

Because a set of p orbitals transforms in the same way as the coordinate axes, this relationship is needed to answer the problem left unresolved in Section 3.2, the transformation of a complete set of p orbitals referred to arbitrary axes under the operations of the C_{2v} point group.

Problem A2.8   Derive equation (A2.7).

Problem A2.9   Consider the following conversion of an axis set x, y, z into the general positions occupied by the set x'', y''' and z' (Figure A2.2). Take z and rotate it about a suitable axis such that it is coincident with z'. Now rotate the other two axes around z' so that they coincide with x'' and y''' (x'' and y''' are in the plane perpendicular to z'). Does this sequence mean that it is possible to relate the two axis sets by just two angles rather than the three of equation (A2.7)? If not, why not?

In the section above, the particular concern has been with sets of matrices which are usually reducible representations. However, similar considerations apply to irreducible representations. That is, there exist sets of irreducible matrices for each group. As will be seen in Section A2.4, it is always possible to manipulate a set of matrices which form a basis for a reducible representation in such a way that they can be re-written as a sum of the irreducible matrices.

As an example of the irreducible matrix representations of a group those for the C_{3v} point group are given in Table A2.2. This table should be compared with the C_{3v} character table given in Table 6.1 (and also in Appendix 3).

Comparison of Table A2.2 with the C_{3v} character table reveals two important things. First, whereas individual operations are listed separately in Table A2.2, in the character table they are grouped into classes. Second, for a given irreducible representation, the irreducible matrices of all operations in any one class have the same character and this is the character listed for the class in the character table. The *rapprochement* between the 'individual operation' and 'classes' presentations is provided by the class algebra which was introduced in

Table A2.2

C_{3v}	E	C_3^+	C_3^-	$\sigma_v(1)$	$\sigma_v(2)$	$\sigma_v(3)$
A_1	(1)	(1)	(1)	(1)	(1)	(1)
A_2	(1)	(1)	(1)	(-1)	(-1)	(-1)
E	$\begin{bmatrix} 1 & 0 \\ 0 & 1 \end{bmatrix}$	$\begin{bmatrix} -\frac{1}{2} & -\frac{\sqrt{3}}{2} \\ \frac{\sqrt{3}}{2} & -\frac{1}{2} \end{bmatrix}$	$\begin{bmatrix} -\frac{1}{2} & \frac{\sqrt{3}}{2} \\ -\frac{\sqrt{3}}{2} & -\frac{1}{2} \end{bmatrix}$	$\begin{bmatrix} -1 & 0 \\ 0 & 1 \end{bmatrix}$	$\begin{bmatrix} \frac{1}{2} & -\frac{\sqrt{3}}{2} \\ -\frac{\sqrt{3}}{2} & -\frac{1}{2} \end{bmatrix}$	$\begin{bmatrix} \frac{1}{2} & \frac{\sqrt{3}}{2} \\ \frac{\sqrt{3}}{2} & -\frac{1}{2} \end{bmatrix}$

Appendix 1.4. There are some applications of group theory where it is necessary to use the complete matrix representations of groups; for instance in some applications of the direct product which will be met in the next section, although these applications will not be included in that section.

A2.2 DIRECT PRODUCTS

In the main text three different uses of the phrase 'direct product' were met. First, the operations of some groups were said to be the direct product of the operations of two other groups. An example is the D_{2h} group, discussed in Section 4.3. Each individual operation of the D_{2h} point group may be regarded as a product of an individual operation of the D_2 group with an individual operation of the C_i group. For such cases the character table of the product group was also said to be the direct product of those of the other two groups. The phrase 'direct product' was also used to describe the multiplication together of two representations of a group, a topic which was discussed at some length in Chapter 10 (where the symbol $\otimes$ was used to denote this particular application of the direct product). Clearly, the concept of a direct product is one of wide applicability in group theory; it also is an important one.

At the end of the previous section it was noted that there exists a close connection between the characters in a character table and sets of irreducible matrices. Just like the case of the (reducible) matrix representations which were discussed in that section in some detail, the multiplication of irreducible matrices is isomorphic to the multiplication of the group operations. Because of this isomorphism and because direct products of group operations can be formed, we would expect there to be, correspondingly, a direct product of matrices. At the beginning of the previous section (A2.1) one way of multiplying two matrices was described; but this was such that the size of the product matrix is often the same as the size of the matrices from which it was formed (e.g. when square matrices are multiplied together). A characteristic of direct products is that an *increase* in size is the norm—the D_{2h} group is larger than D_2 and C_2. This suggests that the direct product of the matrices involves a second form of matrix multiplication. This is not a unique situation. For instance, there are two different ways of combining—multiplying—vectors together.

The *direct product of two matrices* is obtained by individually and separately multiplying every element of each of the two matrices together. Thus, the direct product of the matrices

$$\begin{bmatrix} 3 & 2 \\ 4 & -1 \end{bmatrix} \text{ and } \begin{bmatrix} a & b \\ d & e \end{bmatrix} \text{ is}$$

$$\begin{bmatrix} 3a & 3b & 2a & 2b \\ 3d & 3e & 2d & 2e \\ 4a & 4b & -a & -b \\ 4d & 4e & -d & -e \end{bmatrix}$$

If a general element of the *matrix A* is a_{ij} (i labelling the row and j the column in which a_{ij} occurs in A) and a typical element of B is b_{km} (kth row, mth column), then the general element of the matrix C which is the direct product of A and B is:

$$a_{ij} \cdot b_{km} = c_{ij,km} \tag{A2.8}$$

Of course, the general element $c_{ij,km}$ could simply be labelled according to the row and column in which it occurs in C. To do this, however, would be to lose sight of its origins; the more explicit, although apparently more unwieldy, expression in (A2.8) is therefore preferred. An example of the reason for this follows.

For the matrices themselves,

$$A \otimes B = C \tag{A2.9}$$

where $\otimes$ again indicates that a direct product is being formed.

There are several ways in which the matrix C may be written; a convenient one is

$$C = \begin{bmatrix} a_{11}B & a_{12}B & \cdot & \cdot & \cdot \\ a_{21}B & a_{22}B & \cdot & \cdot & \cdot \\ a_{31}B & a_{32}B & \cdot & \cdot & \cdot \\ \cdot & \cdot & \cdot & \cdot & \cdot \\ \cdot & \cdot & \cdot & \cdot & \cdot \end{bmatrix} \tag{A2.10}$$

where $a_{11}B$ means that each element of the matrix B is multiplied, in order, by a_{11}.

Problem A2.10   Fill in the blanks in the following matrix equation (it may be helpful to regard the 4×4 matrix as consisting of four 2×2's, corresponding to B above).

$$\begin{bmatrix} 1 & 2 \\ 0 & . \end{bmatrix} \otimes \begin{bmatrix} . & . \\ -1 & . \end{bmatrix} = \begin{bmatrix} 3 & 6 & 0 & 0 \\ . & -3 & 0 & 0 \\ . & . & -4 & -8 \\ . & . & . & . \end{bmatrix} \qquad (A2.10)$$

The elements of the group D_{3d} are formed as the direct product of the elements of the C_{3v} group with the elements of the C_i group. The relationship between the operations of D_{3d}, C_{3v}, and C_i is indicated in the table below in the form

	Operations of C_{3v}
Operations of C_i	Operations of D_{3d}

	E	C_3^+	C_3^-	$\sigma_v(1)$	$\sigma_v(2)$	$\sigma_v(3)$
E	E	C_3^+	C_3^-	$\sigma_v(1)$	$\sigma_v(2)$	$\sigma_v(3)$
i	i	S_6^-	S_6^+	$C_1(1)$	$C_2(2)$	$C_2(3)$

A precisely parallel relationship exists between the irreducible matrix representations of the D_{3d}, C_{3v} and C_i groups, a relationship detailed below. First, however, note that the group D_{3d} is also the direct product of D_3 with C_i, and it is this latter product which is conventionally taken to determine the labels of the irreducible representations of D_{3d}.

Problem A2.11   Show that $D_{3d} = D_3 \otimes C_i$.

The matrix representations of the C_i group are:

C_i	E	i
A_g	(1)	(1)
A_u	(1)	(-1)

so that the direct product with the C_{3v} matrix representations given in Table A2.2 leads to the irreducible matrix representations for the D_{3d} group given in

Table A2.3

D_{3d}	E	C_3^+	C_3^-	$\sigma_v(1)$	$\sigma_v(2)$	$\sigma_v(3)$	i	S_6^-	S_6^+	$C_2(1)$	$C_2(2)$	$C_2(3)$
A_{1g}	(1)	(1)	(1)	(1)	(1)	(1)	(1)	(1)	(1)	(1)	(1)	(1)
A_{2g}	(1)	(1)	(1)	(−1)	(−1)	(−1)	(1)	(1)	(1)	(−1)	(−1)	(−1)
E_g	$\begin{bmatrix}1&0\\0&1\end{bmatrix}$	$\begin{bmatrix}-\frac{1}{2}&-\frac{\sqrt{3}}{2}\\\frac{\sqrt{3}}{2}&-\frac{1}{2}\end{bmatrix}$	$\begin{bmatrix}-\frac{1}{2}&\frac{\sqrt{3}}{2}\\-\frac{\sqrt{3}}{2}&-\frac{1}{2}\end{bmatrix}$	$\begin{bmatrix}-1&0\\0&1\end{bmatrix}$	$\begin{bmatrix}\frac{1}{2}&-\frac{\sqrt{3}}{2}\\-\frac{\sqrt{3}}{2}&-\frac{1}{2}\end{bmatrix}$	$\begin{bmatrix}\frac{1}{2}&\frac{\sqrt{3}}{2}\\\frac{\sqrt{3}}{2}&-\frac{1}{2}\end{bmatrix}$	$\begin{bmatrix}1&0\\0&1\end{bmatrix}$	$\begin{bmatrix}-\frac{1}{2}&-\frac{\sqrt{3}}{2}\\\frac{\sqrt{3}}{2}&-\frac{1}{2}\end{bmatrix}$	$\begin{bmatrix}-\frac{1}{2}&\frac{\sqrt{3}}{2}\\-\frac{\sqrt{3}}{2}&-\frac{1}{2}\end{bmatrix}$	$\begin{bmatrix}-1&0\\0&1\end{bmatrix}$	$\begin{bmatrix}\frac{1}{2}&-\frac{\sqrt{3}}{2}\\-\frac{\sqrt{3}}{2}&-\frac{1}{2}\end{bmatrix}$	$\begin{bmatrix}\frac{1}{2}&\frac{\sqrt{3}}{2}\\\frac{\sqrt{3}}{2}&-\frac{1}{2}\end{bmatrix}$
A_{1u}	(1)	(1)	(1)	(1)	(1)	(1)	(−1)	(−1)	(−1)	(−1)	(−1)	(−1)
A_{2u}	(1)	(1)	(1)	(−1)	(−1)	(−1)	(−1)	(−1)	(−1)	(1)	(1)	(1)
E_u	$\begin{bmatrix}1&0\\0&1\end{bmatrix}$	$\begin{bmatrix}-\frac{1}{2}&-\frac{\sqrt{3}}{2}\\\frac{\sqrt{3}}{2}&-\frac{1}{2}\end{bmatrix}$	$\begin{bmatrix}-\frac{1}{2}&\frac{\sqrt{3}}{2}\\-\frac{\sqrt{3}}{2}&-\frac{1}{2}\end{bmatrix}$	$\begin{bmatrix}-1&0\\0&1\end{bmatrix}$	$\begin{bmatrix}\frac{1}{2}&-\frac{\sqrt{3}}{2}\\-\frac{\sqrt{3}}{2}&-\frac{1}{2}\end{bmatrix}$	$\begin{bmatrix}\frac{1}{2}&\frac{\sqrt{3}}{2}\\\frac{\sqrt{3}}{2}&-\frac{1}{2}\end{bmatrix}$	$\begin{bmatrix}-1&0\\0&-1\end{bmatrix}$	$\begin{bmatrix}\frac{1}{2}&\frac{\sqrt{3}}{2}\\-\frac{\sqrt{3}}{2}&\frac{1}{2}\end{bmatrix}$	$\begin{bmatrix}\frac{1}{2}&-\frac{\sqrt{3}}{2}\\\frac{\sqrt{3}}{2}&\frac{1}{2}\end{bmatrix}$	$\begin{bmatrix}1&0\\0&-1\end{bmatrix}$	$\begin{bmatrix}-\frac{1}{2}&\frac{\sqrt{3}}{2}\\\frac{\sqrt{3}}{2}&\frac{1}{2}\end{bmatrix}$	$\begin{bmatrix}-\frac{1}{2}&-\frac{\sqrt{3}}{2}\\-\frac{\sqrt{3}}{2}&\frac{1}{2}\end{bmatrix}$

Table A2.3. It is entirely reasonable that this isomorphism should exist between multiplication of operations and multiplication of matrices in this application of the direct product. The isomorphism exists in each of the individual groups involved in the direct product.

Problem A2.12 Show that $D_{3d} = C_{3v} \otimes C_i$ (see above); then check Table A2.3.

The definition of a direct product given by equations (A2.8) and (A2.9) and the convention given by (A2.10) is, of course, also applicable to the direct products formed between two representations of the same group. Thus the direct product matrices $A_2 \otimes E$ of the C_{3v} point group are (from Table A2.2)

$$
\begin{array}{cccccc}
E & C_3^+ & C_3^- & \sigma_v(1) & \sigma_v(2) & \sigma_v(3)
\end{array}
$$

$$
A_2 \otimes E = \begin{bmatrix} 1 & 0 \\ 0 & 1 \end{bmatrix}
\begin{bmatrix} -\frac{1}{2} & -\frac{\sqrt{3}}{2} \\ \frac{\sqrt{3}}{2} & -\frac{1}{2} \end{bmatrix}
\begin{bmatrix} -\frac{1}{2} & \frac{\sqrt{3}}{2} \\ -\frac{\sqrt{3}}{2} & -\frac{1}{2} \end{bmatrix}
\begin{bmatrix} 1 & 0 \\ 0 & -1 \end{bmatrix}
\begin{bmatrix} -\frac{1}{2} & \frac{\sqrt{3}}{2} \\ \frac{\sqrt{3}}{2} & \frac{1}{2} \end{bmatrix}
\begin{bmatrix} -\frac{1}{2} & -\frac{\sqrt{3}}{2} \\ -\frac{\sqrt{3}}{2} & \frac{1}{2} \end{bmatrix}
$$

Examination of these matrices shows that the characters of the direct product matrices are the same as those obtained by the method described in the text— using a character table and multiplying the characters of the two irreducible representations together. The technique of multiplying characters to obtain direct product characters although simple, ignores the subtle changes that have taken place in the corresponding matrices, particularly in those representing the σ_v operations (for which multiplying by the character 0 might well have appeared trivial).

Problem A2.13 Using the C_{3v} character table, form the direct product of the A_2 and E irreducible representations; compare the answer with the matrix form given above.

As a final example we consider the direct product $E \otimes E$ in C_{3v} but confine the discussion to just two of the product matrices. The two direct products which will be evaluated are those direct product matrices corresponding to $\sigma_v(1)$ and C_3^+ which arise from the $E \otimes E$ direct product. For the first of these, expression in the form given by (A2.10) leads to

$$
\left\{
\begin{array}{cc}
-1 \begin{bmatrix} -1 & 0 \\ 0 & 1 \end{bmatrix} & 0 \begin{bmatrix} -1 & 0 \\ 0 & 1 \end{bmatrix} \\[12pt]
0 \begin{bmatrix} -1 & 0 \\ 0 & 1 \end{bmatrix} & 1 \begin{bmatrix} -1 & 0 \\ 0 & 1 \end{bmatrix}
\end{array}
\right\}
$$

which, on expansion gives

$$\begin{bmatrix} 1 & 0 & 0 & 0 \\ 0 & -1 & 0 & 0 \\ 0 & 0 & -1 & 0 \\ 0 & 0 & 0 & 1 \end{bmatrix}$$

a matrix with a character of 0, the same character as obtained working with the C_{3v} character table.

The second direct product, that corresponding to C_3^+, involves more work. In the form of (A2.10) the product is:

$$\left\{ -\frac{1}{2}\begin{bmatrix} -\frac{1}{2} & -\frac{\sqrt{3}}{2} \\ \frac{\sqrt{3}}{2} & -\frac{1}{2} \end{bmatrix} \quad -\frac{\sqrt{3}}{2}\begin{bmatrix} -\frac{1}{2} & -\frac{\sqrt{3}}{2} \\ \frac{\sqrt{3}}{2} & -\frac{1}{2} \end{bmatrix} \right.$$
$$\left. \frac{\sqrt{3}}{2}\begin{bmatrix} -\frac{1}{2} & -\frac{\sqrt{3}}{2} \\ \frac{\sqrt{3}}{2} & -\frac{1}{2} \end{bmatrix} \quad -\frac{1}{2}\begin{bmatrix} -\frac{1}{2} & -\frac{\sqrt{3}}{2} \\ \frac{\sqrt{3}}{2} & -\frac{1}{2} \end{bmatrix} \right\}$$

leading to

$$\begin{bmatrix} \frac{1}{4} & \frac{\sqrt{3}}{4} & \frac{\sqrt{3}}{4} & \frac{3}{4} \\ -\frac{\sqrt{3}}{4} & \frac{1}{4} & -\frac{3}{4} & \frac{\sqrt{3}}{4} \\ -\frac{\sqrt{3}}{4} & -\frac{3}{4} & \frac{1}{4} & \frac{\sqrt{3}}{4} \\ \frac{3}{4} & -\frac{\sqrt{3}}{4} & -\frac{\sqrt{3}}{4} & \frac{1}{4} \end{bmatrix}$$

and the expected character of 1.

Because these direct product matrices are 4×4 it is clear that they must describe the transformation of four quantities—basis functions—which must themselves be related to the basis functions for the E irreducible representation. The exploration and exploitation of such relationships are an important aspect of advanced group theory but the full development of which is beyond the scope of the present text, although a start has been made in Section 10.2.

Problem A2.14   Evaluate the direct product matrices of $E \otimes E$ in C_{3v} for the operations C_3^- and $\sigma_v(2)$.

A2.3   THE ORTHONORMALITY RELATIONSHIPS

In Section 5.3 group orthogonality relationships were discussed in a general fashion. The object of the present section is to provide a firmer mathematical basis for these relationships. Theorems 2–5 of Section 5.3 were all expressed in terms of the characters of irreducible representations. But it has been seen earlier in this appendix, in Section A2.1, that these characters are derived from irreducible matrices, such as those in Table A2.2. It is through the latter that the relationships of Section 5.3 are derived. The first step is to derive some orthogonality relationships which relate to the irreducible matrices and then to express them in a single comprehensive equation. The theorems of Section 5.3 derive from these matrix element orthogonality relationships. Finally, a proof of the comprehensive equation mentioned above will be indicated. In this section the arguments and notation used by Eyring, Walter and Kimball in Chapter 10 of their book *Quantum Chemistry* (Wiley, 1944) will be used. In this book the argument is more general and rigorous than that given here.

   If Table A2.2 is carefully examined it will be found that if corresponding matrix elements in the same irreducible representation are squared and added together a rather simple result is obtained. Thus, if the 1,2 (top right-hand— first row, second column) elements of the E irreducible matrices are squared and added we obtain:

$$0 + 3/4 + 3/4 + 0 + 3/4 + 3/4 + 0 = 3$$

The answer, 3, is equal to

$$\frac{\text{The number of operations in the group (the order of the group)}}{\text{The dimension of the irreducible representation in the present example}} = \frac{6}{2}$$

Problem A2.15   Check that 3 is the result when the above procedure is applied to the 1,1, to the 2,1 and to the 2,2 elements of the E irreducible matrices in Table A2.2.

This result is perfectly general and may be written as

$$\sum_R \Gamma_i(R)_{mn} \Gamma_i(R)_{mn} = \frac{h}{l_i} \tag{A2.11}$$

where i labels the irreducible representation chosen, $\Gamma_i(R)_{mn}$ indicating the mnth element of the irreducible matrix corresponding to the operation R. The order of the group is denoted by h and l_i is the dimension of the ith irreducible representation. The symbol Γ is that commonly used to indicate a matrix— which may be reducible or irreducible—which transforms isomorphically to a symmetry operation.

   If, on the other hand, instead of squaring matrix elements and then adding, two *different* elements from within the same matrix are multiplied and added then the result is 0. Thus, for the 1,1 and 1,2 elements of the E matrices in Table A2.1 we have

$$0 + \frac{\sqrt{3}}{4} - \frac{\sqrt{3}}{4} + 0 - \frac{\sqrt{3}}{4} + \frac{\sqrt{3}}{4} = 0$$

In general,

$$\sum_R \Gamma_i(R)_{mn} \Gamma_j(R)_{m'n'} = 0 \qquad (A2.12)$$

where either $m \neq m'$ and/or $n \neq n'$. Similarly, for any pair of matrix elements of any two *different* irreducible representations i and j (as long as the matrices are associated with the same operation, R).

$$\sum_R \Gamma_i(R)_{mn} \Gamma_j(R)_{m'n'} = 0 \qquad (A2.13)$$

where $i \neq j$ and where there is no restriction on mn and $m'n'$ as long as they exist. Thus for the 1,1 (the only!) element of the A_2 matrices and the 2,1 element of the E matrices in Table A2.1 we have

$$0 + \frac{\sqrt{3}}{2} - \frac{\sqrt{3}}{2} + 0 - \frac{\sqrt{3}}{2} + \frac{\sqrt{3}}{2} = 0 \qquad (A2.13)$$

Equations (A2.11), (A2.12) and (A2.13), and the restrictions on the quantities within them may be combined into the general—and therefore important—relationship (A2.14), a relationship often called *the great orthogonality theorem*:

$$\sum_R \Gamma_i(R)_{mn} \Gamma_j(R)_{m'n'} = \frac{h}{\sqrt{l_i l_j}} \delta_{ij} \delta_{mm'} \delta_{nn'} \qquad (A2.14)$$

where $\delta_{ij} = 1$ if $i = j$ but 0 if $i \neq j$. Similarly $\delta_{mm'} = 1$ if $m = m'$ but 0 if $m \neq m'$ and $\delta_{nn'} = 1$ if $n = n'$ but 0 if $n \neq n'$.

The appearance of a square root term in (A2.14) is deceptive. Because of the δ_{ij} term this term will only be of importance when $i = j$ and then $\sqrt{l_i l_j} = l_i = l_j$, so we could just as well have put one of these on the right-hand side of (A2.14). The square root is included so that (A2.14) is symmetric in i and j.

In order to relate (A2.14) to the theorems of Section 5.3 it has to be adapted to refer to the characters (= sum of elements on the leading diagonal) of the irreducible matrices. Whereas $\Gamma_i(R)_{mn}$ is a general matrix element, diagonal matrix elements will have $m = n$ and so be of the form $\Gamma_i(R)_{mm}$. For these diagonal elements (A2.14) assumes the form

$$\sum_R \Gamma_i(R)_{mn} \Gamma_j(R)_{m'm'} = \frac{h}{l_j} \delta_{ij} \delta_{mm'} \qquad (A2.15)$$

Because, by definition, the ith irreducible representation is of dimension l_i, there will be l_i terms $\Gamma_i(R)_{mm}$ which have to be added to give the character of the ith irreducible matrix appropriate to the operation R. That is, m assumes values from 1 through to l_i (if different from 1). Similarly, m' runs from 1 to l_j.

Writing, as is conventionally done, $\chi_i(R)$ for the character (the sum of elements along the leading diagonal) of the ith irreducible representation under the operation R we have

$$\chi_i(R) = \sum_m \Gamma_i(R)_{mm} \tag{A2.16}$$

and

$$\chi_j(R) = \sum_{m'} \Gamma_j(R)_{m'm'} \tag{A2.17}$$

Summing each side of (A2.15) over m and m' and substituting (A2.16) and (A2.17) gives

$$\sum_R \Gamma_i(R)\Gamma_j(R) = \frac{h}{l_i} \delta_{ij} \sum_{m=1}^{l_i} \sum_{m'=1}^{l_j} \delta_{mm'} \tag{A2.18}$$

Problem A2.16 Derive equation (A2.18) from (A2.15), (A2.16) and (A2.17).

Because of the $\delta_{mm'}$ term on the right-hand side of (A2.18) ($\delta_{mm'} = 0$ if $m \neq m'$ but 1 if $m = m'$) there will only be a contribution from the two summations when $m = m'$ and then the contribution will be 1. The right-hand side becomes

$$\frac{h}{l_i} \delta_{ij} \sum_{m'=1}^{l_j} 1$$

The meaning of the summation in this expression is 'every term in the summation from 1 through to l_j contributes 1 to the total'—and so the value of the sum is l_j. The summation is not to be confused with $\sum_{m'=1}^{l_j} m'$. This means that (A2.18) can be rewritten

$$\sum_R \chi_i(R)\chi_j(R) = h\delta_{ij} \tag{A2.19}$$

In this expression each operation appears separately. Because for a given irreducible representation the character of each matrix is the same for all operations in the same class their contributions to the left-hand side of (A2.19) can be added together. If there are g_ρ operations in the class ρ:

$$\sum_R \chi_i(R)\chi_j(R) = \sum_\rho \chi_i(R_\rho)\chi_j(R_\rho)g_\rho$$

where $\chi_i(R_\rho)$ is the character of the ith irreducible representation under the class R_ρ (i.e. the quantity given in the group character table). Of course, $\chi_i(R_\rho)$ is numerically equal to $\chi_i(R)$. With this substitution (A2.19) becomes

$$\sum_\rho \chi_i(R_\rho)\chi_j(R_\rho)g_\rho = h\delta_{ij} \tag{A2.20}$$

All of the Theorems 2–5 given in Section 5.3 are contained within (A2.20).

Problem A2.17   Show that Theorems 2–5 of Section 5.3 are contained within (A2.20).

Because the above argument was built on Table A2.2—which contains only real quantities, equation (A2.20) contains the hidden assumption that all characters are real. As Appendix 3 (and Chapter 11) show, this assumption is not valid—some character tables contain complex quantities. Fortunately, the generalization of (A2.20) to include these cases is simple; the general expression is

$$\sum_\rho \chi_i^*(R_\rho)\chi_j(R_\rho)g_\rho = h\delta_{ij} \tag{A2.21}$$

where the * means that one uses not the quantity given in a character table as $\chi_i(R_\rho)$ but, instead, its complex conjugate. An example of this usage is given in some detail in Chapter 11.

Problem A2.18   Show that in the C_{2v} point group there is an infinite number of functions of general form $x^m y^m$ which transform as the totally symmetric (A_1) irreducible representation.

The next task is the derivation of equation (A2.14). Again, the derivation to be given is closely related to that of Eyring, Walter and Kimball in their book *Quantum Chemistry*, in Appendix VI, and uses the same notation but is less rigorous and is structured rather differently.

Consider a function F, which, for generality, is expressed as a sum

$$F = \sum_{k=1}^{l} x_k y_k \tag{A2.22}$$

and which is invariant under all operations of the group—it transforms as the totally symmetric irreducible representation. That is, any operation, R, of the group operating on F gives F

$$RF = F$$

The variables $x_1, \ldots, x_k$ (there are usually no more than three members of the set) themselves form a basis for the representation Γ_x of the group, with matrices $\Gamma_x(R)$. Similarly, the y_k form a basis for Γ_y with matrices $\Gamma_y(R)$. It is to be noted that the functions which here are called $x_k y_k$ may be identified with the functions $x^n y^m$ of Problem A2.18—there is nothing in the following derivation which requires that $n = m = 1$ (be careful not to confuse subscripts with superscripts).

The explicit expression for RF is

$$RF = R\left[\sum_{k=1}^{l} x_k y_k\right] = F$$

Because R operates on each function individually, this is

$$RF = \sum_{k=1}^{l} Rx_k \cdot Ry_k = F$$

Now sum over all operations of the group; the result is

$$\sum_R RF = \sum_R \sum_{k=1}^{l} Rx_k \cdot Ry_k = hF \qquad \text{(A2.23)}$$

there being h operations in the group.

Consider Rx_k; it is evident from equation (A2.6) that, for example,

$$Rx_k = \sum_{s=1}^{n} \Gamma_x(R)_{sk} x_s \qquad \text{(A2.24a)}$$

where $\Gamma_x(R)_{sk}$ is the skth element of the matrix $\Gamma_x(R)$. Similarly,

$$Ry_k = \sum_{t=1}^{m} \Gamma_y(R)_{tk} y_t \qquad \text{(A2.24b)}$$

It will later be shown that the invariance of F requires that

$$\Gamma_x(R)_{ik} = \Gamma_y(R)_{ik} \qquad \text{(A2.25)}$$

so the x, y, suffixes on the Γ's will be dropped. Additionally, set $m = n = l$ in equations (A2.24) and (A2.26). Substituting these two expressions into (A2.23) gives

$$\sum_R RF = \sum_R \sum_{k=1}^{l} \sum_{s=1}^{l} \sum_{t=1}^{l} \Gamma(R)_{sk} \Gamma(R)_{tk} y_s y_t = hF \qquad \text{(A2.26)}$$

Problem A2.19   Derive equation (A2.26).

Because equation (A2.22) contains only terms $x_k y_k$ so too must (A2.26), that is, $s = t$. So, in (A2.26), all terms with $s \neq t$ must equal zero. Thus,

$$\sum_R \Gamma(R)_{sk} \Gamma(R)_{tk} = 0 \qquad (s \neq t) \qquad \text{(A2.27)}$$

Because $\Sigma_R RF$ in (A2.26) spans all the R in the group and because each R has a unique inverse, R^{-1} (Section A1.1), $\Sigma_R R^{-1} F$ also spans all operations of the group. It follows that

$$\sum_R R^{-1} F = \sum_R RF$$

i.e. the sums over R and R^{-1} merely give the same operations in a different order.

In this appendix the relationship between the matrices associated with R and with R^{-1} has not been investigated but it is a simple one—when all the elements of the matrices are real, the case to which the present discussion is limited, one simply interchanges columns and rows of R to obtain R^{-1} and vice versa.

Problem A2.20 Use the above technique to obtain the inverses of the matrices in Table A2.1. Check your answers by reference to Table 8.2.

Thus, for example, the inverse of the matrix associated with $R_z(\alpha)$ in equation (A2.6) is simply

$$R_z^{-1}(\alpha) = \begin{bmatrix} \cos \alpha & -\sin \alpha & 0 \\ \sin \alpha & \cos \alpha & 0 \\ 0 & 0 & 1 \end{bmatrix}$$

Problem A2.20 Show, using the above 3×3 matrix and the one given in equation (A2.6), that

$$\Gamma R_z(\alpha) \Gamma R_z^{-1}(\alpha) = \Gamma R_z^{-1}(\alpha) \Gamma R_z(\alpha) = \Gamma E$$

Note that this relationship is obtained from equation (A1.4) by the substitution of a matrix for the corresponding operation.

Returning to equation (A2.26), it is clear from the above that on the left-hand side $\Sigma_R RF$ can be replaced by $\Sigma_R R^{-1} F$. However, when this substitution is made, on the right-hand side those matrix elements associated with R must be replaced by those associated with R^{-1}. This is done, as has been seen, by interchanging rows and columns—that is, by reversing the subscripts on the matrix elements (this causes no problems because we are dealing with square matrices).

It follows that

$$\sum_R R^{-1}F = \sum_R \sum_{k=1}^{l} \sum_{s=1}^{l} \sum_{t=1}^{l} \Gamma(R)_{ks}\Gamma(R)_{kt}x_s y_t = hF \qquad (A2.28)$$

Problem A2.21   Check equation (A2.28).

Therefore, for $s \neq t$ it follows that (using the discussion under equation (A2.26)):

$$\sum_R \Gamma(R)_{ks}\Gamma(R)_{kt} = 0 \qquad s \neq t \qquad (A2.29)$$

Having exploited the $s \neq t$ situation to the full, now set $s = t = j$ in (A2.26) and thus obtain

$$\sum_R RF = \sum_R \sum_{k=1}^{l} \sum_{j=1}^{l} \Gamma(R)_{jk}\Gamma(R)_{jk}x_j y_j = hF \qquad (A2.30)$$

Comparing this with (A2.22) we deduce that

$$\sum_R \sum_{k=1}^{l} \Gamma(R)_{jk}\Gamma(R)_{jk} = h \qquad \text{i.e. } j \text{ runs from 1 to } l \qquad (A2.31)$$

Similarly, setting $s = t = j$ in (A2.28) gives

$$\sum_R \sum_{k=1}^{l} \Gamma(R)_{kj}\Gamma(R)_{kj} = h \qquad j = 1, ..., l \qquad (A2.32)$$

Problem A2.22   Check the derivation of equations (A2.31) and (A2.32).

The right-hand sides of (A2.31) and (A2.32) are both independent of j and k, no matter the order in which these dummy suffixes appear in the left-hand side expressions. It follows that each term in the summation over k makes an equal contribution to the sum. The contribution of each term is therefore given by

$$\sum_R \Gamma(R)_{jk}\Gamma(R)_{jk} = \frac{h}{l} \qquad (A2.33)$$

Finally, combine equations (A2.27), (A2.29) and (A2.33). None of these expressions contain any suffixes on the Γ's but we use (A2.24) to justify the requirement that the suffixes be identical, although, strictly, this should be formally proved. This combination gives, with minor changes in notation, the expression

$$\sum_R \Gamma_i(R)_{mn} \Gamma_j(R)_{m'n'} = \frac{h}{\sqrt{l_i l_j}} \delta_{ij} \delta_{mm'} \delta_{nn'} \qquad (A2.14)$$

which, as indicated, is identical to (A2.14).

Problem A2.23 Detail the derivation of (A2.14) along the lines outlined in the text.

The derivation of this equation was the object of the present section and many readers may wish to stop at this point. There is, however, one loose end in the derivation—equation (A2.25) was not proven

$$\Gamma_x(R)_{ik} = \Gamma_y(R)_{ik} \qquad (A2.25)$$

The validity of this equation was essential to the derivation of equation (A2.14). For completion, the somewhat lengthy, proof follows.

Start by operating on each side of (A2.22) with an operation, R, of the group. That is, start with

$$RF = R \sum_{k=1}^{r} x_k y_k$$

where, for convenience the l of equation (A2.22) has been replaced by r.

$$= \sum_{k=1}^{r} R(x_k y_k)$$

$$= \sum_{k=1}^{r} R(x_k) \cdot R(y_k)$$

$$= F \text{(see the discussion associated with equation (A2.22))}.$$

That is, $F = \sum_{k=1}^{r} R(x_k) . R(y_k)$ for all R. It is now convenient to write (A2.22) in expanded form:

$$F = \sum_{k=1}^{r} x_k y_k = x_1 y_1 + x_2 y_2 + \cdots + x_r y_r \qquad (A2.34)$$

Let R operate on just the x_k in (A2.22); we have

$$F' = \sum_{k=1}^{r} R(x_k) \cdot y_k \qquad (A2.35)$$

or, in expanded form:

$$F' = R(x_1) y_1 + R(x_2) y_2 + \cdots + R(x_r) y_r$$

Now, for an individual x_k,

$$Rx_k = \sum_{s=1}^{n} \Gamma_x(R)_{sk} x_s \tag{A2.36}$$

where $\Gamma_n(R)_{sk}$ is the skth element of $\Gamma_n(R)$. So, (A2.35) becomes

$$F' = y_1 \sum_{s=1}^{n} \Gamma_x(R)_{s1} x_s + y_2 \sum_{s=1}^{n} \Gamma_x(R)_{s2} + \cdots + y_r \sum_{s=1}^{n} \Gamma_x(R)_{sr} x_s \tag{A2.37}$$

Note that within each summation term, each function $x_1, x_2, \ldots, x_k$ may appear. Regroup the terms in (A2.37) according to these x's; each x may be multiplied by any of the y's. Note that in each $\Gamma_x(R)_{sk}$ term in (A2.37), the s goes with the x and the k with the y. The rearranged expression is

$$F' = x_1 \sum_{k=1}^{r} \Gamma_x(R)_{1k} y_k + x_2 \sum_{k=1}^{r} \Gamma_x(R)_{2k} y_k + \cdots + x_n \sum_{k=1}^{r} \Gamma_x(R)_{nk} y_k$$

If we now operate on the y's on the right-hand side of this expression with R then F' becomes equal to F and we have

$$F = x_1 \sum_{k=1}^{r} \Gamma_x(R)_{1k} Ry_k + \cdots + x_n \sum_{k=1}^{r} \Gamma_x(R)_{nk} Ry_k$$

By comparison with (A2.34) we have

$$\sum_{k=1}^{r} \Gamma_x(R)_{ik} Ry_k = y_i \tag{A2.38}$$

The inverse of the operation R is R^{-1}; operating with it on both sides of (A2.38) gives

$$R^{-1} \sum_{k=1}^{r} \Gamma_x(R)_{ik} Ry_k = R^{-1} y_i$$

Remembering that $\Gamma_x(R)_{ik}$ is a number, this is

$$\sum_{k=1}^{r} \Gamma_x(R)_{ik} R^{-1} Ry_k = R^{-1} y_i$$

that is, by equation (A1.4),

$$\sum_{k=1}^{r} \Gamma_x(R)_{ik} y_k = R^{-1} y_i \tag{A2.39}$$

Now a general expression for Ry_i, analogous to (A2.25), is

$$Ry_i = \sum_{k=1}^{r} \Gamma_y(R)_{ki} y_k$$

so that $R^{-1}y_i$ is given by

$$R^{-1}y_i = \sum_{k=1}^{r} \Gamma_y(R)_{ik} y_k.$$

Comparison with equation (A2.39) requires that

$$\sum_{k=1}^{r} \Gamma_x(R)_{ik} = \sum_{k=1}^{r} \Gamma_y(R)_{ik}$$

We have thus found a relationship between the elements of $\Gamma_x(R)$ and $\Gamma_y(R)$. But this relationship holds for all choice of R and all ik—we placed no restrictions on either. This generality only arises when corresponding individual terms in the two summations are themselves equal. That is,

$$\Gamma_x(R)_{ik} = \Gamma_y(R)_{ik}$$

which is equation (A2.25).

A2.4 THE REDUCTION OF REDUCIBLE REPRESENTATIONS

An extremely important property of the matrices which multiply isomorphically to group operations is the fact that 'their characters are invariant to a similarity transformation', a phrase which will be found in almost every book on the subject. What does the phrase in parentheses mean? This will be explained shortly but first, a digression. Consider a set of matrices $A, B, C \ldots$ which form a representation of the group. From these, other matrices can be generated which also form a representation of the group, as will now be shown. The importance of this step is that the new matrices may be chosen so that they are more convenient to work with than the starting set. In particular, the new matrices can be cunningly engineered to be a sum of irreducible matrices.

Suppose that the multiplication of the original matrices is such that

$$AB = C$$

it has to be shown that a similar relationship holds for the new matrices. If this can be done then they, indeed, multiply correctly. The engineering process consists of selecting another matrix, call it M, not a member of the set $A, B, C \ldots$, but one which is of the same dimension—so, if A, B, C are all

6×6 matrices, so too must be M. For the moment M will not be specified further beyond requiring that it has inverse, M^{-1}. Form the products

$$M^{-1}AM = A'$$
$$M^{-1}BM = B'$$
$$M^{-1}CM = C'$$
$$\begin{matrix} . & . \\ . & . \end{matrix}$$

where the product matrices A', B', C' ... will be of the same dimension as A, B, C, ... The set of matrices A', B', C' ... are a new set of representation matrices of the same group. This can be shown by considering the product $A'B'$:

$$A'B' = M^{-1}AMM^{-1}BM = M^{-1}ABM = M^{-1}CM = C'$$

That is $A'B' = C'$ whenever $AB = C$ and so A', B', C' ... multiply in the same way as (isomorphically with) A, B, C ....

The matrices A and A' are said to be 'related by a similarity transformation' as are B and B', C and C', etc. It is the property of matrices related in this way that they have the same characters—hence the phrase 'their characters are invariant to a similarity transformation'. This relationship between the characters is not difficult to prove. The character of the matrix A, $\chi(A)$, is given by

$$\chi(A') = \sum_{i} A_{ii}$$

which, written in terms of the elements of product $M^{-1}AM$ is

$$\chi(A') = \sum_{i} \sum_{j} \sum_{k} M_{ik}^{-1} A_{kj} M_{ji}$$

But each term on the right-hand side is a number and the same result is obtained no matter in which order they are multiplied together. It is therefore permissible to rewrite this expression as

$$\chi(A') = \sum_{i} \sum_{j} \sum_{k} A_{kj} M_{ji} M_{ik}^{-1}$$

$$= \sum_{j} \sum_{k} A_{kj} \sum_{i} M_{ji} M_{ik}^{-1}$$

But the final summation is that required for matrix multiplication (see equation (A2.1)), so

$$\chi(A') = \sum_{j} \sum_{k} A_{kj} (MM^{-1})_{jk}$$

But $MM^{-1} = E$, a unit matrix with 1's on the diagonal and zeros elsewhere, so

$$\chi(A') = \sum_j \sum_k A_{kj}\delta_{jk} \quad \delta_{jk} = 1 \quad \text{if} \quad j = k$$
$$= 0 \quad \text{if} \quad j \neq k$$
$$= \sum_k A_{kk}$$

We conclude that

$$\chi(A') = \chi(A)$$

which demonstrates the invariance of the character under a similarity transformation.

The transformation of reducible matrix representations into sums of irreducible matrix representations is of vital importance. Although it is rarely necessary to go through the formal procedure it is implicitly involved, for example, in the formation of the symmetry-adapted linear combinations of orbitals used in the discussions of molecular bonding in this book. The procedure involved may be seen in the schematic equation given below; by an educated choice of M, the general matrix A is converted to the matrix A' which has irreducible matrices (cross-hatched) strung along its leading diagonal, all other elements being zero.

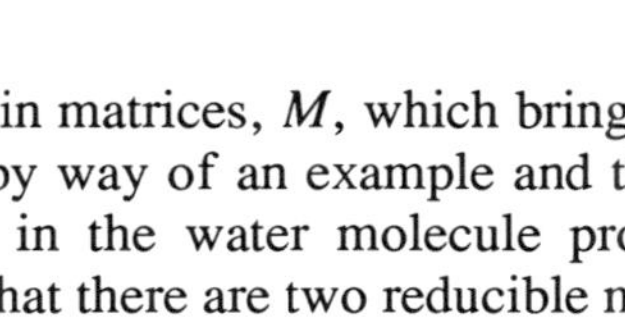

$$M^{-1}[A]M = \begin{bmatrix} \boxed{\diagdown} & 0 & 0 & 0 \\ 0 & \boxed{\diagdown} & 0 & 0 \\ 0 & 0 & \boxed{\diagdown} & \\ 0 & 0 & & \boxed{\diagdown} \end{bmatrix} \tag{A2.40}$$

Figure A2.3

The next step is to obtain matrices, M, which bring about this simplification. It is simplest to proceed by way of an example and this we do by returning to the two-hydrogen atoms in the water molecule problem (Section 3.2), for which Table A2.1 shows that there are two reducible matrices to consider,

$$\begin{bmatrix} 1 & 0 \\ 0 & 1 \end{bmatrix} \quad \text{and} \quad \begin{bmatrix} 0 & 1 \\ 1 & 0 \end{bmatrix}$$

The first of these is diagonal (only zeros off the leading diagonal) so the similarity transformation has to lead to the retention of this characteristic. The second matrix is non-diagonal and has to be transformed to diagonal form. It will shortly be shown how the matrix, M, is obtained but, for the moment, it is simply given and shown that it leads to the desired result.

In the present example the matrix M is

$$\begin{bmatrix} \frac{1}{\sqrt{2}} & \frac{1}{\sqrt{2}} \\ \frac{1}{\sqrt{2}} & -\frac{1}{\sqrt{2}} \end{bmatrix}$$

so that M^{-1} is

$$\begin{bmatrix} \frac{1}{\sqrt{2}} & \frac{1}{\sqrt{2}} \\ \frac{1}{\sqrt{2}} & -\frac{1}{\sqrt{2}} \end{bmatrix}$$

i.e. M is self-inverse. Next, evaluate the two products of the form $M^{-1}AM$, working from the left. For the first:

$$\begin{bmatrix} \frac{1}{\sqrt{2}} & \frac{1}{\sqrt{2}} \\ \frac{1}{\sqrt{2}} & -\frac{1}{\sqrt{2}} \end{bmatrix}\begin{bmatrix} 1 & 0 \\ 0 & 1 \end{bmatrix}\begin{bmatrix} \frac{1}{\sqrt{2}} & \frac{1}{\sqrt{2}} \\ \frac{1}{\sqrt{2}} & -\frac{1}{\sqrt{2}} \end{bmatrix} = \begin{bmatrix} \frac{1}{\sqrt{2}} & \frac{1}{\sqrt{2}} \\ \frac{1}{\sqrt{2}} & -\frac{1}{\sqrt{2}} \end{bmatrix}\begin{bmatrix} \frac{1}{\sqrt{2}} & \frac{1}{\sqrt{2}} \\ \frac{1}{\sqrt{2}} & -\frac{1}{\sqrt{2}} \end{bmatrix} = \begin{bmatrix} 1 & 0 \\ 0 & 1 \end{bmatrix}$$

Since this case A was an identity matrix (1's for all diagonal elements and zeros elsewhere) this product was really

$$M^{-1}EM = M^{-1}M = E$$

so it could have been anticipated that A would be left unchanged by the similarity transformation. The second case is less trivial.

$$\begin{bmatrix} \frac{1}{\sqrt{2}} & \frac{1}{\sqrt{2}} \\ \frac{1}{\sqrt{2}} & -\frac{1}{\sqrt{2}} \end{bmatrix}\begin{bmatrix} 0 & 1 \\ 1 & 0 \end{bmatrix}\begin{bmatrix} \frac{1}{\sqrt{2}} & \frac{1}{\sqrt{2}} \\ \frac{1}{\sqrt{2}} & -\frac{1}{\sqrt{2}} \end{bmatrix} = \begin{bmatrix} \frac{1}{\sqrt{2}} & \frac{1}{\sqrt{2}} \\ -\frac{1}{\sqrt{2}} & \frac{1}{\sqrt{2}} \end{bmatrix}\begin{bmatrix} \frac{1}{\sqrt{2}} & \frac{1}{\sqrt{2}} \\ \frac{1}{\sqrt{2}} & -\frac{1}{\sqrt{2}} \end{bmatrix} = \begin{bmatrix} 1 & 0 \\ 0 & -1 \end{bmatrix}$$

As required, this matrix has been transformed into a diagonal form. Both of the original 2×2 matrices have now been block-diagonalized into the form indicated on the right-hand side of equation (A2.40). The transformed matrices are (with the blocking indicated as sub-matrices):

$$\begin{bmatrix} (1) & 0 \\ 0 & (1) \end{bmatrix} \quad \text{and} \quad \begin{bmatrix} (1) & 0 \\ 0 & (-1) \end{bmatrix}$$

Remembering that, as (A2.2)–(A2.5) show, each of the original matrices has to be considered twice, write separately the sub-matrices (here 1×1, i.e. numbers) that appear on the diagonals of the transformed matrices. We obtain:

Operation	E	C_2	σ_v	σ_v'	
First sub-matrix	(1)	(1)	(1)	(1)	A_1
Second sub-matrix	(1)	(−1)	(1)	(−1)	B_2

The characters of these matrices are equal to the matrices themselves and so it is evident that the A_1 and B_1 representations of Chapter 3 (Section 3.3) have been generated.

Problem A2.24   Show that the four 4×4 matrices given on pages 320 and 321 are simultaneously reduced to a diagonal form by a similarity transformation using for M:

$$\begin{bmatrix} \frac{1}{2} & \frac{1}{2} & \frac{1}{2} & \frac{1}{2} \\ \frac{1}{2} & \frac{1}{2} & -\frac{1}{2} & -\frac{1}{2} \\ \frac{1}{2} & -\frac{1}{2} & \frac{1}{2} & -\frac{1}{2} \\ \frac{1}{2} & -\frac{1}{2} & -\frac{1}{2} & \frac{1}{2} \end{bmatrix}$$

Next, the key problem; how does one obtain the matrix which reduces a set of reducible matrices to block diagonal form? The answer is indicated by comparing the matrix used in the example detailed above with the two symmetry-adapted A_1 and B_1 functions obtained in Chapter 3 (at the end of Section 3.4). These were:

$$\psi(A_1) = \frac{1}{\sqrt{2}} (h_1 + h_2)$$

$$\psi(B_1) = \frac{1}{\sqrt{2}} (h_1 - h_2)$$

If the coefficients which multiply h_1 and h_2 in these expressions are written in matrix form we obtain:

$$\begin{bmatrix} \frac{1}{\sqrt{2}} & \frac{1}{\sqrt{2}} \\ \frac{1}{\sqrt{2}} & -\frac{1}{\sqrt{2}} \end{bmatrix}$$

which is the matrix M used above. The result is general, the matrix which diagonalizes a reducible matrix is obtained by forming a matrix from the coefficients with which the basis functions appear in the symmetry-adapted functions. The method of obtaining these symmetry adapted functions has been described in Chapter 4 (Section 4.6) extended in Chapters 5 (Section 5.5) and 6 (Section 6.2) and simplified for difficult problems in Appendix 4. There is one point at which care is needed. This is that it is important to make sure that the listing of the basis functions used in obtaining a reducible matrix (and in real-life problems this is the way such matrices are usually obtained) is identical to that implied in the matrix obtained from the coefficients in the symmetry-adapted functions. Thus, in the example above the functions $\psi(A_1)$ and $\psi(B_1)$ are obtained by multiplying out the product

$$\begin{bmatrix} \frac{1}{\sqrt{2}} & \frac{1}{\sqrt{2}} \\ \frac{1}{\sqrt{2}} & -\frac{1}{\sqrt{2}} \end{bmatrix} \begin{bmatrix} h_1 \\ h_2 \end{bmatrix}$$

and here the listing of h_1 and h_2 coincides with that of (A2.2)–(A2.5). It so happens that in this case we would have escaped penalty had h_1 and h_2 have been listed in the incorrect order and thus obtained for M the matrix

$$\begin{bmatrix} \frac{1}{\sqrt{2}} & \frac{1}{\sqrt{2}} \\ -\frac{1}{\sqrt{2}} & \frac{1}{\sqrt{2}} \end{bmatrix}$$

but, in general, one could not expect to be so fortunate.

Problem A2.25   Show that the alternative form of M given above leads to the same results as that form used in the text.

This appendix is concluded with the derivation of the algebraic expression that is the basis of the recipe frequently used in the main text to obtain the irreducible components of a reducible representation. As the discussion in this section has shown, a reducible matrix representation of a group can be reduced to a matrix which contains only irreducible matrices along its leading diagonal by a suitable similarity transformation. Further, the character of the original and transformed matrices are identical. If $\chi(R)$ is written for the character of a reducible matrix under the operation R then:

$$\chi(R) = \sum_{j=1}^{k} a_j \chi_j(R) \tag{A2.41}$$

where $\chi_j(R)$ is the character of the jth irreducible representation under the same operation (there are k irreducible representations in all) and a_j is either zero or a positive integer, the number of times the jth irreducible representation occurs in the reducible representation. It is these a_j's that we seek to determine. Equation (A2.19) is

$$\sum_{R} \chi_i(R)\chi_j(R) = h\delta_{ij} \tag{A2.19}$$

where $\delta_{ij} = 1$ if $i = j$ but 0 if $i \neq j$. Now multiply each side of (A2.19) by a_j and sum over all j. We thus obtain, using (A2.41),

$$\sum_{R} \chi_i(R) \sum_{j=1}^{k} a_j \chi_j(R) = \sum_{R} \chi_i(R)\chi(R) = ha_i \tag{A2.42}$$

Only a_j appears on the right-hand side of this expression because all other a_j's have been multiplied by 0 because of the δ_{ij}. Rearranging (A2.42) equation (A2.43) is obtained

$$a_i = \frac{1}{h} \sum_R \chi(R)\chi_i(R) \tag{A2.43}$$

the equation implicitly used in the text.

Appendix 3
Character Tables of the more Important Point Groups

At the right of each character table in this compilation two columns of bases are given for irreducible representations, Rotations and Translations, given in the first column, are needed for vibrational analyses (see Chapter 9) and for some forms of spectroscopy (see Chapter 10). The second column is useful for other spectroscopies (Chapter 10) and for discussions of molecular bonding. Invariably, x^2, y^2 and z^2 in some combination or independently transform as the totally symmetric irreducible representation. It follows that any linear combination of the functions so transforming and, in particular, $x^2 + y^2 + z^2 = r^2$, transform under this irreducible representation. The function r^2 is spherically symmetrical and so is associated with the s orbital of an atom.

'Note' comments have usually either been repeated where relevant or cross references given. However, the reader encountering problems should scan the notes for related character tables, where he or she may well find helpful comments.

Whenever possible the direct product nature of a character table has been indicated by divisions within the character table itself. It is often possible to simplify a problem by working in a subgroup instead of the full group and the divisions within the character table in this appendix are intended to facilitate this.

1 THE ICOSAHEDRAL GROUPS

I_h: The icosahedral group.

I_h	E	$12C_5$	$12C_5^2$	$20C_3$	$15C_2$	i	$12S_{10}$	$12S_{10}^3$	$20S_6$	15σ		
A_g	1	1	1	1	1	1	1	1	1	1		$x^2+y^2+z^2$
T_{1g}	3	$-2\cos 144$	$-2\cos 72$	0	-1	3	$-2\cos 144$	$-2\cos 72$	0	-1	(R_x, R_y, R_z)	
T_{2g}	3	$-2\cos 72$	$-2\cos 144$	0	-1	3	$-2\cos 72$	$-2\cos 144$	0	-1		
G_g	4	-1	-1	1	0	4	-1	-1	1	0		
H_g	5	0	0	-1	1	5	0	0	-1	1		$(\frac{1}{\sqrt6}[2z^2-x^2-y^2], \frac{1}{\sqrt2}[x^2-y^2], xy, yz, zx)$
A_u	1	1	1	1	1	-1	-1	-1	-1	-1		
T_{1u}	3	$-2\cos 144$	$-2\cos 72$	0	-1	-3	$2\cos 144$	$2\cos 72$	0	1	(T_x, T_y, T_z)	(x, y, z)
T_{2u}	3	$-2\cos 72$	$-2\cos 144$	0	-1	-3	$2\cos 72$	$2\cos 144$	0	1		
G_u	4	-1	-1	1	0	-4	1	1	-1	0		
H_u	5	0	0	-1	1	-5	0	0	1	-1		

Notes: (1) This character table is the direct product of I with C_i, indicated by lines.
(2) When working with groups containing a C_5 axis the following relationships will be found useful.

$$-2\cos 72° = -0.61803 = \tfrac{1}{2}(1 - 5^{1/2}) = \alpha$$
$$-2\cos 144° = 1.61803 = \tfrac{1}{2}(1 + 5^{1/2}) = \beta$$
$$\alpha^2 = 1 + \alpha; \quad \beta^2 = 1 + \beta; \quad \alpha\beta = -1; \quad \alpha + \beta = 1$$

(3) The icosahedral groups are the only ones which contain fourfold (G) and fivefold (H) degenerate representations. In some compilations these are given the alternative labels of U (fourfold) and V (fivefold). Both the icosahedral and cubic groups contain triply degenerate representations; in this compilation they are denoted by T; in some others, and more seldom, the symbol F is used.

Example: The icosahedron shown in Figure 7.31.

I: The group of pure rotations of an icosahedron.

I	E	$12C_5$	$12C_5^2$	$20C_3$	$15C_2$		
A	1	1	1	1	1		$x^2 + y^2 + z^2$
T_1	3	$-2\cos 144$	$-2\cos 72$	0	-1	$(R_x, R_y, R_z), (T_x, T_y, T_z)$	
T_2	3	$-2\cos 72$	$-2\cos 144$	0	-1	(x, y, z)	
G	4	-1	-1	1	0		
H	5	0	0	-1	1		$\left(\frac{1}{\sqrt{6}}[2z^2 - x^2 - y^2], \frac{1}{\sqrt{2}}[x^2 - y^2], xy, yz, zx\right)$

Note: See comments under the I_h group.

Example: The easiest way of obtaining a figure of I symmetry is to take one edge of the icosahedron shown in Figure 7.31 and make it a zigzag (the C_2 axis passing through the mid-point of this edge must be preserved). The pure rotation operations are then used to generate this zigzag in each of the other edges. A figure of I symmetry results. Faces that are planar in the icosahedron become fluted under the above procedure.

2 CUBIC POINT GROUPS

The two most important cubic point groups are O_h and T_d.

O_h	E	$8C_3$	$6C_4$	$3C_2$	$6C_2'$	i	$8S_6$	$6S_4$	$3\sigma_h$	$6\sigma_d$		
A_{1g}	1	1	1	1	1	1	1	1	1	1		$x^2+y^2+z^2$
A_{2g}	1	1	-1	1	-1	1	1	-1	1	-1		
E_g	2	-1	0	2	0	2	-1	0	2	0		$\left(\frac{1}{\sqrt6}[2z^2-x^2-y^2],\frac{1}{\sqrt2}[x^2-y^2]\right)$
T_{1g}	3	0	1	-1	-1	3	0	1	-1	-1	(R_x, R_y, R_z)	
T_{2g}	3	0	-1	-1	1	3	0	-1	-1	1		(xy, yz, zx)
A_{1u}	1	1	1	1	1	-1	-1	-1	-1	-1		
A_{2u}	1	1	-1	1	-1	-1	-1	1	-1	1		
E_u	2	-1	0	2	0	-2	1	0	-2	0		
T_{1u}	3	0	1	-1	-1	-3	0	-1	1	1	(T_x, T_y, T_z)	(x, y, z)
T_{2u}	3	0	-1	-1	1	-3	0	1	1	-1		

Notes: (1) In some texts inorganic chemists refer to one set of the d orbitals on a transition metal atom at the centre of an octahedral complex as d_{γ} (which are e_g above) and d_{ϵ} (which are t_{2g}).

(2) The O_h group is a direct product of O and C_i. This is indicated by the lines in the character table.

(3) Although the σ_h mirror planes also satisfy the definition for σ_d here and later (for the D_{nh} groups) the h subscript is given precedence.

(4) The $1/\sqrt6$ and $1/\sqrt2$ in the expressions for e_g basis functions normalize each function to unity. In the text the $1/\sqrt2$ on x^2-y^2 was not used (to keep the discussion as simple as possible) and $1/\sqrt3$ therefore appeared in front of $2z^2-x^2-y^2$.

Examples: (a) A cube, an octahedron (Figure 7.1). (b) SF_6 (Figure 7.2 when $M = S$, $L = F$).

351

T_d	E	$8C_3$	$3C_2$	$6S_4$	$6\sigma_d$		
A_1	1	1	1	1	1		$x^2+y^2+z^2$
A_2	1	1	1	-1	-1		
E	2	-1	2	0	0		$(\frac{1}{\sqrt{6}}[2z^2-x^2-y^2],\frac{1}{\sqrt{2}}[x^2-y^2])$
T_1	3	0	-1	1	-1	(R_x, R_y, R_z)	
T_2	3	0	-1	-1	1	(T_x, T_y, T_z)	$(x, y, z);\ (xy, yz, zx)$

Examples: (a) A tetrahedron (Figure 7.1). (b) CH_4, P_4, $Ni(CO)_4$, $C(CH_3)_4$, in its most symmetrical configuration.

Other cubic point groups

O: The group of the pure rotation operations of the octahedron.

O	E	$8C_3$	$3C_2$	$6C_4$	$6C_2'$		
A_1	1	1	1	1	1		$x^2+y^2+z^2$
A_2	1	1	1	-1	-1		
E	2	-1	2	0	0		$(\frac{1}{\sqrt{6}}[2z^2-x^2-y^2],\frac{1}{\sqrt{2}}[x^2-y^2])$
T_1	3	0	-1	1	-1	$(T_x, T_y, T_z)(R_x, R_y, R_z)$	(x, y, z)
T_2	3	0	-1	-1	1		(xy, yz, zx)

Example: To obtain a figure of O symmetry follow the instructions given under the I character table, replacing 'Figure 7.30' by 'Figure 7.1', 'icosahedron' and 'octahedron' and 'I' by 'O'.

T: The group of the pure rotation operations of the tetrahedron.

T	E	$4C_3$	$4C_3^2$	$3C_2$		
A	1	1	1	1		$x^2+y^2+z^2$
E	$\begin{cases}1\\1\end{cases}$	$\begin{matrix}\varepsilon\\\varepsilon^2\end{matrix}$	$\begin{matrix}\varepsilon^2\\\varepsilon\end{matrix}$	$\begin{matrix}1\\1\end{matrix}$		$\left(\dfrac{1}{\sqrt{6}}[2z^2-x^2-y^2],\dfrac{1}{\sqrt{2}}[x^2-y^2]\right)$
T	3	0	0	-1	$(T_x, T_y, T_z)(R_x, R_y, R_z)$	$(x, y, z); (xy, yz, zx)$

Note: $\varepsilon\left(\varepsilon=\exp\left[\dfrac{2\pi i}{3}\right]\right)$ and $\varepsilon^2\left(\varepsilon^2=\exp\dfrac{4\pi i}{3}=\exp\dfrac{-2\pi i}{3}\right)$ are complex conjugates. See also Chapter 11 and the notes under the C_3 group below.

Example: To obtain a figure of T symmetry follow the instructions given under the I character table, replacing 'Figure 7.3' by 'Figure 7.1', 'icosahedron' by 'tetrahedron' and 'I' by 'T'.

T_h	E	$4C_3$	$4C_3^2$	$3C_2$	i	$4S_6^5$	$4S_6$	$3\sigma_h$		
A_g	1	1	1	1	1	1	1	1		$x^2+y^2+z^2$
E_g	$\begin{cases}1\\1\end{cases}$	$\begin{matrix}\varepsilon\\\varepsilon^2\end{matrix}$	$\begin{matrix}\varepsilon^2\\\varepsilon\end{matrix}$	$\begin{matrix}1\\1\end{matrix}$	$\begin{matrix}1\\1\end{matrix}$	$\begin{matrix}\varepsilon\\\varepsilon^2\end{matrix}$	$\begin{matrix}\varepsilon^2\\\varepsilon\end{matrix}$	$\begin{matrix}1\\1\end{matrix}$		$\left(\dfrac{1}{\sqrt{6}}[2z^2-x^2-y^2],\dfrac{1}{\sqrt{2}}[x^2-y^2]\right)$
T_g	3	0	0	-1	3	0	0	-1	(R_x, R_y, R_z)	(xy, yz, zx)
A_u	1	1	1	1	-1	-1	-1	-1		
E_u	$\begin{cases}1\\1\end{cases}$	$\begin{matrix}\varepsilon\\\varepsilon^2\end{matrix}$	$\begin{matrix}\varepsilon^2\\\varepsilon\end{matrix}$	$\begin{matrix}1\\1\end{matrix}$	$\begin{matrix}-1\\-1\end{matrix}$	$\begin{matrix}-\varepsilon\\-\varepsilon^2\end{matrix}$	$\begin{matrix}-\varepsilon^2\\-\varepsilon\end{matrix}$	$\begin{matrix}-1\\-1\end{matrix}$		
T_u	3	0	0	-1	-3	0	0	1	(T_x, T_y, T_z)	(x, y, z)

Notes: (1) The T_h character table is the direct product of T with C_i. This is indicated by the lines in the character table.

(2) For the meaning of ε see under the T character table above.

Example: An 'octahedral' complex $[M(OH_2)_6]$ in which all the atoms of *trans* H_2O ligands lie in a σ_h plane.

3   THE GROUPS D_{nh}

A regular, planar polygon with n sides has D_{nh} symmetry. So, an equilateral triangle has D_{3h} symmetry, a square has D_{4h} symmetry. The label D arises because of the presence of twofold axes (Dihedral axes) perpendicular to a C_n axis. There are n of these twofold axes. The subscript h means that all of the groups have a unique mirror plane perpendicular to the C_n axis (if this axis is vertical then the mirror plane is horizontal). Although these groups all have σ_d operations, the σ_h takes precedence in the labelling, hence D_{nh}. Also to avoid possible confusion with the D_{nd} groups many authors label as σ_v some or all of the σ_d mirror planes.

D_{2h}	E	$C_2(z)$	$C_2(y)$	$C_2(x)$	i	$\sigma(xy)$	$\sigma(zx)$	$\sigma(yz)$		
A_g	1	1	1	1	1	1	1	1		z^2, x^2, y^2
B_{1g}	1	1	-1	-1	1	1	-1	-1	R_z	xy
B_{2g}	1	-1	1	-1	1	-1	1	-1	R_y	zx
B_{3g}	1	-1	-1	1	1	-1	-1	1	R_x	yz
A_u	1	1	1	1	-1	-1	-1	-1		xyz
B_{1u}	1	1	-1	-1	-1	-1	1	1	T_z	z
B_{2u}	1	-1	1	-1	-1	1	-1	1	T_y	y
B_{3u}	1	-1	-1	1	-1	1	1	-1	T_x	x

Notes: (1) Because there are three mutually perpendicular C_2 axes the choice of x, y and z is arbitrary. A relabelling of these axes will lead to an interchange of the labels B_1, B_2 and B_3. Similarly, the h, v subscript notation on the mirror planes is unhelpful and so the mirror planes and the corresponding operations are defined by the Cartesian axes that lie in them.

(2) The D_{2h} group is a direct product of D_2 and C_i. This is indicated by the lines in the character table.

Examples: C_2H_4; B_2H_6 (see Chapter 4).

D_{3h}	E	$2C_3$	$3\sigma_v$	σ_h	$2S_3$	$3\sigma_d$		
A_1'	1	1	1	1	1	1		$z^2;\, x^2+y^2$
A_2'	1	1	-1	1	1	-1	R_z	
E'	2	-1	0	2	-1	0	$(\bar{T}_x, T_y)$	$(x,y);\ (\tfrac{1}{\sqrt{2}}(x^2-y^2), xy)$
A_1''	1	1	1	-1	-1	-1		
A_2''	1	1	-1	-1	-1	1	T_z	z
E''	2	-1	0	-2	1	0	(R_y, R_x)	(zx, yz)

Notes: (1) The D_{3h} group is a direct product of D_3 and C_s. This is indicated by the lines in the character table. Irreducible representations symmetric with respect to reflection in the σ_h mirror plane are denoted by ′, while antisymmetry is denoted by ″.

(2) The $1/\sqrt{2}$ factor on (x^2-y^2) as an E' basis function means that, like xy, it is normalized to unity.

Example: A triangular prism (Figure A3.1).

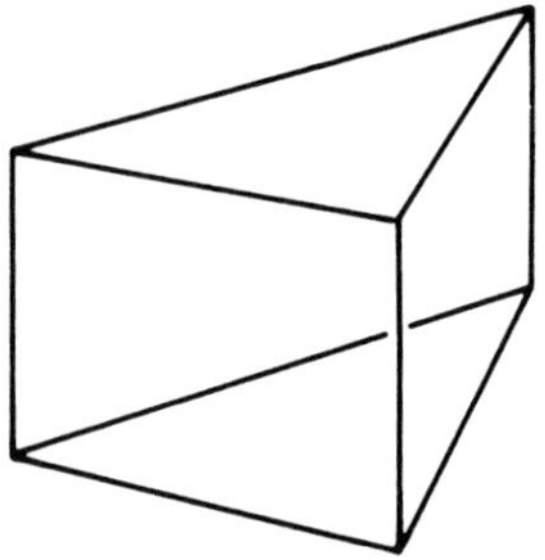

Figure A3.1 A triangular prism, of D_{3h} symmetry.

D_{4h}	E	$2C_4$	C_2	$2C_2'$	$2C_2''$	i	$2S_4$	σ_h	$2\sigma_d$	$2\sigma_d'$		
A_{1g}	1	1	1	1	1	1	1	1	1	1		z^2, x^2+y^2
A_{2g}	1	1	1	-1	-1	1	1	1	-1	-1	R_z	
B_{1g}	1	-1	1	1	-1	1	-1	1	1	-1		x^2-y^2
B_{2g}	1	-1	1	-1	1	1	-1	1	-1	1		xy
E_g	2	0	-2	0	0	2	0	-2	0	0	(R_x, R_y)	(zx, yz)
A_{1u}	1	1	1	1	1	-1	-1	-1	-1	-1		
A_{2u}	1	1	1	-1	-1	-1	-1	-1	1	1	T_z	z
B_{1u}	1	-1	1	1	-1	-1	1	-1	-1	1		
B_{2u}	1	-1	1	-1	1	-1	1	-1	1	-1		
E_u	2	0	-2	0	0	-2	0	2	0	0	(T_x, T_y)	(x, y)

Notes: (1) The D_{4h} group is a direct product of D_4 and C_i. This is indicated by the lines in the character table.

(2) The choice between which pair of C_2 axes (and operations) are labelled C_2' and those which are labelled C_2'' is arbitrary. A redefinition will interchange B_{1g} with B_{2g} and B_{1u} with B_{2u}. Similarly, the choice of the vertical planes σ_d and σ_d' is arbitrary but the σ_d planes must contain the C_2' axes and the σ_d' planes must contain the C_2''.

(3) In this character table the strict definition of σ_d has been followed but many authors label the mirror planes containing a C_2' axis as σ_v and those containing a C_2'' axis as σ_d.

Examples: A square prism (Figure A3.2).

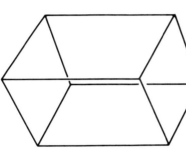

Figure A3.2 A square prism, of D_{4h} symmetry.

D_{5h}	E	$2C_5$	$2C_5^2$	$5C_2$	σ_h	$2S_5$	$2S_5^2$	$5\sigma_d$		
A_1'	1	1	1	1	1	1	1	1		x^2+y^2, z^2
A_2'	1	1	1	-1	1	1	1	-1	R_z	
E_1'	2	$2\cos 72$	$2\cos 144$	0	2	$2\cos 72$	$2\cos 144$	0	(T_x, T_y)	(x, y)
E_2'	2	$2\cos 144$	$2\cos 72$	0	2	$2\cos 144$	$2\cos 72$	0		$\left(\frac{1}{\sqrt{2}}[x^2-y^2], xy\right)$
A_1''	1	1	1	1	-1	-1	-1	-1		
A_2''	1	1	1	-1	-1	-1	-1	1	T_z	z
E_1''	2	$2\cos 72$	$2\cos 144$	0	-2	$-2\cos 72$	$-2\cos 144$	0	(R_x, R_y)	(xz, yz)
E_2''	2	$2\cos 144$	$2\cos 72$	0	-2	$-2\cos 144$	$-2\cos 72$	0		

Notes: (1) The D_{5h} group is a direct product of D_5 and C_s. This is indicated by the lines in the character table. Irreducible representations symmetric with respect to reflection in the σ_h mirror plane are denoted by the superscript $'$ and antisymmetry is denoted by $''$.

(2) See the notes under the I_h character table.

Examples: A regular pentagonal prism (Figure A3.3) and eclipsed ferrocene (Figure A3.4).

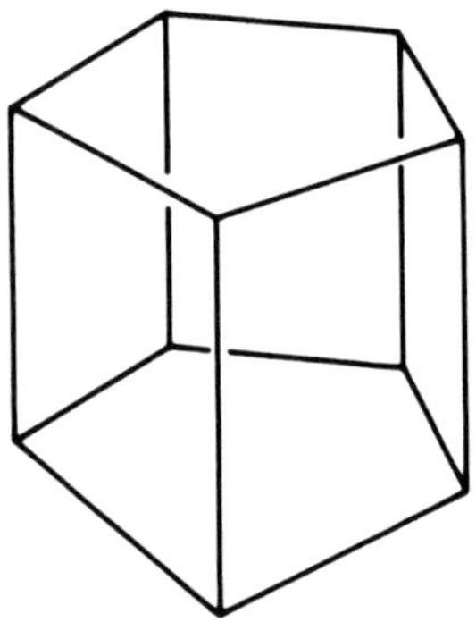

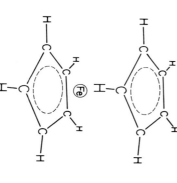

Figure A3.3 A pentagonal prism, of D_{5h} symmetry.

Figure A3.4 Ferrocene, $Fe(C_5H_5)_2$, in the eclipsed configuration, of D_{5h} symmetry.

D_{6h}	E	$2C_6$	$2C_3$	C_2	$3C_2'$	$3C_2''$	i	$2S_3$	$2S_6$	σ_h	$3\sigma_d$	$3\sigma_d'$		
A_{1g}	1	1	1	1	1	1	1	1	1	1	1	1		$x^2+y^2,\ z^2$
A_{2g}	1	1	1	1	-1	-1	1	1	1	1	-1	-1	R_z	
B_{1g}	1	-1	1	-1	1	-1	1	-1	1	-1	1	-1		
B_{2g}	1	-1	1	-1	-1	1	1	-1	1	-1	-1	1		
E_{1g}	2	1	-1	-2	0	0	2	1	-1	-2	0	0	(R_x, R_y)	$(xz,\ yz)$
E_{2g}	2	-1	-1	2	0	0	2	-1	-1	2	0	0		$(\frac{1}{\sqrt2}[x^2-y^2],\ xy)$
A_{1u}	1	1	1	1	1	1	-1	-1	-1	-1	-1	-1		
A_{2u}	1	1	1	1	-1	-1	-1	-1	-1	-1	1	1	T_z	z
B_{1u}	1	-1	1	-1	1	-1	-1	1	-1	1	-1	1		
B_{2u}	1	-1	1	-1	-1	1	-1	1	-1	1	1	-1		
E_{1u}	2	1	-1	-2	0	0	-2	-1	1	2	0	0	(T_x, T_y)	(x, y)
E_{2u}	2	-1	-1	2	0	0	-2	1	1	-2	0	0		

Notes: (1) The D_{6h} character table is a direct product of D_6 and C_i. This is indicated by the lines in the character table.

(2) The choice between which pair of C_2 axes (and operations) are labelled C_2' and those which are labelled C_2'' is arbitrary. A redefinition will interchange B_{1g} with B_{2g} and B_{1u} with B_{2u}. Similarly the choice of the vertical planes σ_d and σ_d' is arbitrary but the σ_d planes must contain the C_2' axes and the σ_d' planes must contain the C_2''.

Examples: A regular hexagonal prism (Figure A3.5) and the benzene molecule (Figure A3.6).

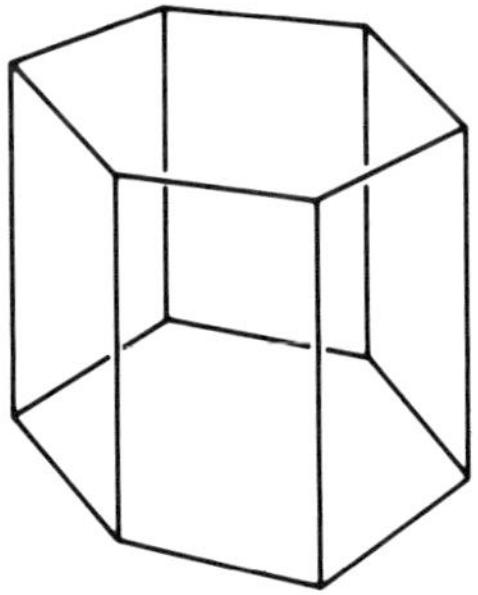

Figure A3.5 A hexagonal prism, of D_{6h} symmetry.

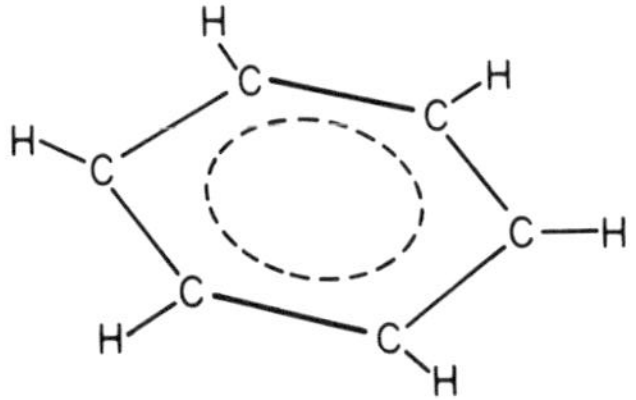

Figure A3.6 The benzene molecule, of D_{6h} symmetry.

4   THE GROUPS D_{nd}

These groups do not have the σ_h mirror plane of the D_{nh} groups. Objects with D_{nd} symmetry typically have two similar halves, staggered with respect to each other. Thus solid objects of D_{nd} symmetry are called 'antiprisms', a term which indicates the staggering.

When n is odd, the groups D_{nd} are direct products of the groups D_n and C_i (when n is even, the direct products $D_n \times C_i$ are D_{nh} groups).

D_{2d}	E	$2S_4$	C_2	$2C_2'$	$2\sigma_d$		
A_1	1	1	1	1	1		x^2+y^2, z^2
A_2	1	1	1	-1	-1	R_z	
B_1	1	-1	1	1	-1		x^2-y^2
B_2	1	-1	1	-1	1	T_z	z; xy
E	2	0	-2	0	0	(T_x, T_y); (R_x, R_y)	(x, y); (xz, yz)

Notes: (1) The x axis is taken as coincident with one C_2'.

(2) Singly degenerate irreducible representations which are symmetric with respect to an S_4 operation are indicated by an A label; antisymmetry is indicated by a B label.

(3) This group is sometimes, but increasingly rarely, called V_d.

(4) Many people find the $2C_2'$ axes and $2\sigma_d$ mirror planes difficult to locate in this group. Time spent with the examples would be time well spent.

Examples: A triangular dodecahedron (Figure A3.7) and the molecule spiropentane (Figure A3.8).

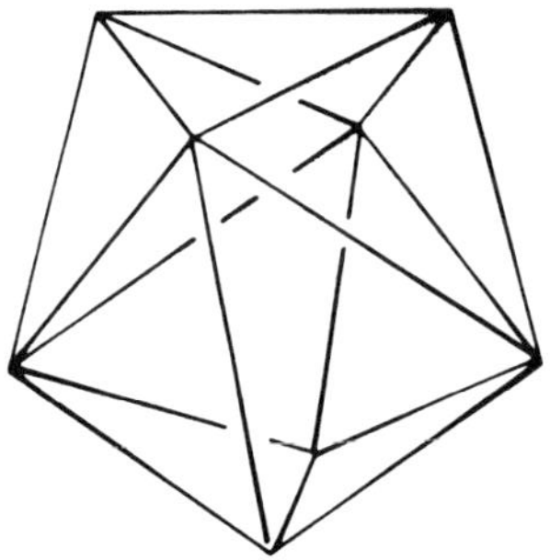

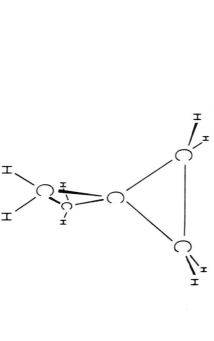

Figure A3.7 The triangular dodecahedron, of D_{2d} symmetry. Note that all of the apices of this figure lie in one of two mutually perpendicular planes.

Figure A3.8 The molecular spiropentane, C_5H_8, of D_{2d} symmetry. The 'outer' carbon atoms in this figure lie at four of the eight apices of the triangular dodecahedron shown in Figure A3.7.

D_{3d}	E	$2C_3$	$3C_2$	i	$2S_6$	$3\sigma_d$		
A_{1g}	1	1	1	1	1	1		x^2+y^2, z^2
A_{2g}	1	1	-1	1	1	-1	R_z	
E_g	2	-1	0	2	-1	0	(R_x, R_y)	$\left(\frac{1}{\sqrt{2}}[x^2-y^2], xy\right); (xz, yz)$
A_{1u}	1	1	1	-1	-1	-1		
A_{2u}	1	1	-1	-1	-1	1	T_z	z
E_u	2	-1	0	-2	1	0	(T_x, T_y)	(x, y)

Notes: (1) This group is a direct product of D_3 with C_i, indicated by the lines in the character table.

(2) The C_3 and i operations may be considered as derived from the S_6 because $S_6^2 = C_3$ and $S_6^3 = i$.

Example: The staggered ethane molecule (Figure A3.9).

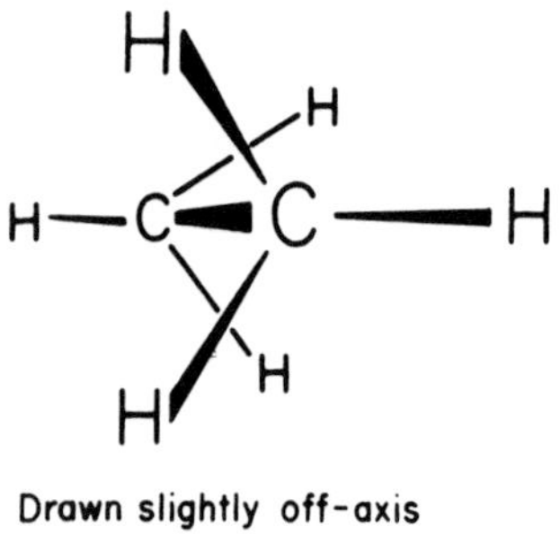

Drawn slightly off-axis

Figure A3.9 The staggered ethane molecule, of D_{3d} symmetry. This symmetry is best seen if the molecule is viewed along the C–C bond but it is difficult to draw it adequately in this orientation.

D_{4d}	E	$2S_8$	$2C_4$	$2S_8^3$	C_2	$4C_2'$	$4\sigma_d$		
A_1	1	1	1	1	1	1	1		$x^2+y^2;\ z^2$
A_2	1	1	1	1	1	-1	-1	R_z	
B_1	1	-1	1	-1	1	1	-1		
B_2	1	-1	1	-1	1	-1	1	T_z	z
E_1	2	$\sqrt{2}$	0	$-\sqrt{2}$	-2	0	0	(T_x, T_y)	(x, y)
E_2	2	0	-2	0	2	0	0		$(\tfrac{1}{\sqrt{3}}[x^2-y^2], xy)$
E_3	2	$-\sqrt{2}$	0	$\sqrt{2}$	-2	0	0	(R_y, R_x)	(xz, yz)

Note: The unique C_2 and the C_4 operations may be considered to be derived from the S_8 because $S_8^2 = C_4$; $S_8^4 = C_2$.

Example: The square antiprism (Figure A3.10).

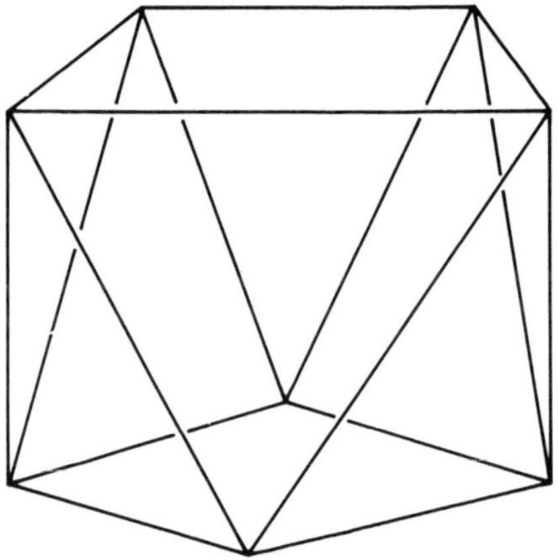

Figure A3.10 The square antiprism, of D_{4d} symmetry.

D_{5d}	E	$2C_5$	$2C_5^2$	$5C_2$	i	$2S_{10}^3$	$2S_{10}$	$5\sigma_d$		
A_{1g}	1	1	1	1	1	1	1	1		$x^2+y^2;\ z^2$
A_{2g}	1	1	1	-1	1	1	1	-1	R_z	
E_{1g}	2	$2\cos 72$	$2\cos 144$	0	2	$2\cos 72$	$2\cos 144$	0	(R_x, R_y)	(xz, yz)
E_{2g}	2	$2\cos 144$	$2\cos 72$	0	2	$2\cos 144$	$2\cos 72$	0		$\left(\frac{1}{\sqrt{2}}[x^2-y^2], xy\right)$
A_{1u}	1	1	1	1	-1	-1	-1	-1		
A_{2u}	1	1	1	-1	-1	-1	-1	1	T_z	z
E_{1u}	2	$2\cos 72$	$2\cos 144$	0	-2	$-2\cos 72$	$-2\cos 144$	0	(T_x, T_y)	(x, y)
E_{2u}	2	$2\cos 144$	$2\cos 72$	0	-2	$-2\cos 144$	$-2\cos 72$	0		

Notes: (1) This group is the direct product of C_5 and C_i, indicated by the lines in the character table.
(2) Before working with this group refer to the notes under the I_h character table.
(3) Many of the operations of the group may be considered to be derived from the S_{10} because $S_{10}^2 = C_5$; $S_{10}^4 = C_5^2$; $S_{10}^5 = i$.

Examples: The pentagonal antiprism (Figure A3.11), staggered ferrocene (Figure A3.12).

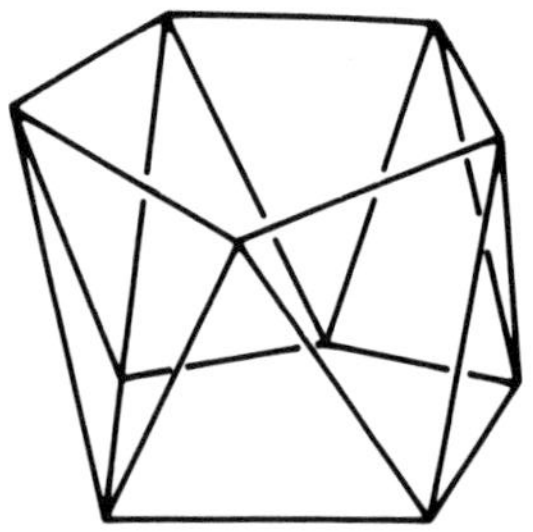

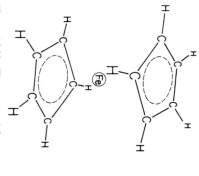

Figure A3.11 The pentagonal antiprism, of D_{5d} symmetry.

Figure A3.12 The staggered ferrocene molecule, $Fe(C_5H_5)_2$, of D_{5d} symmetry.

D_{6d}	E	$2S_{12}$	$2C_6$	$2S_4$	$2C_3$	$2S_{12}^5$	C_2	$6C_2'$	$6\sigma_d$		
A_1	1	1	1	1	1	1	1	1	1		$x^2+y^2;\ z^2$
A_2	1	1	1	1	1	1	1	-1	-1	R_z	z
B_1	1	-1	1	-1	1	-1	1	1	-1		
B_2	1	-1	1	-1	1	-1	1	-1	1		
E_1	2	$\sqrt{3}$	1	0	-1	$-\sqrt{3}$	-2	0	0	(T_x, T_y)	(x, y)
E_2	2	1	-1	-2	-1	1	2	0	0		$\left(\frac{1}{\sqrt{2}}(x^2-y^2),\ xy\right)$
E_3	2	0	-2	0	2	0	-2	0	0		
E_4	2	-1	-1	2	-1	-1	2	0	0		
E_5	2	$-\sqrt{3}$	1	0	-1	$\sqrt{3}$	-2	0	0	(R_y, R_x)	(yz, zx)

Note: Many of the operations of the group may be derived from S_{12} because $S_{12}^2 = C_6$; $S_{12}^3 = S_4$; $S_{12}^4 = C_3$; $S_{12}^6 = C_2$.

Examples: A hexagonal antiprism (Figure A3.13), staggered dibenzene-chromium (Figure A3.14).

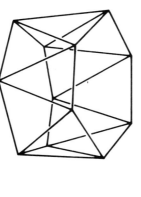

Figure A3.13 The hexagonal antiprism, of D_{6d} symmetry.

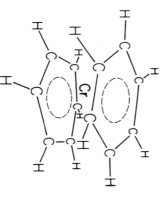

Figure A3.14 The staggered dibenzenechromium molecule, $Cr(C_6H_6)_2$, of D_{6d} symmetry.

5   THE GROUP D_n

These are groups of proper (i.e. pure) rotations corresponding to bodies in which there are n C_2 axes perpendicular to a principal C_n axis. To obtain solid figures of these geometries it is simplest to take a polyhedron shown for a D_{nh} or D_{nd} symmetry and to systematically introduce zigzag edges such as used to derive figures for the groups O and T. Molecules of D_{nh} or D_{nd} symmetries drop to D_n symmetry when the 'top' and 'bottom' parts of the molecule are given small, arbitrary, twists in opposite directions about the z axis.

D_{nh} and D_{nd} (n odd) group are direct products of D_n with either C_i or C_s. For problems in these groups it is often simplest to work in D_n symmetry and move to the full group at a later stage.

D_2	E	$C_2(z)$	$C_2(y)$	$C_2(x)$		
A	1	1	1	1		$x^2;\ y^2,\ z^2$
B_1	1	1	-1	-1	$T_z;\ R_z$	$z;\ xy$
B_2	1	-1	1	-1	$T_y;\ R_y$	$y;\ zx$
B_3	1	-1	-1	1	$T_x;\ R_x$	$x;\ yz$

Notes: (1) Because there are three mutually perpendicular C_2 axes the choice of x, y and z is arbitrary. A relabelling of these axes will lead to an interchange of the labels B_1, B_2 and B_3.

(2) Because x^2 and y^2 transform, separately, as A, it follows that the function $x^2 - y^2$ also has A symmetry.

Example: Ethene in which the two CH_2 groups have been made non-coplanar by a counter-rotation of these two units about the C–C axis (Figure A3.15).

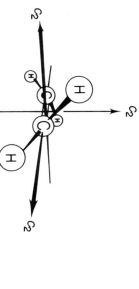

Figure A3.15   A slightly twisted ethene molecule, of D_2 symmetry.

D_3	E	$2C_3$	$3C_2$		
A_1	1	1	1		z^2, x^2+y^2
A_2	1	1	-1	$T_z; R_z$	z
E	2	-1	0	$(T_x, T_y); (R_y, R_x)$	$(x,y); (\frac{1}{\sqrt{2}}(x^2-y^2), xy)$; (zx, yz)

Example: Ethane in which the two CH_3 units have been counter-rotated about the C—C axis so that the molecule is neither eclipsed nor staggered (Figure A3.16).

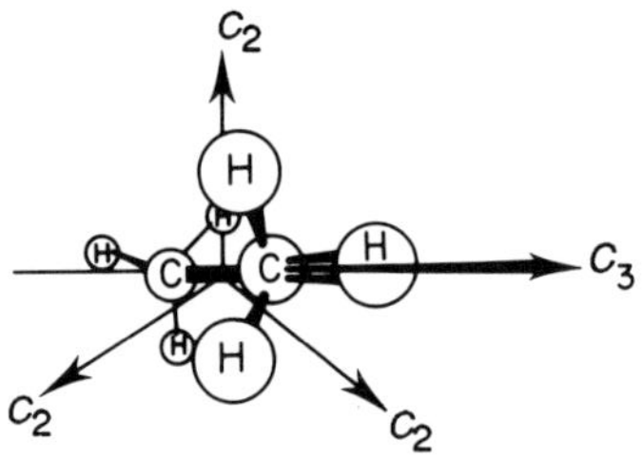

Figure A3.16   An ethane molecule which is neither eclipsed nor staggered , of D_3 symmetry.

D_4	E	$2C_4$	C_2	$2C_2'$	$2C_2''$		
A_1	1	1	1	1	1		$z^2, x^2 + y^2$
A_2	1	1	1	-1	-1	$T_z; R_z$	z
B_1	1	-1	1	1	-1		$x^2 - y^2$
B_2	1	-1	1	-1	1		xy
E	2	0	-2	0	0	$(T_x, T_y); (R_x, R_y)$	$(x, y); (zx, yz)$

Notes: (1) The x axis has been taken as coincident with one C_2'.
(2) The choice between which set of two C_2 axes is called $2C_2'$ and which is called $2C_2''$ is arbitrary. If the choice opposite to that above is taken then the labels B_1 and B_2 have to be interchanged (B_1 has a character of 1 under the C_2' operations).
Example: A twisted cube (Figure A3.17).

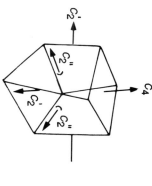

Figure A3.17   A slightly twisted cube, of D_4 symmetry.

369

D_5	E	$2C_5$	$2C_5^2$	$5C_2$		
A_1	1	1	1	1		$x^2+y^2; z^2$
A_2	1	1	1	−1	$T_x; R_z$	z
E_1	2	2 cos 72	2 cos 144	0	$(T_x, T_y); (R_x, R_y)$	$(x, y); (xz, yz)$
E_2	2	2 cos 144	2 cos 72	0		$\left(\frac{1}{\sqrt{2}}[x^2-y^2], xy\right)$

Note: Before working with this group refer to Note (2) under the D_{5h} character table.
Example: Ferrocene when the carbon atoms in opposite rings are neither staggered nor eclipsed (Figure A3.18).

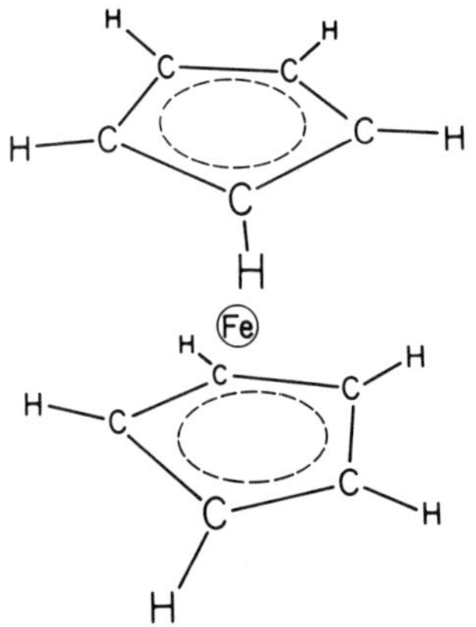

Figure A3.18 A ferrocene molecule, $Fe(C_5H_5)_2$, in which the two rings are neither staggered nor eclipsed, of D_5 symmetry.

D_6	E	$2C_6$	$2C_3$	C_2	$3C_2'$	$3C_2''$		
A_1	1	1	1	1	1	1		$x^2 + y^2, z^2$
A_2	1	1	1	1	-1	-1	$T_z; R_z$	z
B_1	1	-1	1	-1	1	-1		
B_2	1	-1	1	-1	-1	1		
E_1	2	1	-1	-2	0	0	$(T_x, T_y); (R_x, R_y)$	(xz, yz)
E_2	2	-1	-1	2	0	0		$\left(\frac{1}{\sqrt{2}}[x^2 - y^2], xy\right)$

Notes: (1) The x axis has been taken as coincident with one C_2'.

(2) The choice between which set of three C_2 axes is called $3C_2'$ and which is called $3C_2''$ is arbitrary. If the choice opposite to that above is taken then the labels B_1 and B_2 on functions will have to be interchanged (B_1 has a character of 1 under the C_2' operations).

Example: Dibenzenechromium when the carbon atoms in opposite rings are neither staggered nor eclipsed (Figure A3.19).

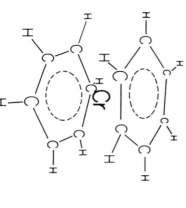

Figure A3.19   A dibenzenechromium molecule, $Cr(C_6H_6)_2$, in which the two rings are neither staggered nor eclipsed, of D_6 symmetry.

6 THE GROUPS C_{nv}

This set of character tables is of considerable importance since the groups involved are of common occurrence. One problem that frequently occurs with them is that of ambiguity about the choice of x and y (z presents no problems). Thus, for C_{2v}, a change of choice can change the meaning of labels B_1 and B_2. Related problems arise for all C_{nv} groups with n even. Thus for C_{4v}, as B_1 and B_2 change with choice, so too do the characters describing the functions $x^2 - y^2$ and xy (most evident when the choice of x and y lies between placing them in the $2\sigma_v$ and $2\sigma_v'$). This has been discussed in more detail in the relevant chapters in the book.

C_{2v}	E	C_2	σ_v	σ_v'		
A_1	1	1	1	1	T_z	$z; z^2, x^2, y^2$
A_2	1	1	-1	-1	R_z	xy
B_1	1	-1	1	-1	$T_y; R_x$	$y; zx$
B_2	1	-1	-1	1	$T_x; R_y$	$x; yz$

Notes: (1) x is taken as lying in σ_v'.

(2) Interchange of the labels σ_v and σ_v' or (equivalently) interchange of choice of direction of x and y axes (one lies in each mirror plane) interchanges the labels B_1 and B_2 (B_1 has a character of 1 under the σ_v operation).

Example: The water molecule. See Chapters 2 and 3.

C_{3v}	E	$2C_3$	$3\sigma_v$		
A_1	1	1	1	T_z	$z; z^2, x^2 + y^2$
A_2	1	1	-1	R_z	
E	2	-1	0	$(T_x, T_y); (R_y, R_x)$	$(x, y); (zx, yz); (xy, \frac{1}{\sqrt{2}}[x^2 - y^2])$

Example: The ammonia molecule. See Chapter 6.

C_{4v}	E	$2C_4$	C_2	$2\sigma_v$	$2\sigma_v'$		
A_1	1	1	1	1	1	T_z	z; z^2, x^2+y^2
A_2	1	1	1	−1	−1	R_z	
B_1	1	−1	1	1	−1		x^2-y^2
B_2	1	−1	1	−1	1		xy
E	2	0	−2	0	0	(T_x, T_y); (R_x, R_y)	(x, y); (zx, yz)

Notes: (1) x is taken as lying in one σ_v plane.

(2) Interchange of the labels σ_v and σ_v' (and the choice is arbitrary) leads to an interchange of the labels B_1 and B_2 (B_1 has a character of 1 under the σ_v operations).

Example: The BrF$_5$ molecule (Chapter 5).

C_{5v}	E	$2C_5$	$2C_5^2$	$5\sigma_v$		
A_1	1	1	1	1	T_z	$z;\ x^2+y^2,\ z^2$
A_2	1	1	1	-1	R_z	
E_1	2	$2\cos 72$	$2\cos 144$	0	$(T_y,\ T_x);\ (R_x,\ R_y)$	$(x,\ y);\ (xz,\ yz)$
E_2	2	$2\cos 144$	$2\cos 72$	0		$\left(\frac{1}{\sqrt{2}}[x^2-y^2],\ xy\right)$

Note: Before working with this group refer to Note (2) under the D_{5h} character table.
Example: η^5-cyclopentadienecarbonylnickel (Figure A3.20).

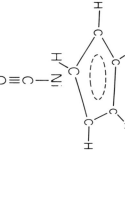

Figure A3.20 The η^5-cyclopentadiene-carbonylnickel molecule, $Ni(C_5H_5)CO$, of C_{5v} symmetry.

C_{6v}	E	$2C_6$	$2C_3$	C_2	$3\sigma_v$	$3\sigma_v'$		
A_1	1	1	1	1	1	1		z; x^2+y^2, z^2
A_2	1	1	1	1	-1	-1	T_z R_z	
B_1	1	-1	1	1	1	1		
B_2	1	-1	1	1	-1	-1		
E_1	2	1	-1	-2	0	0	(T_x, T_y); (R_y, R_x)	(x, y); (xz, yz)
E_2	2	-1	-1	2	0	0		$\left(\frac{1}{\sqrt{2}}[x^2-y^2], xy\right)$

Note: Interchange of the labels σ_v and σ_v' (and the choice is arbitrary) leads to an interchange of the labels B_1 and B_2 on functions (B_1 has a character of 1 under the σ_v operations).

Example: The compound η^6-hexamethylbenzene, η^6-benzenechromium (Figure A3.21).

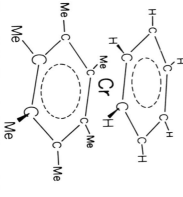

Figure A3.21 The molecule η^6-hexamethylbenzene,η^6-benzenechromium, $Cr(C_6Me_6)(C_6H_6)$, in which the two rings are eclipsed, of C_{6v} symmetry.

7   THE GROUPS C_{nh}

These groups have a derivation similar to that of the D_{nh} group—they are direct products of C_n with either C_i (n even) or C_s (n odd). The only one which has been found to be of real chemical importance is C_{2h}; however, C_{3h} is included to give an example of the n odd case.

C_{2h}		E	C_2	i	σ_h		
A_g		1	1	1	1	R_z	x^2, y^2, z^2, xy
B_g		1	-1	1	-1	$R_x; R_y$	$yz; zx$
A_u		1	1	-1	-1	T_z	z
B_u		1	-1	-1	1	$T_x; T_y$	$z; y$

Note: This group is a direct product of the C_2 and C_i groups, indicated in the character table by lines:

C_{3h}		E	C_3	C_3^2	σ_h	S_3	S_3^5		
A'		1	1	1	1	1	1	R_z	x^2+y^2, z^2
E'	$\begin{cases} \\ \end{cases}$	$\begin{matrix}1\\1\end{matrix}$	$\begin{matrix}\varepsilon\\\varepsilon^2\end{matrix}$	$\begin{matrix}\varepsilon^2\\\varepsilon\end{matrix}$	$\begin{matrix}1\\1\end{matrix}$	$\begin{matrix}\varepsilon\\\varepsilon^2\end{matrix}$	$\begin{matrix}\varepsilon^2\\\varepsilon\end{matrix}$	(T_x, T_y)	$(x, y); (\frac{1}{\sqrt{2}}[x^2-y^2], xy)$
A''		1	1	1	-1	-1	-1	T_z	z
E''	$\begin{cases} \\ \end{cases}$	$\begin{matrix}1\\1\end{matrix}$	$\begin{matrix}\varepsilon\\\varepsilon^2\end{matrix}$	$\begin{matrix}\varepsilon^2\\\varepsilon\end{matrix}$	$\begin{matrix}-1\\-1\end{matrix}$	$\begin{matrix}-\varepsilon\\-\varepsilon^2\end{matrix}$	$\begin{matrix}-\varepsilon^2\\-\varepsilon\end{matrix}$	(R_y, R_x)	(yz, zx)

Notes: (1) See the notes on the C_3 group for the meaning of ε and ε^2

(2) This group is a direct product of the C_3 and C_s group, indicated in the character table by the lines.

8   THE GROUPS C_n

These are cyclic groups with character tables that look rather strange when compared with most of those encountered earlier in this appendix. They only look strange when compared with other point groups. For many other groups—for instance, in the translation groups encountered in theories of crystal structure, but not discussed in Chapters 12 and 13—the appearance of complex numbers is the norm.

C_2	E	C_2		
A	1	1	$T_z; R_z$	$z; z^2, y^2, x^2; xy$
B	1	-1	$T_x; T_y; R_x; R_y$	$x; y; z; yz, xz$

Note: In this and two of the next three tables the notation that

$$\varepsilon = \exp\left(\frac{2\pi i}{n}\right) = \cos\left(\frac{2\pi}{n}\right) + i \sin\left(\frac{2\pi}{n}\right)$$

$$\varepsilon^2 = \exp\left(\frac{-2\pi i}{n}\right) = \cos\left(\frac{2\pi}{n}\right) + i \sin\left(\frac{2\pi}{n}\right)$$

is used, so that here $(n = 3)$.

$$\varepsilon = \cos 120 + i \sin 120 = -\frac{1}{2} + i\sqrt{\frac{3}{2}}$$

$$\varepsilon^2 = \cos 120 + i \sin 120 = -\frac{1}{2} - i\sqrt{\frac{3}{2}}$$

C_3	E	C_3	C_3^2		
A	1	1	1	$T_z; R_z$	$z; x^2+y^2, z^2$
E	$\left\{\begin{matrix}1\\1\end{matrix}\right.$	$\begin{matrix}\varepsilon\\\varepsilon^2\end{matrix}$	$\begin{matrix}\varepsilon^2\\\varepsilon\end{matrix}$	$(T_x, T_y); (R_x, R_y)$	$(x, y); (\frac{1}{\sqrt{2}}[x^2-y^2], xy) : (yz, zx)$

Note that ε and ε^2 are complex conjugates. That is, in the case of $n = 3$, $\varepsilon^2 = \varepsilon^*$; the ε^2 notation has been used in the C_3, T, T_h and C_{3h} character tables.

C_4	E	C_4	C_2	C_4^3		
A	1	1	1	1	$T_z; R_z$	$z; x^2 + y^2, z^2$
B	1	-1	1	-1		
E	$\begin{cases} 1 \\ 1 \end{cases}$	$\begin{matrix} i \\ -i \end{matrix}$	$\begin{matrix} -1 \\ -1 \end{matrix}$	$\begin{matrix} -i \\ i \end{matrix}$	$(T_x, T_y); (R_x, R_y)$	$(x, y); (yz, zx)$

Note: It is easy to show by substitution in the equations given above that for $n = 4$, $\exp(2\pi i/n) = i$
Chapter 11 gives detailed examples of working with the C_4 group.

C_5	E	C_5	C_5^2	C_5^3	C_5^4		
A	1	1	1	1	1	$T_z; R_z$	$z; x^2 + y^2, z^2$
E_1	$\begin{cases} 1 \\ 1 \end{cases}$	$\begin{matrix} \varepsilon \\ \varepsilon^* \end{matrix}$	$\begin{matrix} \varepsilon^2 \\ \varepsilon^{2*} \end{matrix}$	$\begin{matrix} \varepsilon^{2*} \\ \varepsilon^2 \end{matrix}$	$\begin{matrix} \varepsilon^* \\ \varepsilon \end{matrix}$	$(T_x, T_y); (R_x, R_y)$	$(x, y); (yz, zx)$
E_2	$\begin{cases} 1 \\ 1 \end{cases}$	$\begin{matrix} \varepsilon^2 \\ \varepsilon^{2*} \end{matrix}$	$\begin{matrix} \varepsilon^* \\ \varepsilon \end{matrix}$	$\begin{matrix} \varepsilon \\ \varepsilon^* \end{matrix}$	$\begin{matrix} \varepsilon^2 \\ \varepsilon^2 \end{matrix}$		$\left(\frac{1}{\sqrt{2}}[x^2 - y^2], xy\right)$

Note: For a definition of ε etc., see under C_3.

C_6	E	C_6	C_3	C_2	C_3^2	C_6^5		
A	1	1	1	1	1	1	$T_z; R_z$	$z; x^2+y^2, z^2$
B	1	−1	1	−1	1	−1		
E_1	$\left\{\begin{array}{l}1\\1\end{array}\right.$	$\begin{array}{l}\varepsilon\\\varepsilon^*\end{array}$	$\begin{array}{l}-\varepsilon^*\\-\varepsilon\end{array}$	$\begin{array}{l}-1\\-1\end{array}$	$\begin{array}{l}-\varepsilon\\-\varepsilon^*\end{array}$	$\begin{array}{l}\varepsilon^*\\\varepsilon\end{array}$	$(T_x, T_y); (R_x, R_y)$	$(x, y); (yz, zx)$
E_2	$\left\{\begin{array}{l}1\\1\end{array}\right.$	$\begin{array}{l}-\varepsilon^*\\-\varepsilon\end{array}$	$\begin{array}{l}-\varepsilon\\-\varepsilon^*\end{array}$	$\begin{array}{l}1\\1\end{array}$	$\begin{array}{l}-\varepsilon^*\\-\varepsilon\end{array}$	$\begin{array}{l}-\varepsilon\\-\varepsilon^*\end{array}$		$\left(\frac{1}{\sqrt{2}}[x^2-y^2], xy\right)$

Note: For a definition of ε, etc., see under C_3.

9   THE GROUPS S_n (n EVEN) (INCLUDES C_i)

Another set of cyclic groups is denoted S_n. These only exist for n even because odd values of n do not satisfy the requirement $(S_n)^n = E$. The S_2 group is usually labelled C_i because the operations S_2 and i are identical (this is demonstrated in Figure 7.29).

C_i	E	i		
A_g	1	1	R_z; R_x; R_y	z^2; y^2; x^2; xy; yz, zx
A_u	1	-1	T_z; T_x; T_y	x; y; z

Note: C_i often forms a direct product with another group. In this case the g and u suffixes of the C_i group are carried into the labels of the direct product group.

S_4	E	S_4	C_2	S_4^3		
A	1	1	1	1	R_z	x^2+y^2, z^2
B	1	-1	1	-1	T_z; z	
E	$\left\{\begin{matrix}1\\1\end{matrix}\right.$	$\begin{matrix}i\\-i\end{matrix}$	$\begin{matrix}-1\\-1\end{matrix}$	$\begin{matrix}-i\\i\end{matrix}$	(T_x, T_y); (R_x, R_y)	$\left(\frac{1}{\sqrt{2}}[x^2-y^2], xy\right)(x, y)$; (yz, zx)

Note: See the note under the C_4 group.

10   THE GROUP C_s AND THE TRIVIAL GROUP C_i

The group C_s, like the group C_i, often participates in a direct product group. In the case of C_s it is the post-superscript primes which carry over into the labels of the irreducible representations of the product group.

C_s	E	σ		
A'	1	1	R_z; T_x; T_y	x; y; z^2; y^2; x^2; xy
A''	1	-1	T_z; R_x; R_y	z; yz; zx

Note: The group C_1 is trivial because it is the symmetry of an object which has no symmetry! The only symmetry operation is the identity.

C_1	E
A	1

In this group no bases are listed —all bases give rise to the A irreducible representation!

11   THE INFINITESIMAL ROTATION (LINEAR) GROUPS $C_{\infty v}$ AND $D_{\infty h}$

Molecules in which all atoms lie on a common axis demand special attention because a rotation of any magnitude about this axis is a symmetry operation. The attack which proves profitable on this problem is to regard all such rotations to be (very large) multiples of an infinitesimally small rotation. That is, there is a C_∞ axis and associated operations. The character table gives the character for the operation of rotation by an arbitrary angle ϕ denoted C_∞^ϕ. Not only is there an infinite number of operations based on C_∞ there is also an infinite number of σ_v mirror planes. Fortunately, they all fall into a single class. If the linear molecule has no centre of symmetry then the appropriate group is $C_{\infty v}$. With a centre of symmetry the group is the direct product of $C_{\infty v}$ with C_i and is denoted $D_{\infty h}$. Because the groups are infinite, the usual method of reducing a reducible representation will not work. However, reduction by inspection is usually possible. Appendix 5 discusses this problem in more detail. The alternative labels for irreducible representations for $C_{\infty v}$ and for $D_{\infty h}$ antedate the system used in this book. It is the Σ, Π, Δ system which is the more commonly used.

The $D_{\infty h}$ point group is the direct product of $C_{\infty v}$ and C_i; it would therefore be expected that all terms in the irreducible representations carrying a g suffix would appear with the same coefficients as their counterparts in $C_{\infty v}$. Similarly, the u-suffix representations should carry the coefficients with changed signs. Inspection of the character table of $D_{\infty h}$ reveals that these expectations do not appear to be obeyed. For instance, E_1 in $C_{\infty v}$ has the character 2 cos ϕ under $2C_\infty^\phi$, whereas E_{1g} in $D_{\infty h}$ has a character of -2 cos ϕ under $2S_\infty^\phi$. The reason for this apparent anomaly lies in the nature of S_n operations. Consider C_3. The combination $i.C_3$ corresponds to an S_6 operation, not to a S_3—it is $i.C_6$ which corresponds to an S_3. (See, for instance, Figure 7.5(b); there is no C_6 axis in an octahedron, only C_3.) In general, corresponding to a pure rotation operation involving a rotation of ϕ will be a rotation-inversion operation involving a rotation of $-(180 - \phi)$. But

$$\cos\{-(180 - \phi)\} = -\cos \phi$$

which explains the apparent anomalies in the $D_{\infty h}$ character table.

In previous character tables a prime (and double prime) notation has been used to indicate symmetry behaviour with respect to a mirror plane. These mirror planes were all of the σ_h type. In $C_{\infty v}$ and $D_{\infty h}$ a $+,-$ notation has been used but this is because the mirror planes are of the σ_v type.

$C_{\infty v}$

$C_{\infty v}$	E	$2C_\infty^\phi$	...	$\infty\sigma_v$		
$A_1 \equiv \Sigma^+$	1	1	...	1	T_z	$z; x^2+y^2, z^2$
$A_2 \equiv \Sigma^-$	1	1	...	-1	R_z	
$E_1 \equiv \Pi$	2	$2\cos\phi$	...	0	$(T_x, T_y); (R_x, R_y)$	$(x, y); (zx, yz)$
$E_2 \equiv \Delta$	2	$2\cos 2\phi$	...	0		$\frac{1}{\sqrt2}[x^2 - y^2], xy$
$E_3 \equiv \Phi$	2	$2\cos 3\phi$	...	0		
...						

$D_{\infty h}$

$D_{\infty h}$	E	$2C_\infty^\phi$	...	$\infty\sigma_v$	i	$2S_\infty^\phi$	...	∞C_2		
$A_{1g} \equiv \Sigma_g^+$	1	1	...	1	1	1	...	1		x^2+y^2, z^2
$A_{2g} \equiv \Sigma_g^-$	1	1	...	-1	1	1	...	-1	R_z	
$E_{1g} \equiv \Pi_g$	2	$2\cos\phi$	...	0	2	$-2\cos\phi$	...	0	(R_x, R_y)	(zx, yz)
$E_{2g} \equiv \Delta_g$	2	$2\cos 2\phi$	...	0	2	$2\cos 2\phi$	...	0		$\frac{1}{\sqrt2}[x^2-y^2], xy$
$E_{3g} \equiv \Phi_g$	2	$2\cos 3\phi$	...	0	2	$-2\cos 3\phi$	...	0		
...										
$A_{1u} \equiv \Sigma_u^+$	1	1	...	1	-1	-1	...	-1	T_z	z
$A_{2u} \equiv \Sigma_u^-$	1	1	...	-1	-1	-1	...	1		
$E_{1u} \equiv \Pi_u$	2	$2\cos\phi$	...	0	-2	$2\cos\phi$	...	0	(T_x, T_y)	(x, y)
$E_{2u} \equiv \Delta_u$	2	$2\cos 2\phi$	...	0	-2	$-2\cos 2\phi$	...	0		
$E_{3u} \equiv \Phi_u$	2	$2\cos 3\phi$	...	0	-2	$2\cos 3\phi$	...	0		
...										

Appendix 4
The Fluorine Group Orbitals of π Symmetry in SF$_6$

It is inevitable that in the application of group theory to chemistry some short-cuts exist—and are exploited—which circumvent tedious or difficult mathematics. The experienced worker can often astonish the inexperienced by their ability to write down the correct linear combinations for a new problem—with no apparent work. In this appendix, an attempt is made to reveal some of the tricks. Thus, although at first sight it seems an advantage to have high symmetry this is sometimes not the case when carrying out a detailed calculation—for instance, there would be a considerable number of different interactions possible between two sets of triply degenerate orbitals in a bonding problem. In such a case it may help to pretend that the symmetry is lower than is in fact the case because the consequent reduced degeneracy forces a pairing between individual members of each set, thus reducing the number of interactions to be considered. Having paired the orbitals by this device, the low symmetry geometry can be forgotten and the correct point group used.

It is a similar trick which provides an alternative to the projection operator method of obtaining linear combinations of orbitals (Sections 4.6, 5.5 and 6.2) and which proves to be easier to use in high symmetry cases. It uses knowledge of the correct combinations in a lower symmetry case to obtain those of a higher symmetry molecule, the lower symmetry group being a subgroup of the higher. There is no unique path in this approach—different workers might choose different low symmetry groups. For a given choice of subgroup there may be several equally valid ways of proceeding. Those experienced in the art develop a 'nose' which is based on a mixture of experience and the ability to anticipate problems that will be encountered along each alternative path. Something of this 'nose' will be evident in the next section where an attempt has been made to give the reasons for expecting a particular approach to be fruitful (or not, as the case may be).

In tackling the problem of generating the fluorine group orbitals of π symmetry in SF$_6$—of O_h symmetry—a choice of lower symmetry group must first be made. It is usually sensible to choose the subgroup of the highest symmetry for which detailed results are available. In the present case this suggests that the C_{4v} subgroup of O_h be chosen because some ligand group orbitals for a molecule of this symmetry—BrF$_5$—were obtained in Chapter 5. It is true that in that chapter only Br–F σ-bonding interactions were considered but perhaps they can be used as a base from which to obtain the π combinations. In Chapter 5 it was explicitly recognized that the fluorine σ orbitals would be mixtures of s and p atomic orbitals; for simplicity they were there drawn as pure s orbitals. In Figure A4.1 they are drawn again, but this time as pure p orbitals; the C_{4v} symmetry labels are included. Suppose the p orbitals are tilted out of the plane, as shown in Figure A4.2. The symmetry labels of Figure A4.1 remain appropriate as do the linear combinations. This is really an indication that in BrF$_5$ there is no symmetry-dictated requirement that the Br–F σ bonding orbitals have their maxima in the plane defined by the fluorine atoms. If the tilting processes is now completed Figure A4.3 is obtained, which shows that the π orbital combinations have been obtained—starting from the σ! This method could be used because there is no operation in C_{4v} which interchanges—and thus compares—the 'top' with the 'bottom' of each p orbital in Figure A4.3. The trick could not have been used in the D_{4h} subgroup because the σ_h mirror plane in that group gives this comparison and so distinguishes between σ and π orbitals. None the less, the combinations shown in Figure A4.3 remain correct in D_{4h} because C_{4v} is also a subgroup of D_{4h}—but the symmetry labels would have to be changed.

The next step is that of recognizing that the twelve p$_\pi$ orbitals of the fluorine atoms in SF$_6$ can be obtained from those of the three planes of four atoms shown in Figure A4.4. Recall that the twelve π orbitals of Figure A4.4

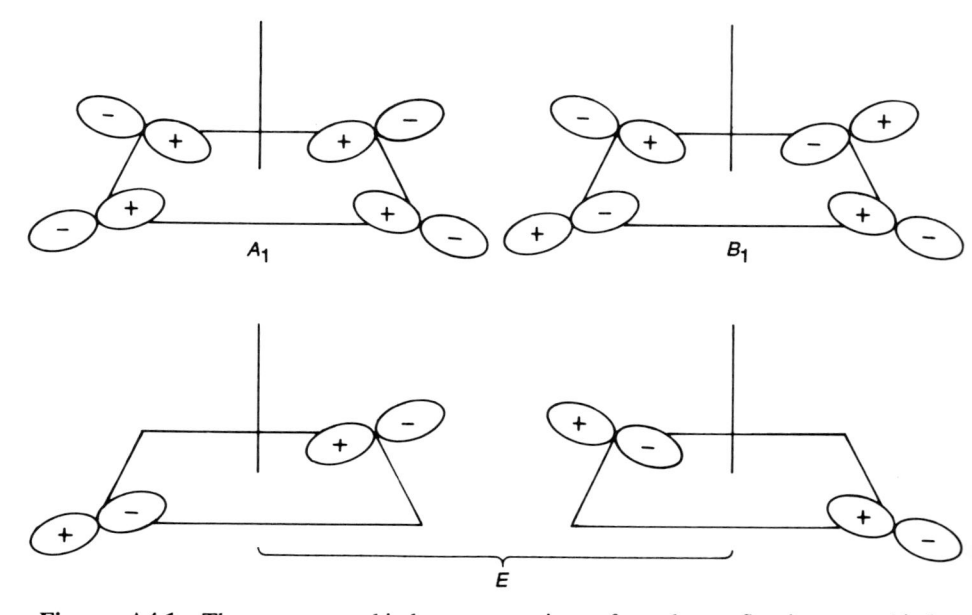

Figure A4.1 The pure p orbital representation of coplanar fluorine σ orbital symmetry-adapted combinations in BrF$_5$.

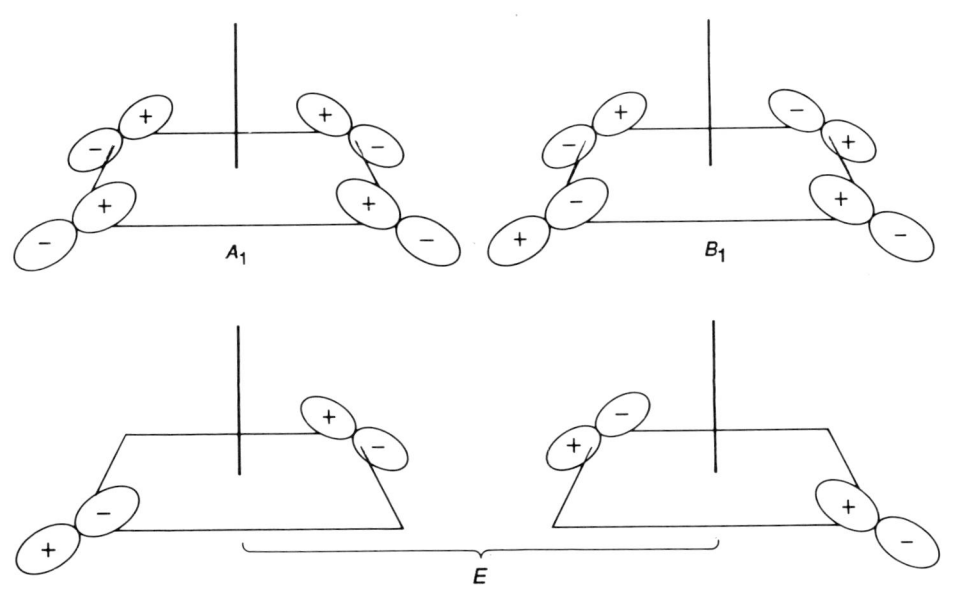

Figure A4.2 The same p orbitals as in Figure A4.1 but with each tilted out of the plane in the direction of the apical fluorine in BrF$_5$.

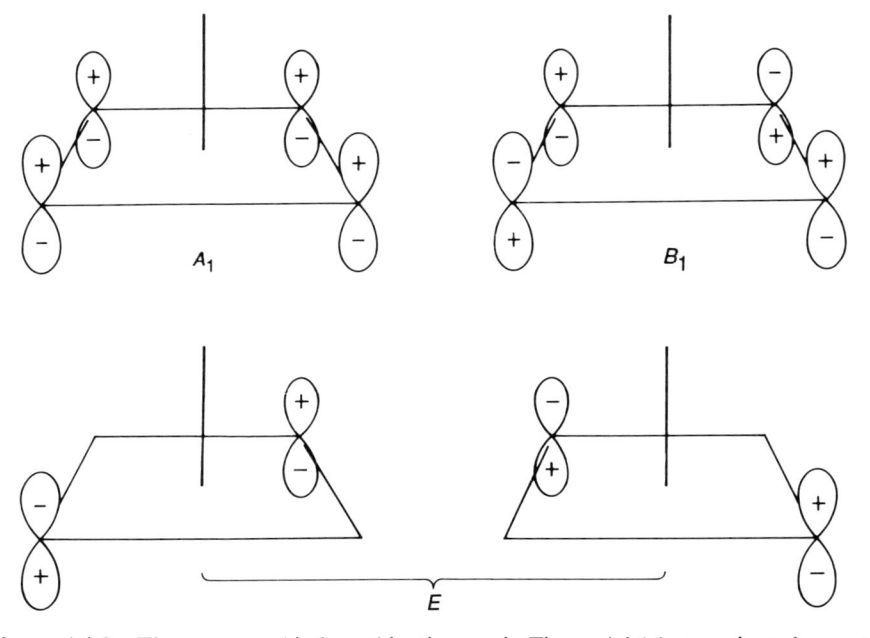

Figure A4.3 The same p orbital combinations as in Figure A4.1 but reoriented so as to be perpendicular to the original set.

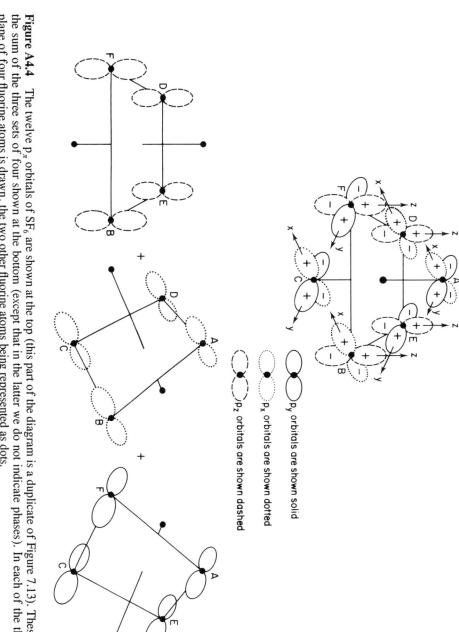

Figure A4.4 The twelve p_π orbitals of SF_6 are shown at the top (this part of the diagram is a duplicate of Figure 7.13). These twelve are the sum of the three sets of four shown at the bottom (except that in the latter we do not indicate phases). In each of the three sets the plane of four fluorine atoms is drawn, the two other fluorine atoms being represented as dots.

p_z orbitals are shown dashed

p_x orbitals are shown dotted

p_y orbitals are shown solid

transform as $T_{1g} + T_{1u} + T_{2g} + T_{2u}$; i.e. four different sets of triply degenerate orbitals. This triple degeneracy neatly matches the three planes and associated sets of orbitals shown in Figure A4.4. If this is exploited and the three orbitals, one from each plane, which correspond to the A_1 combination in Figure A4.3 are collected together:

$$\tfrac{1}{2}[p_z(B) + p_z(D) + p_z(E) + p_z(F)] \quad \leftarrow$$
$$\tfrac{1}{2}[p_y(A) + p_y(C) + p_y(E) + p_y(F)]$$
$$\tfrac{1}{2}[p_x(A) + p_x(B) + p_x(C) + p_x(D)]$$

the T_{1u} set of ligand π orbitals in O_h is obtained, as may be checked by considering their transformations as a set. The combination shown in Figure 7.15 is indicated by an arrow. Similarly, the three combinations corresponding to the B_1 in Figure A4.3 are:

$$\tfrac{1}{2}[p_z(B) + p_z(D) - p_z(E) - p_z(F)] \quad \leftarrow$$
$$\tfrac{1}{2}[p_y(A) + p_y(C) - p_y(E) - p_y(F)]$$
$$\tfrac{1}{2}[p_x(A) + p_x(B) - p_x(C) - p_x(D)]$$

which is the T_{2u} set of ligand π orbitals in O_h, the combination shown in Figure 7.17 being arrowed.

It is at this point that anticipation cautions against continuing on and finishing the problem. There are two indicators that should cause us to pause. First, the next step would have involved three pairs of orbitals (the three sets, one from each plane, corresponding to the degenerate E set in Figure A4.3). It seems that six orbitals, apparently all degenerate, would be obtained. However, we are looking for two sets of three (T_{1g} and T_{2g}) and the two sets are not expected to be degenerate. Second, some arbitrariness has been exercised in the procedure that has been followed. In particular, when working with the planes shown in Figures A4.4 only p_π orbitals which have their maximum amplitude perpendicular to this plane have been considered. Equally, the choice could have been made to work with the p_π orbitals which 'lie in' the plane, as shown in Figure A4.5, although it is not immediately clear how to proceed had this alternative choice been made. Experience suggests that when six apparently degenerate orbitals are obtained, as seems to be the case here, this is because symmetry-distinct combinations have been mixed together. After all, this is always mathematically possible even if it is a step which would not be made from choice. Further, experience is that because a choice exists between 'perpendicular p_π orbitals' and 'coplanar p_π orbitals' each alternative must be expected to appear to an equal extent in the answer.

The way to extract symmetry-distinct combinations from sets in which they have been mixed together is to take suitable linear combinations of members of

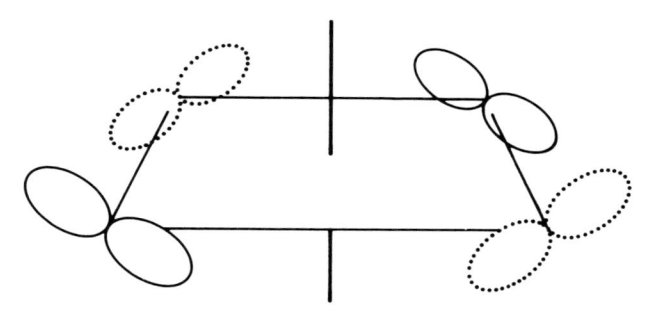

Figure A4.5 A set of four fluorine p$_\pi$ orbitals which lie in the plane of the fluorine atoms (cf. Figure A4.3 and the second part of Figure A4.4, where the sets of fluorine p$_\pi$ orbitals shown are all perpendicular to the plane of the four fluorine atoms).

the mixed-up sets (a set of six orbitals in the present case). These six (un-normalized) are:

$$\psi_1 = [p_z(E) - p_z(F)] \qquad \psi_2 = [p_z(B) - p_z(D)]$$
$$\psi_3 = [p_y(A) - p_y(C)] \qquad \psi_4 = [p_y(E) - p_y(F)]$$
$$\psi_5 = [p_x(A) - p_x(C)] \qquad \psi_6 = [p_x(B) - p_x(D)]$$

Which orbitals should be combined together? It is here that the expectation of a 'coplanar' set of p$_\pi$ orbitals comes to our aid. Note that the set of 'coplanar' p$_\pi$ orbitals shown in Figure A4.5 contain contributions from

$$p_y(F), \ p_x(B), \ p_y(E) \text{ and } p_x(D)$$

i.e. those orbitals contained in ψ_4 and ψ_6 in the list above. Clearly, then, we have to combine these two. Because ψ_4 and ψ_6 are symmetry-equivalent (a C_4 rotation turns ψ_4 into ψ_6) they must be expected to contribute equally to the combinations. The only way for this to occur is to combine them first with the same and then with opposite signs. The result is (giving the final combinations in normalized form):

$$\psi_6 + \psi_4: \quad \tfrac{1}{2}[p_x(B) - p_x(D) + p_x(E) - p_x(F)]$$
$$\psi_6 - \psi_4: \quad \tfrac{1}{2}[p_x(B) - p_x(D) - p_x(E) + p_x(F)]$$

Having thus discovered a way forward with one pair, it makes sense to apply the same procedure to the pairs:

$$\psi_1 \text{ and } \psi_5$$
$$\psi_2 \text{ and } \psi_3$$

and thus obtain the complete sets:

$$T_{1g}: \quad \tfrac{1}{2}[p_x(A) - p_x(C) + p_z(E) - p_z(F)] \quad \leftarrow$$
$$\tfrac{1}{2}[p_y(A) - p_y(C) - p_z(B) + p_z(D)]$$
$$\tfrac{1}{2}[p_x(B) - p_x(D) + p_y(E) - p_y(F)]$$

$$T_{2g}: \quad \tfrac{1}{2}[p_x(A) - p_x(C) - p_z(E) + p_z(F)] \quad \leftarrow$$
$$\tfrac{1}{2}[p_y(A) - p_y(C) + p_z(B) - p_z(D)]$$
$$\tfrac{1}{2}[p_x(B) - p_x(D) - p_y(E) + p_y(F)]$$

The combinations illustrated in Figure 7.14 (T_{1g}) and 7.16 (T_{2g}) are indicated by arrows.

The reader may well object that while a method has been given for obtaining combinations, it has not been shown that they are the ones that are

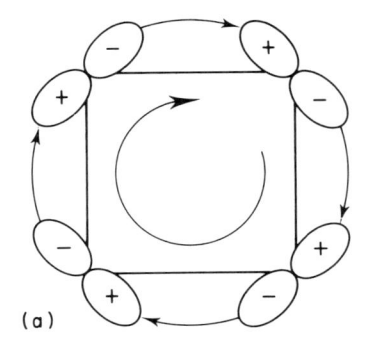

(a)

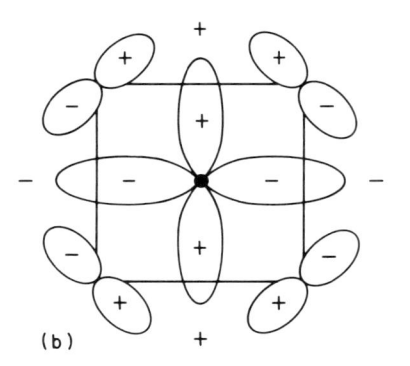

(b)

Figure A4.6 (a) A T_{1g} function compared to a rotation (central arrow). Note the arrows drawn between lobes of adjacent p_π orbitals.
(b) Comparison with a typical T_{2g} function (xy, d_{xy}, etc.)—at the centre—with the nodal pattern of the corresponding T_{2g} combination of p_π orbitals.

required. Put it another way; how were the six combinations allocated correctly between the T_{1g} and T_{2g} sets? Formally, of course, the answer is 'by considering their transformations', but, fortunately, experience relieves us of the tedium of this step. The character table for the O_h point group given in Appendix 3 shows that T_{1g} functions have the characteristic of a rotation while T_{2g} functions behave like products of coordinate axes. Figure A4.6 shows how these two observations may be used as a yardstick to discriminate between T_{1g} and T_{2g} functions.

This method of assignment of functions generated by a building-up procedure appears to break down when there is no basis function listed against an irreducible representation in a character table—for instance in O_h there is nothing listed against A_{1u}. This does not mean that no basis functions exist—they always do. Instead, they tend to be rather complicated, containing many nodes (thus, in Problem 4.2 the f_{xyz} orbital had to be invoked in order to find a function which spanned one particular irreducible representation). Such high nodality is usually enough to identify functions transforming under such an irreducible representation—it seldom happens that one is interested in more than one of this type at a time.

In this appendix an attempt has been made to give some insight into the way that experienced practitioners tackle some group theoretical problems. The approach used is complementary to more formal short-cut treatments which can be given, one of which is described in the reprint of an article in the *Journal of Chemical Education* which follows. The case of the fluorine π orbitals in SF$_6$ is not included in this article and the reader may find it of value to extend the treatment to include it.

LIGAND GROUP ORBITALS OF COMPLEX IONS†

Several articles which discuss ligand field theory reflect the growing interest in this refinement of simple crystal field theory.[1-5] The reasons why covalency needs to be introduced into the latter theory are too well known to need elaboration here. Rather, we shall discuss a problem which arises in teaching the theory to undergraduate classes. As we have pointed out, the derivation of the 'correct linear combinations of ligand orbitals' in ligand field theory is a step almost invariably omitted in expositions of the subject suitable for undergraduates.[5] The reason is simple: the derivation is difficult. The derivation in the most important case—that of an octahedral complex—using a group theoretical approach has been discussed.[5] In the present article an alternative derivation is given of the form of ligand group orbitals (l.g.o.'s) which is more

---

† Reprinted with minor alterations from an article by S. F. A. Kettle in the *Journal of Chemical Education*, **46** (1966), 652. Copyright (1966) by the Division of Chemical Education, American Chemical Society, and reprinted with permission.

suitable for undergraduate tuition. In particular, no use is made of detailed group theory.

The method which we use may be termed the method of 'ascent in symmetry'. The l.g.o.'s of a complicated molecule are derived from those of simpler 'molecules' which are fragments of the complicated one, i.e. vectors appropriate to any point group are derived as linear combinations of the vectors of its subgroup, implicit use being made of the group correlation tables. The method is one of considerable power; e.g. one may obtain the l.g.o.'s for an icosahedral arrangement of equivalent σ type orbitals in relatively simple form by this method. The standard group theoretical procedure is most unwieldy in this case, making it necessary to resort to a Schmidt orthogonalization procedure.

The method may conveniently be based on three axioms:

Axiom 1 The l.g.o.'s of a complicated molecule are related to those of its fragments by the condition that only sets with non-zero overlap may interact.

Axiom 2 Two equivalent orbitals are properly considered in in-phase and out-of-phase combinations. So, the correct combination of two localized orbitals σ_1 and σ_2 are, with neglect of overlap,†

$$\psi_s = \frac{1}{\sqrt{2}} (\sigma_1 + \sigma_2)$$

$$\psi_a = \frac{1}{\sqrt{2}} (\sigma_1 - \sigma_2)$$

Axiom 3 If a set of l.g.o.'s is d-fold degenerate and is formed from n equivalent orbitals then the sum of squares of coefficients with which each equivalent orbital appears in the set is d/n. This axiom may often be used in the simpler form that for every set of n equivalent orbitals $(\sigma_1, \sigma_2, ..., \sigma_n)$ there is always a totally symmetric combination:

$$\frac{1}{\sqrt{n}} (\sigma_1 + \sigma_2 + \cdots + \sigma_n)$$

We now illustrate the use of these axioms by deriving the l.g.o.'s appropriate to five different stereochemistries. In all cases we include group theoretical labels although these are not essential to the argument.

---

† For consistency, a positive phase is assigned to each localized orbital.

Example 1   The σ l.g.o.'s of a planar AB_3 molecule (D_{3h} symmetry). Label the σ orbitals σ_1, σ_2 and σ_3. Consider σ_1 and σ_2. From Axiom 2 the correct combinations, neglecting overlap, are:

$$\psi_s = \frac{1}{\sqrt{2}} (\sigma_1 + \sigma_2)$$

$$\psi_a = \frac{1}{\sqrt{2}} (\sigma_1 - \sigma_2)$$

By Axiom 1, of these only ψ_s can interact with σ_3 (the nodal plane implicit in ψ_a bisects σ_3). We have, then:

$$\psi_1 = \frac{1}{\sqrt{1+\lambda^2}} (\psi_s + \lambda\sigma_3)$$

and

$$\psi_2 = \frac{1}{\sqrt{1+\lambda^2}} (\lambda\psi_s - \sigma_3)$$

where the constant λ has to be determined. Now, from Axiom 3 the first of these combinations must be, in expanded form,

$$\psi_1 = \frac{1}{\sqrt{3}} (\sigma_1 + \sigma_2 + \sigma_3)$$

so, by comparison of coefficients, $\lambda = 1/\sqrt{2}$ (we need only consider the positive root). It follows that

$$\psi_2 = \frac{1}{\sqrt{6}} (\sigma_1 + \sigma_2 - 2\sigma_3).$$

ψ_1 is of A_1' symmetry and ψ_a and ψ_2 together transform as E'.

Example 2   The σ l.g.o.'s of a planar AB_4 molecule (D_{4h} symmetry). Label the σ orbitals cyclically σ_1, σ_2, σ_3 and σ_4. Consider the pairs σ_1 and σ_3; σ_2 and σ_4. By Axiom 2 we have the combinations:

$$\psi_1 = \frac{1}{\sqrt{2}} (\sigma_1 + \sigma_3) \qquad \psi_2 = \frac{1}{\sqrt{2}} (\sigma_1 - \sigma_3)$$

$$\psi_3 = \frac{1}{\sqrt{2}} (\sigma_2 + \sigma_4) \qquad \psi_4 = \frac{1}{\sqrt{2}} (\sigma_2 - \sigma_4)$$

The nodal plane implicit in ψ_2 contains atoms 2 and 4. Similarly, the ψ_4 nodal plane contains atoms 1 and 3. It follows, from Axiom 1, that only ψ_1 and ψ_3 interact. From Axiom 3, one combination is

$$\psi_5 = (\sigma_1 + \sigma_2 + \sigma_3 + \sigma_4) \quad \text{i.e.} \quad \frac{1}{\sqrt{2}} (\psi_1 + \psi_3),$$

so the other must be

$$\psi_6 = (\sigma_1 - \sigma_2 + \sigma_3 - \sigma_4) \quad \text{i.e.} \quad \frac{1}{\sqrt{2}} (\psi_1 - \psi_3),$$

ψ_5 is of A_{1g} symmetry, ψ_2 and ψ_4 together transform under the E_u irreducible representation and ψ_6 is of B_{2g} symmetry.

Example 3 The σ l.g.o.'s of a tetrahedral AB$_4$ molecule (T_d symmetry). The derivation in this case is identical to that in Example 2. ψ_5 transforms as A_1 and ψ_2, ψ_4 and ψ_6 as T_2. In this T_2 set the Cartesian coordinates onto which ψ_2, ψ_4 and ψ_6 have a one-to-one mapping are not equivalently orientated. Two, those which map onto ψ_2 and ψ_4, pass through the edges of the cube corresponding to the tetrahedron, but the third passes through the midpoint of faces. The T_2 l.g.o. set which maps onto the usual choice of Cartesian axes for the tetrahedron is:

$$\frac{1}{\sqrt{2}} (\psi_2 + \psi_4) = \tfrac{1}{2} (\sigma_1 + \sigma_2 - \sigma_3 - \sigma_4)$$

$$\frac{1}{\sqrt{2}} (\psi_2 - \psi_4) = \tfrac{1}{2} (\sigma_1 - \sigma_2 - \sigma_3 + \sigma_4)$$

and

$$\psi_6 = \tfrac{1}{2}(\sigma_1 - \sigma_2 + \sigma_3 - \sigma_4)$$

Example 4 The σ l.g.o.'s of an octahedral AB$_6$ molecule (O_h symmetry). We isolate four ligand σ orbitals in a plane and label them cyclically σ_1, σ_2, σ_3, and σ_4. The correct combinations for this set are given in Example 2. Above and below this plane, respectively, lie the orbitals σ_5 and σ_6. We consider the combinations:

$$\psi_1 = \tfrac{1}{2}(\sigma_1 + \sigma_2 + \sigma_3 + \sigma_4)$$

$$\psi_2 = \frac{1}{\sqrt{2}} (\sigma_1 - \sigma_3) \qquad \psi_5 = \frac{1}{\sqrt{2}} (\sigma_5 + \sigma_6)$$

$$\psi_3 = \frac{1}{\sqrt{2}} (\sigma_2 - \sigma_4) \qquad \psi_6 = \frac{1}{\sqrt{2}} (\sigma_5 - \sigma_6)$$

$$\psi_4 = \tfrac{1}{2}(\sigma_1 - \sigma_2 + \sigma_3 - \sigma_4)$$

Axiom 1, applied by the 'nodal plane' criterion, shows that only ψ_1 and ψ_5 are non-orthogonal. The combination

$$\frac{1}{\sqrt{1+\lambda^2}} (\psi_1 + \lambda\psi_5) \text{ leads to (Axiom 3)}$$

$$\psi_7 = \frac{1}{\sqrt{6}} (\sigma_1 + \sigma_2 + \sigma_3 + \sigma_4 + \sigma_5 + \sigma_6) = \sqrt{\frac{2}{3}}\, \psi_1 + \frac{1}{\sqrt{3}}\, \psi_5$$

It follows, by comparison of coefficients, that $\lambda = (1/\sqrt{2})$ so that the combination:

$$\frac{1}{\sqrt{1+\lambda^2}} (\lambda\psi_1 - \psi_5)$$

is

$$\psi_8 = \frac{1}{\sqrt{12}} (\sigma_1 + \sigma_2 + \sigma_3 + \sigma_4 - 2\sigma_5 - 2\sigma_6).$$

ψ_7 is of A_{1g} symmetry, ψ_4 and ψ_8 transform as E_g and ψ_2, ψ_3 and ψ_6 as T_{1u}.

Example 5　The σ l.g.o.'s of an AB_8 Archimedean antiprismatic molecule (D_{4d} symmetry).

This example again uses the results of Example 2 by considering the allowed combinations between two square planar arrangements of ligand orbitals, rotated with respect to one another by 45°. In order to use Axiom 1 the nodal planes of the two sets must be brought into coincidence. This involves the rotation of coordinate axes as discussed in Example 3. Label the ligand orbitals cyclically $\sigma_1, \ldots, \sigma_8$, those of one plane being $\sigma_1, \ldots, \sigma_4$ and those of the other $\sigma_5, \ldots, \sigma_8$. σ_5 is positioned so that viewed down the fourfold rotation axis it appears to lie between σ_1 and σ_2.

Appropriate combinations are:

$$\psi_1 = \tfrac{1}{2}(\sigma_1 + \sigma_2 + \sigma_3 + \sigma_4) \qquad \psi_5 = \tfrac{1}{2}(\sigma_5 + \sigma_6 + \sigma_7 + \sigma_8)$$

$$\psi_2 = \frac{1}{\sqrt{2}}(\sigma_1 - \sigma_3) \qquad \psi_6 = \tfrac{1}{2}(\sigma_5 - \sigma_6 - \sigma_7 + \sigma_8)$$

$$\psi_3 = \frac{1}{\sqrt{2}}(\sigma_2 - \sigma_4) \qquad \psi_7 = \tfrac{1}{2}(\sigma_5 + \sigma_6 - \sigma_7 - \sigma_8)$$

$$\psi_4 = \tfrac{1}{2}(\sigma_1 - \sigma_2 + \sigma_3 - \sigma_4) \qquad \psi_8 = \tfrac{1}{2}(\sigma_5 - \sigma_6 + \sigma_7 - \sigma_8)$$

Application of Axiom 1 shows that we must consider further combinations between the pairs:

$$\psi_1 \quad \text{and} \quad \psi_5$$
$$\psi_2 \quad \text{and} \quad \psi_2$$
$$\psi_3 \quad \text{and} \quad \psi_7$$

The first pair gives

$$\psi_9 = \frac{1}{\sqrt{8}} (\sigma_1 + \sigma_2 + \sigma_3 + \sigma_4 + \sigma_5 + \sigma_6 + \sigma_7 + \sigma_8)$$

$$\psi_{10} = \frac{1}{\sqrt{8}} (\sigma_1 + \sigma_2 + \sigma_3 + \sigma_4 - \sigma_5 - \sigma_6 - \sigma_7 - \sigma_8)$$

while use of Axiom 3, in its more detailed form, shows that the correct combinations of ψ_2 and ψ_6 are

$$\psi_{11} = \frac{1}{2}(\sigma_1 - \sigma_3) + \frac{1}{2\sqrt{2}}(\sigma_5 - \sigma_6 - \sigma_7 + \sigma_8)$$

$$\psi_{12} = \frac{1}{2}(\sigma_1 - \sigma_3) - \frac{1}{2\sqrt{2}}(\sigma_5 - \sigma_6 - \sigma_7 + \sigma_8)$$

and of ψ_3 and ψ_7

$$\psi_{13} = \frac{1}{2}(\sigma_2 - \sigma_4) + \frac{1}{2\sqrt{2}}(\sigma_5 + \sigma_6 - \sigma_7 - \sigma_8)$$

$$\psi_{14} = \frac{1}{2}(\sigma_2 - \sigma_4) - \frac{1}{2\sqrt{2}}(\sigma_5 + \sigma_6 - \sigma_7 - \sigma_8)$$

since it is evident that ψ_{11} and ψ_{13} must be degenerate as must also be ψ_{12} and ψ_{14}. The symmetries of these combinations are

$$\psi_4 \text{ and } \psi_5 : E_2$$
$$\psi_9 : A_1$$
$$\psi_{10} : B_2$$
$$\psi_{11} \text{ and } \psi_{13} : E_1$$
$$\psi_{12} \text{ and } \psi_{14} : E_3$$

Example 6 As an example of the application of the method to combinations of ligand orbitals of diatomic π symmetry we consider the l.g.o.'s of π symmetry in a tetrahedral complex. Although the standard technique can be used to obtain the correct combinations, in practice the calculation is rather difficult.

We choose axes and orientations as shown in Figure A4.7. One set of ligand π orbitals (labelled α) is 'coplanar' with the z axis. The other set (labelled β) lies in planes perpendicular to the z axis. If the x and y axes are chosen as shown some slight simplification results. Our basic combinations are

$$\psi_1 = \frac{1}{\sqrt{2}}(\alpha_1 + \alpha_2) \qquad \psi_2 = \frac{1}{\sqrt{2}}(\alpha_1 - \alpha_2)$$

$$\psi_3 = \frac{1}{\sqrt{2}}(\alpha_3 + \alpha_4) \qquad \psi_4 = \frac{1}{\sqrt{2}}(\alpha_3 - \alpha_4)$$

$$\psi_5 = \frac{1}{\sqrt{2}}(\beta_1 + \beta_2) \qquad \psi_6 = \frac{1}{\sqrt{2}}(\beta_1 - \beta_2)$$

$$\psi_7 = \frac{1}{\sqrt{2}}(\beta_3 + \beta_4) \qquad \psi_8 = \frac{1}{\sqrt{2}}(\beta_3 - \beta_4)$$

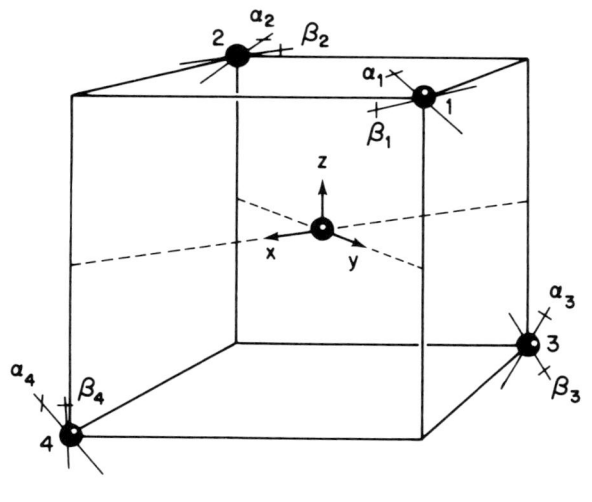

Figure A4.7 Cartesian axes and orientation of ligand π orbitals in a tetrahedral complexion.

Following the usual procedure it is readily seen that

$$\psi_9 = \frac{1}{\sqrt{2}}\,(\psi_1 + \psi_3) = \tfrac{1}{2}(\alpha_1 + \alpha_2 + \alpha_3 + \alpha_4)$$

$$\psi_{10} = \frac{1}{\sqrt{2}}\,(\psi_1 - \psi_3) = \tfrac{1}{2}(\alpha_1 + \alpha_2 - \alpha_3 - \alpha_4)$$

are orthogonal to all other combinations. It follows that they must be members of degenerate sets for they contain no β component (cf. Axiom 3). Consider ψ_9. This obviously transforms like the z axis and so will be a member of a triply degenerate set of which the other components transform as x and y. It follows that $d/n = 3/8$. However, the coefficient of the α's, squared, in ψ_9 is $\tfrac{1}{4}$ so there must be an α component in the x and y transforming members. Evidently, these components are derived from $\psi_2(y)$ and $\psi_4(x)$, each of which must appear with a coefficient of $\tfrac{1}{2}$ ($\tfrac{3}{8} - \tfrac{1}{4} = \tfrac{1}{8} = [\tfrac{1}{2}(1/\sqrt{2})]^2$). Now, ψ_6 transforms like x and ψ_8 like y so we are evidently seeking combinations of ψ_6 with ψ_4 and of ψ_8 with ψ_2. The correct combinations are

$$\psi_{11} = \frac{\sqrt{3}}{2}\,\psi_6 - \tfrac{1}{2}\psi_4 = \frac{1}{2\sqrt{2}}\left[\sqrt{3}\,(\beta_1 - \beta_2) - \alpha_3 + \alpha_4\right]$$

and

$$\psi_{12} = \frac{\sqrt{3}}{2}\,\psi_8 - \tfrac{1}{2}\psi_2 = \frac{1}{2\sqrt{2}}\left[\sqrt{3}\,(\beta_3 - \beta_4) - \alpha_1 + \alpha_2\right]$$

where we have been careful to make sure that the phases of ψ_6 and ψ_4 mapping onto x and of ψ_8 and ψ_2 onto y are identical.

Combinations of ψ_6 and ψ_4 and of ψ_8 and ψ_2 orthogonal to ψ_{11} and ψ_{12} are

$$\psi_{13} = \tfrac{1}{2}\psi_6 + \frac{\sqrt{3}}{2}\psi_4 = \frac{1}{2\sqrt{2}}\left[\beta_1 - \beta_2 + \sqrt{3}\,(\alpha_3 - \alpha_4)\right]$$

and

$$\psi_{14} = \tfrac{1}{2}\psi_8 + \frac{\sqrt{3}}{2}\psi_2 = \frac{1}{2\sqrt{2}}\left[\beta_3 - \beta_4 + \sqrt{3}\,(\alpha_1 - \alpha_2)\right]$$

We have only to deal with ψ_5 and ψ_7, which are not orthogonal; the correct combinations are

$$\psi_{16} = \tfrac{1}{2}(\beta_1 + \beta_2 + \beta_3 + \beta_4)$$
$$\psi_{17} = \tfrac{1}{2}(\beta_1 + \beta_2 - \beta_3 - \beta_4)$$

The symmetries of these combinations are

$$\psi_9, \psi_{11}, \psi_{12}: T_2$$
$$\psi_{13}, \psi_{14}, \psi_{16}: T_1$$
$$\psi_{10}, \psi_{17}: E$$

REFERENCES

1. W. Manch and W. C. Fernelius, *J. Chem. Educ.*, **38** (1961), 192.
2. A. D. Liehr, *J. Chem. Educ.*, **39** (1962), 135.
3. F. A. Cotton, *J. Chem. Educ.*, **41** (1964), 466.
4. H. B. Gray, *J. Chem. Educ.*, **41** (1966), 2.
5. S. F. A. Kettle, *J. Chem. Educ.*, **43** (1966), 21.

Appendix 5
The C$_{\infty v}$ and D$_{\infty h}$ Point Groups

The method developed in the text for the reduction of reducible representations does not work for infinite groups. This is because in the usual reduction formula (equation (A2.43))

$$a_i = \frac{1}{h} \sum_R \chi(R)\chi_i(R)$$

the quantity h is infinity, and the set of R over which the summation has to be done is also infinite. Instead, it is simplest to work the equation from which this is derived (equation (A2.41)) and in which a_i appears on the right-hand-side.

$$\chi(R) = \sum_i a_i\chi_i(R) \tag{A5.1}$$

There have been several papers in the chemical education literature on the reduction of reducible representations of infinite point groups, most of them being based on equation (A5.1).[1-7] Comparison of them makes interesting reading, the method given in this appendix differs—just a little—from all of them; the reader encountering difficulties has, then, several sources to which to turn for help.

The $C_{\infty v}$ character table is

$C_{\infty v}$	E	$2C_\infty^\phi$	$\ldots$	$\infty\sigma_v$
$A_1 \equiv \Sigma^+$	1	1	$\ldots$	1
$A_2 \equiv \Sigma^-$	1	1	$\ldots$	-1
$E_1 \equiv \Pi$	2	$2\cos\phi$	$\ldots$	0
$E_2 \equiv \Delta$	2	$2\cos 2\phi$	$\ldots$	0
$E_3 \equiv \Phi$	2	$2\cos 3\phi$	$\ldots$	0
$\ldots$	$\ldots$	$\ldots$	$\ldots$	$\ldots$

A general reducible representation (Γ_{red}) may have contributions from all irreducible representations, i.e. be of the form

$$\Gamma_{red} = a_1\Sigma^+ + a_2\Sigma^- + a_3\Pi + a_4\Delta + a_5\Phi + \cdots$$

so the problem is to determine the coefficients a_1, a_2, $a_3 \ldots$, etc. Under the operation E the character of this reducible representation will be:

$$\chi_1 = a_1 + a_2 + 2(a_3 + a_4 + a_5 + a_6 \cdots) \tag{A5.2}$$

because Σ^+ and Σ^- each have characters of unity under this operation and all other irreducible representations a character of 2. Similarly, under the operation C_∞^ϕ, the character of the reducible representation is

$$\chi_2^\phi = a_1 + a_2 + 2(a_3 \cos \phi + a_4 \cos 2\phi + a_5 \cos 3\phi + \phi_6 \cos 4\phi + \cdots).$$

Finally, under the operation σ_v the reducible representation has a character

$$\chi_3 = a_1 - a_2 \qquad (A5.3)$$

We have, then, a series of equations from which the unknown coefficients $a_1, a_2, a_3 \ldots$ can be determined, often by inspection. The fact that there are apparently only three equations but possibly more than three unknowns is no problem because there are really as many equations as needed. This is because it is possible to extract subsidiary equations by equating coefficients of, for example, cos ϕ, of cos 2ϕ, and so on. Thus, if Γ_{red} contains a term $n \cos \phi$ under the C_∞^ϕ operation this can only arise from the Π irreducible representation and it can immediately be concluded that $a_3 = n/2$.

For some cases it may be simpler to allow ϕ to assume specific values; in such cases χ_2^ϕ may be chosen to become selective of the terms within the parentheses. Thus, for $\phi = 90°$, cos $\phi = \cos 3\phi = \cos 5\phi = \cdots = 0$, but cos $2\phi = \cos 6\phi = \cdots = -1$ and cos $4\phi = \cos 8\phi = \cdots = 1$ giving

$$\chi_2^{90} = a_1 + a_2 + 2(-a_4 + a_6 \ldots) \qquad (A5.4)$$

while for $\phi = 180°$, cos $\phi = \cos 3\phi = \cos 5\phi = \cdots = -1$ and cos $2\phi = \cos 4\phi = \cdots 1$, giving

$$\chi_2^{180} = a_1 + a_2 + 2(-a_3 + a_4 - a_5 + a_6 \ldots) \qquad (A5.5)$$

In this way as many simultaneous equations in the unknowns, the a_n's, as are needed to solve the problem can be generated. The second example illustrates the use of this method. In practice, if there are more than a few terms the method becomes cumbersome.

Example Problem: What are the symmetry species of the vibrations of a linear X–Y–Z molecule?
Solution: Following the method developed in Chapter 9, consider the transformation of Cartesian displacement coordinates associated with each atom of the molecule:

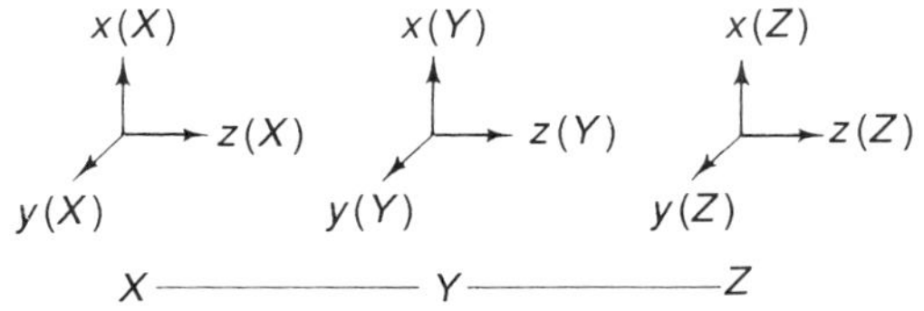

Figure A5.1

and obtain the reducible representation

$$
\begin{array}{ccc}
E & 2C_\infty^\phi & \infty\sigma_v \\
9 & (3+6\cos\phi) & 3
\end{array}
$$

(Under the C_∞^ϕ rotations each of the z displacement coordinates remains itself; each of the x and y are rotated by ϕ and so the projection of the rotated displacement onto the original is $\cos\phi$—see either equation (A2.6) or Section 6.1).

As indicated above, the $\cos\phi$ factor can at once be separated out from this reducible representation; given a character of $6\cos\phi$ under $2C_\infty^\phi$, the characters under E and $\infty\sigma_v$ follow

$$
\begin{array}{ccc}
E & 2C_\infty^\phi & \infty\sigma_v \\
6 & 6\cos\phi & 0
\end{array}
$$

which is 3Π (see the $C_{\infty v}$ character table above). Subtracting these characters from the original reducible representation we are left with

$$
\begin{array}{ccc}
E & C_\infty^\phi & \infty\sigma_v \\
3 & 3 & 3
\end{array}
$$

which is $3\Sigma^+$.

We conclude that the original reducible representation reduces into the components

$$3\Sigma^+ + 3\Pi$$

The next step is to subtract the symmetry species of the translations and rotations. The character table for $C_{\infty v}$ given in Appendix 3 can be used to determine these. However, because the molecule is linear, the $3N-5$ rule (N = the number of atoms) operates and R_z is not included in the rotations. From Appendix 3 we have

Translations: $\Sigma^+ + \Pi$
Rotations: Π

It is concluded that the vibrations of a linear X–Y–Z molecule transform as

$$2\Sigma^+ + \Pi$$

(The two Σ^+ vibrations arise from the X–Y and Y–Z stretching motions; the Π from the X–Y–Z angle deformation.)

Comment: Had a term in $\cos 2\phi$ under $2C_\infty^\phi$ been obtained in this development, a Δ contribution would first have been extracted, a term in $\cos 3\phi$ would have led to a Φ contribution and so on.

Example 2 Problem: The central atom in a linear X–Y–Z molecule is a transition metal atom. Determine the symmetries of its d orbitals.

Comment: Given the ϕ-angular dependence of the d orbitals—and this is explicit in their algebraic forms—the first method above could be used. However, many will prefer to use schematic diagrams of the angular dependence and the present method is well suited to this, more qualitative, description.

Solution: Take the coordinate axes of the central atom to be those given in Problem 1. By drawing the individual d orbitals (and viewing them 'down' the z-axis, as is appropriate for a rotation about this axis), or by inspection characters can be derived. Take specific values of ϕ of 90° and 180° to obtain the characters

	E	$2C_\infty^{90}$	$2C_\infty^{180}$	$\infty\sigma_v$
d_{z^2}	1	1	1	1
d_{xz}, d_{yz}	2	0	-2	0
$d_{x^2-y^2}, d_{xy}$	2	-2	2	0
Γ_{red}	5	-1	1	1

It is, of course, simplest not to work with Γ_{red} but to separately consider each of the three representations of which it is the sum. However, for the purpose of the present example, the more difficult problem of working with Γ_{red} in a systematic manner will be tackled. This is done by the use of equation s(A5.2) through to (A5.4). In these equations a decision must be made about how many of the $a_1, a_2, a_3 \ldots$ to include on the right-hand side. A wrong choice will either lead to a set of equations which are not internally consistent or, the final check, an answer which does lead to the regeneration of the original reducible representation. In the present example only a_1, a_2, a_3 and a_4 on the right-hand side of these equations will be included (there are four characters in the reducible representation and so four unknowns can be determined; if we wished to include a_5 on the right-hand side we would have had to include another rotation, C_∞^{45}, for instance, when generating the reducible representation).

The substitutions give

In (A5.2) (E): $\qquad\qquad 5 = a_1 + a_2 + 2a_3 + 2a_4$

In (A5.4) ($2C_\infty^{90}$): $\qquad -1 = a_1 + a_2 - 2a_4$

In (A5.5) ($2C_\infty^{180}$): $\qquad 1 = a_1 + a_2 - 2a_3 + 2a_4$

In (A5.3) ($\infty\sigma_v$): $\qquad\quad 1 = a_1 - a_2$

By straightforward manipulation of these equations:

$$a_1 = 1, \quad a_2 = 0, \quad a_3 = 1, \quad a_4 = 1$$

i.e. $\Gamma_{red} = \Sigma^+ + \Pi + \Delta$.

Had we worked with the three simpler problems rather than the complicated one we would have identified d_{z^2} as Σ^+, (d_{xz}, d_{yz}) as Π and $(d_{x^2-y^2}, d_{xy})$ as Δ.

Problem A5.1 Show that if an attempt is made to solve the above problem using only a_1, a_2 and a_3 then the equations obtained are not internally consistent.

Problem A5.2 Show that the set of p orbitals on an atom in a molecule of $C_{\infty v}$ symmetry transform as

$$\Sigma^+ + \Pi$$

Use this result to show that the configuration p^1p^1 (where two different sets of p orbitals are involved) on such an atom gives rise to terms which transform as

$$2\Sigma^+ + \Sigma^- + 2\Pi + \Delta$$

(*Hint:* Consider the direct product $(\Sigma^+ + \Pi) \otimes (\Sigma^+ + \Pi)$.)

In this appendix $C_{\infty v}$ has been used as an illustrative infinite point group; the methods described may be immediately extended to $D_{\infty h}$ by including characters appropriate to S_n and C_2 operations.

REFERENCES

1. L. Schäfer and S. J. Cyvin, *J. Chem. Ed.*, **48** (1971), 295.
2. D. P. Strommen and E. P. Lippincott, *J. Chem. Ed.*, **49** (1972), 341.
3. H. P. Fritzer, *Match*, **3** (1977), 21.
4. J. M. Alvariño, *J. Chem. Ed.*, **55** (1978), 307.
5. R. L. Flurry Jr., *J. Chem. Ed.*, **56** (1979), 638.
6. D. P. Strommen, *J. Chem. Ed.*, **56** (1979), 640.
7. J. M. Alvariño and A. Chamorro, *J. Chem. Ed.*, 57 (1980), 785.

Appendix 6
Hermann–Mauguin (H–M) and Schönflies (S) Equivalents

Point group operations

Hermann–Mauguin	Schönflies	Comments
1	E	
$\bar{1}$	i	A bar over a symbol in the H–M notation indicates inversion in a centre of symmetry.
2	C_2	
3	C_3	Note that rotation is in the same sense in the two
3_2	C_3^2	notations. This is particularly important when
4	C_4	S_4 or S_6 axes are present.
4_3	C_4^3	
6	C_6	
6_5	C_6^5	
m	σ	Unlike σ, the m symbol never carries suffixes. It is very important to recognize that the common
$\bar{3}$	S_6^5	practice is to use different definitions for these
$\bar{3}^2$	S_6	operations in the two notations (H–M; rotation +
$\bar{4}$	S_4^3	inversion: S; rotation + reflection) leading to
$\bar{4}^3$	S_4	apparently perverse correspondences. More
$\bar{6}$	S_3^2	comfortable correspondences can be obtained by defining rotation to be in opposite senses in the
$\bar{6}^5$	S_3	two notations (not done here).

Point groups
Although extension to other cases is straightforward, in practice, use of the Hermann–Mauguin notation is normally confined to the 32 crystallographic point groups. Only these are given here.

Hermann–Mauguin	Schönflies	Comments
1	C_1	
$\bar{1}$	C_i	
2	C_2	
m	C_s	
$\dfrac{2}{m}$	C_{2h}	
222	D_2	
$mm2$	C_{2v}	In H–M sometimes called mm
$\dfrac{222}{mmm}$	D_{2h}	In H–M sometimes called mmm
4	C_4	
$\bar{4}$	S_4	
$\dfrac{4}{m}$	C_{4h}	
422	D_4	
$4mm$	C_{4v}	
$\bar{4}2m$	D_{2d}	
$\dfrac{444}{mmm}$	D_{4h}	In H–M sometimes called $\dfrac{4}{mmm}$
3	C_3	
$\bar{3}$	S_6	
32	D_3	
$3m$	C_{3v}	Beware confusion with $m3$
$\dfrac{\bar{3}2}{m}$	D_{3d}	In H–M sometimes called $\bar{3}m$
$\dfrac{m}{6}$	C_6	
$\dfrac{\bar{3}}{m}$	C_{3h}	
$\dfrac{6}{m}$	C_{6h}	
622	D_6	
$6mm$	C_{6v}	
$\dfrac{\bar{3}m2}{m}$	D_{3h}	In H–M sometimes called $\bar{6}m2$
$\dfrac{622}{mmm}$	D_{6h}	In H–M sometimes called $\dfrac{6}{mmm}$
23	T	
$\dfrac{2\bar{3}}{m}$	T_h	In H–M often called $m3$ (beware confusion with $3m$)
432	O	
$\bar{4}3m$	T_d	
$\dfrac{4}{m}\bar{3}\dfrac{2}{m}$	O_h	In H–M often called $m3m$

Further Reading

The subjects of symmetry and group theory cover an enormous range. At one extreme are popular treatments of manifestations in art and science (recent articles in the chemical press have covered subjects as diverse as Hungarian needlework and motor-car hub covers). At the other extreme are mathematical discussions which appear to have no relationship with the content of the present book (which treats only a small and specialist part of the subject). For most readers the sensible way to progress would be to read a text covering the material contained in this book but in which the mathematics is fully integrated. Progress and understanding should both be rapid. There is a timelessness about the subject so that the fact that a book is relatively old should be regarded as no disadvantage. The following are listed in an approximate order of mathematical difficulty, although the subjectiveness of any assessment has to be recognized. The list is not comprehensive and a browse through the shelves of a library may well be productive.

Symmetry Discovered by J. Rosen (Cambridge University Press, Cambridge, 1975). Not mathematical at all, but beautiful.
Symmetry in Chemistry by H. H. Jaffé and M. Orchin (Wiley, New York, 1965). Little more mathematical than the present book but more 'conventional' in that symmetry elements, operations and examples are the subject of the first three chapters.
Symmetry and Stereochemistry by J. D. Donaldson and S. D. Ross (Intertext, London, 1972) has an emphasis on point group and space group symmetries.
Molecular Symmetry and Group Theory by A. Vincent (Wiley, London, 1977). A programmed introduction which proceeds gently, each frame providing an answer to the problem set in the previous frame.
Symmetry in Chemical Bonding and Structure by W. E. Hatfield and W. E. Parker (Merrill, Columbus, Ohio, 1974). Not very mathematical.

Now come a series of books which have a full integration of the mathematics and discussion. The choice between them is very personal.

Chemical Applications of Group Theory by F. A. Cotton (Wiley, New York, 1990). Now in its third edition, this is the best known of the books in this category. The treatment of space groups is through the eyes of a crystallographer and so has some differences with the presentation in the present text.

Symmetry, Orbitals and Spectra (S.O.S.) by M. Orchin and H. H. Jaffé (Wiley–Interscience, New York, 1971). The touch of humour in the title is significant. Rather more mathematical than the other book by the same authors (see above).
Symmetry and its Applications in Science by A. D. Boardman, D. E. O'Connor and P. A. Young (McGraw-Hill, London, 1973).
Group Theory and Chemistry by D. M. Bishop (Clarendon, Oxford, 1973).

Now a set that are more mathematical but in being so are more rigorous. Again, much the same material is covered and so the previous study of one of the texts above should make understanding both easier and more rapid.

Introductory Group Theory by J. R. Ferraro and J. S. Ziomek (Plenum, New York, 1969).
Symmetry by R. McWeeny (Pergamon, Oxford, 1963). Perhaps the easiest to follow of this group.
Molecular Symmetry by D. S. Schonland (Van Nostrand, London, 1965).
Molecular Symmetry and Spectroscopy by P. R. Bunker (Academic, New York, 1979).
Group Theory in Quantum Mechanics by V. Heine (Pergamon, London, 1960). This book goes well beyond the content of the present volume; it uses Hermann–Mauguin notation.
Molecular Aspects of Symmetry by R. M. Hochstrasser (W. A. Benjamin, New York, 1966). A notable book; it includes accessible treatment of topics not commonly discussed. Unfortunately, it has its fair share of printer's errors.
Group Theory and its Physical Applications by L. M. Falicov (University of Chicago, Chicago, 1966). A small book that deserves to be better known. Rigorous but relatively easy to read.

Turning to the topic of space groups, the approach followed in the present text is closer to that traditional in physics rather than that common in chemistry.

Space Groups and their Representations by G. F. Koster (Academic, New York, 1957). This could be regarded as essential reading. Perhaps the most important part of this book for the reader of the present volume is the first half—and this is the easiest to read. It is a reprint of an article that appeared in *Solid State Physics*, **5** (1957), 13.
Space Groups for Solid State Scientists by G. Burns and A. M. Glazer (Academic, Boston, 1990). Now in its second edition, this is a very easy to read book that covers the basics of group theory and then immediately applies them to space groups.
International Tables for Crystallography T. Hahn (ed.) (Reidel, Boston, 1983). The earlier edition is *International Tables for X-Ray Crystallography, Vol. 1*, N. F. M. Henry and K. Lonsdale (eds) (Kynoch, Birmingham, 1952). One or other of these is essential for serious work. Both have been reprinted several times

Index